Evolution of Biological Systems in Random Media: Limit Theorems and Stability

MATHEMATICAL MODELLING:
Theory and Applications

VOLUME 18

This series is aimed at publishing work dealing with the definition, development and application of fundamental theory and methodology, computational and algorithmic implementations and comprehensive empirical studies in mathematical modelling. Work on new mathematics inspired by the construction of mathematical models, combining theory and experiment and furthering the understanding of the systems being modelled are particularly welcomed.

Manuscripts to be considered for publication lie within the following, non-exhaustive list of areas: mathematical modelling in engineering, industrial mathematics, control theory, operations research, decision theory, economic modelling, mathematical programmering, mathematical system theory, geophysical sciences, climate modelling, environmental processes, mathematical modelling in psychology, political science, sociology and behavioural sciences, mathematical biology, mathematical ecology, image processing, computer vision, artificial intelligence, fuzzy systems, and approximate reasoning, genetic algorithms, neural networks, expert systems, pattern recognition, clustering, chaos and fractals.

Original monographs, comprehensive surveys as well as edited collections will be considered for publication.

The titles published in this series are listed at the end of this volume.

Evolution of Biological Systems in Random Media: Limit Theorems and Stability

by

Anatoly Swishchuk
Department of Mathematics & Statistics,
York University, Toronto, ON, Canada

and

Jianhong Wu
Department of Mathematics & Statistics,
York University, Toronto, ON, Canada

KLUWER ACADEMIC PUBLISHERS
DORDRECHT / BOSTON / LONDON

A C.I.P. Catalogue record for this book is available from the Library of Congress.

ISBN 978-90-481-6398-4

Published by Kluwer Academic Publishers,
P.O. Box 17, 3300 AA Dordrecht, The Netherlands.

Sold and distributed in North, Central and South America
by Kluwer Academic Publishers,
101 Philip Drive, Norwell, MA 02061, U.S.A.

In all other countries, sold and distributed
by Kluwer Academic Publishers,
P.O. Box 322, 3300 AH Dordrecht, The Netherlands.

Printed on acid-free paper

To our families

TABLE OF CONTENTS

PREFACE

The book is devoted to the study of limit theorems and stability of evolving biological systems of "particles" in random environment. Here the term "particle" is used broadly to include moleculas in the infected individuals considered in epidemic models, species in logistic growth models, age classes of population in demographics models, to name a few. The evolution of these biological systems is usually described by difference or differential equations in a given space X of the following type

$$\mathbf{x}_{n+1} = \mathbf{x_n} + \mathbf{g}(\mathbf{x}_n, y),$$

and

$$d\mathbf{x}_t/dt = \mathbf{g}(\mathbf{x}_t, y),$$

here, the vector x describes the state of the considered system, g specifies how the system's states are evolved in time (discrete or continuous), and the parameter y describes the change of the environment. For example, in the discrete-time logistic growth model

$$N_{n+1} = N_n + r(y)N_n(1 - N_n/K(y)),$$

or the continuous-time logistic growth model

$$dN_t/dt = r(y)N_t(1 - N_t/K(y)),$$

N_n or N_t is the population of the species at time n or t, $r(y)$ is the per capita birth rate, and $K(y)$ is the carrying capacity of the environment, we naturally have $X = R$, $X_n \equiv N_n(X_t \equiv N_t)$, $\mathbf{g}(\mathbf{x}, y) = r(y)x(1 - x/K(y)), \mathbf{x} \in X$. Note that for a predator-prey model and for some epidemic models, we will have that $X = R^2$ and $X = R^3$, respectively. In th case of logistic growth models, parameters $r(y)$ and $K(y)$ normally depend on some random variable y. This is unavoidable due to many sources of noise such as a): random influences of food supply and other factors that have the impact on both growth rate and carrying capacity of the environment, and b): measurements errors as we are unable to make the measurement of infinite precision. Therefore, we may get many different values of r and K for a given biological species even in a closed system, and these values change over the time as well. Consequently, predictions and conclusions based on the above models with deterministic parameter values can hardly fit to the reality in these biological evolutions.

We are thus motivated to discuss the aforementioned models involving some random noise (random process, in general). In the particular case of logistic growth model, we will have $r \equiv r(y_t)$ and $K \equiv K(y_t)$, where y_t is a certain random process. Of course, many of the strong attributes (such as stability) of deterministic systems should persist under smooth perturbations of the system, but may be destroyed by random perturbations.

In this book, we describe and develop some methods to qualitatively analyze the biological systems with parameters depending on random processes. In particular, we assume that the random environment is described by Markov renewal or semi-Markov processes, and that the local characteristics of the evolving biological systems depend on the state of the random environment, but the evolution of the biological systems does not affect the environment. Therefore, the dynamics of such a system is described by two objects: a random process $y(t)$ for the environment, and a nonlinear dynamical system for the interaction of the particles of the population biological systems. As such, the model equations take the form of nonlinear stochastic equations (discrete-time or continuous-time) abstractly formulated as a stochastic operator integral equation in a Banach space. Stochastic models to be examined include an epidemic model; slow genetic selection model; demographic model; branching model; predator-prey model; logistic growth model, with both discrete and continuous time.

The goal of this book is to illustrate recent results and methods in several major areas of nonlinear random evolutional equations such as averaging and merging, diffusion approximation, normal deviations, stability in averaging and diffusion approximation schemes. To achive this goal, we develope some limit theorems for difference and differential equations in Markov renewal and semi-Markov random environments by using the general theory of random evolutions.

We now give a short description of the organization of the book.

Chapter 1 is devoted to the description of random media (RM) in the form of Markov chains, Markov renewal or semi-Markov processes and of their ergodic and martingale properties. Also, we introduce the notions of semigroups and their generators.

In Chapter 2, we consider general limit theorems for random evolutions and their applications to difference and differential equations in Markov and semi-Markov RM in series scheme: averaging, merging, diffusion approximation, normal deviations and stability.

The results of Chapter 2 are applied to the aforementioned biological systems in random media: epidemic model (Chapter 3); slow genetic model (Chapter 4); branching model (Chapter 5); demographic model (Chapter 6); logistic growth model (Chpater 7); predator-prey model (Chapter 8).

In these Chapters we firstly describe deterministic biological systems in details, and then we give the description of these systems in random media, namely, stochastic biological models. Each of these stochastic biological systems may be represented as a random evolution. That is why we can apply all the results from the theory of random evolutions (averaging, merging, diffusion approximation, normal deviations) from Chapter 2 to the systems.

It is hoped that this book should be useful for both experts and newcomers in biomathematics, dynamical systems, stochastic analysis, applied probability, and for those who are interested in applications to evolutionary biology.

We would like to acknowledge the support from Natural Sciences and Engineering Research Council of Canada, Canada Research Chairs Program and the Network of Centers of Excellence Program "Mathematics for Information Technology and Complex Systems".

Anatoly V. Swishchuk & Jianhong Wu
Laboratory for Industrial and Applied Mathematics
York University
Toronto, Canada
May 2003

LIST OF NOTATIONS

$(\Omega, \mathcal{F}, \mathcal{F}_n, \mathcal{P})$	—	probability space Ω with σ-algebra $\mathcal{F}$, filtration $\mathcal{F}_n$ and probability $\mathcal{P}$
E	—	expectation by $\mathcal{P}$
$(Y, \mathcal{Y})$	—	measurable phase space
Z^+	:=	$1, 2, 3, 4, \ldots$
$(y_n)_{n \in Z_+}$	-	Markov chain in Y
R^+	:=	$[0, +\infty)$
$(y_n; \theta_n)_{n \in Z^+}$	—	Markov renewal process, $\theta_k \in R^+$
τ_n	:=	$\sum_{k=0}^{n} \theta_k$
$\nu(t) := \max\{ n : \tau_n \leq t \}$	—	counting process
$y(t) := y_{\nu(t)}$	—	semi-Markov process
$\gamma(t)$	:=	$t - \tau_{\nu(t)}$
$p(dy)$	—	stationary probabilities of a Markov chain $(y_n)_{n \in Z^+}$
$P(y, dz)$	:=	$\mathcal{P}(\omega : y_{n+1} \in dz / y_n = y)$
$G_y(t)$	:=	$\mathcal{P}(\omega : \theta_{n+1} \leq t / y_n = y)$
$m(y)$	:=	$\int_0^\infty t G_y(dt)$
m	:=	$\int_Y p(dy) m(y)$
$\pi(dy)$	:=	$\frac{p(dy) m(y)}{m}$
$\mathbf{R}_0$	—	potential of Markov chain $(y_n)_{n \in Z^+}$
$(V, \mathcal{U})$	—	merged phase space
Y_v	—	vth ergodic component of Y, $\forall\, v \in V$
$p_v(dy)$	—	stationary probabilities in Y_v, $\forall\, v \in V$
$m(v)$	:=	$\int_{Y_v} p_v(dy) m(y)$

$\pi_v(dy)$	:=	$\frac{p_v(dy)m(y)}{m(v)}$
$w(t)$	—	standard Wiener process
$(\mathbf{B}, \mathcal{B}, \|\cdot\|)$	—	separable Banach space
$\mathbf{B}^*$	—	dual space of $\mathbf{B}$
l	—	linear continuous functional, $l \in \mathbf{B}^*$
$\Gamma(t), \Gamma_y(t)$	—	semigroups of operators of $t, \forall y \in Y$
$\Gamma, \Gamma(y)$	—	infinitesimal operators, $\forall\ y \in Y$
$V(t), V^\varepsilon(t), V_\varepsilon(t)$	—	initial random evolutions
$\hat{V}(t), V^0(t), V_0(t)$	—	limiting random evolutions (averaged or merged, etc.)
$(X, \mathcal{X})$	—	a linear phase space
$(X_n)_{n\in Z^+}((X(t), X_{\nu(t)})_{t\in R_+})$	—	state of discrete (continuous) dynamical system in X
$\mathcal{P}_{x,y}(\cdot)$	:=	$\mathcal{P}(\cdot/X_0 = x, y_0 = y)$
$E_{x,y}$	—	expectation by $\mathcal{P}_{x,y}$
$(X_n^\epsilon)_{n\in Z^+}((X^\epsilon(t), X_{\nu(t)}^\epsilon)_{t\in R^+})$	—	discrete (continuous) dynamical systems in series scheme
$(\hat{x}_t, \tilde{x}_t, \bar{x}_t, \hat{x}(t), \tilde{x}(t))_{t\in R^+}$	—	limiting dynamical systems
$(m_n)_{n\in Z^+}((M_n)_{n\in Z^+}(m(t), M(t), M_t)_{t\in R^+})$	—	discrete (continuous) martingales
$(m_\epsilon(t), M_t^\epsilon)_{t\in R^+}$	—	martingales in series scheme
$(S_n, I_n, R_n)_{n\in Z^+}((S(t), I(t), R(t))_{t\in R^+})$	—	susceptible, infectives and removed in epidemic model with discrete (continuous) time
$(S_n^\epsilon, I_n^\epsilon, R_n^\epsilon)_{n\in Z^+}((S^\epsilon(t), I^\epsilon(t), R^\epsilon(t))_{t\in R^+})$	—	susceptible, infectives and removed in epidemic model with discrete (continuous) time in series scheme
$((\hat{S}(t), \hat{I}(t), \hat{R}(t), \tilde{S}(t), \tilde{I}(t), \tilde{R}(t))_{t\in R^+})$	—	limiting epidemic models
$(g_n)_{n\in Z^+}((G(t))_{t\in R^+})$	—	proportion of the gene pool of type A in the n reproduction
$(g_n^\epsilon)_{n\in Z^+}((g_{\nu(t)}^\epsilon, g^\epsilon(t))_{t\in R^+})$	—	discrete (continuous) genetic selection models in series scheme

$(\tilde{g}(t), G_0(t))_{t\in R^+}$	—	limiting genetic selection models
$(Q^\epsilon_n)_{n\in Z^+}((Q^\epsilon_{\nu(t)}, Q^\epsilon(t))_{t\in R^+})$	—	discrete (continuous) generating function for Bellman-Harris branching process in series scheme
$(\tilde{Q}(t), \tilde{Q}_t, \hat{Q}(t), \bar{Q}(t), \hat{Q}(t))_{t\in R^+}$	—	limiting generating functions for branching models
$(\vec{v}_n)_{n\in Z^+}$	—	vector of age profile at the *nth* census in demographic model
$(\vec{v}^{\,\epsilon}_n)_{n\in Z^+}((\vec{v}_{\nu(t)}, \vec{v}\ (t))_{t\in R^+})$	—	discrete (continuous) age profile in demographic models in series scheme
M	—	Fibonacci and Leslie reproduction matrices
$(N_n)_{n\in Z^+}((N(t))_{t\in R^+})$	—	discrete (continuous) logistic growth model (LGM)
$(N^\epsilon_n)_{n\in Z^+}((N^\epsilon_{\nu(t)}, N(t)^\epsilon)_{t\in R^+})$	—	discrete (continuous) LGM in series scheme
$(\tilde{N}(t), \hat{N}_t, \bar{N}(t))_{t\in R^+}$	—	limiting LGM
$(N_n, P_n)_{n\in Z^+}((N(t), P(t))_{t\in R^+})$	—	discrete (continuous) predator-prey model (PPM)
$(N^\epsilon_n, P^\epsilon_n)_{n\in Z^+}((N^\epsilon(t), P^\epsilon(t))_{t\in R^+})$	—	discrete (continuous) PPM in series scheme
$(\tilde{N}(t), \tilde{P}(t)), (\hat{N}_t, \hat{P}_t), (\bar{N}(t), \bar{P}(t))_{t\in R^+}$	—	limiting PPM
$(Z^\epsilon(t))_{t\in R^+}$	—	normal deviated dynamical and biological systems in series scheme
$(\tilde{Z}(t))_{t\in R^+}$	—	limiting normal deviated dynamical and biological systems
$(\mathbf{X}^\epsilon_n)_{n\in Z^+}((\mathbf{X}^\epsilon(t), \mathbf{X}^\epsilon_{\nu(t)})_{t\in R^+})$	—	vector discrete (continuous) dynamical systems in series scheme
$(\hat{\mathbf{x}}_t, \tilde{\mathbf{x}}_t, \bar{\mathbf{x}}_t, \hat{\mathbf{x}}(t), \tilde{\mathbf{Z}}(t))_{t\in R^+}$	—	limiting vector dynamical systems

RE	—	random evolutions
RM	—	random media
MRM	—	Markov random media
SMRM	—	semi-Markov random media
DE	—	difference equations
BS	—	biological systems
EM	—	epidemic model
GSM	—	genetic selection model
BM	—	branching model
DM	—	demographic model
LGM	—	logistic growth model
PPM	—	predator-prey model
DA	—	diffusion approximation
ND	—	normal deviations
SS	—	stochastic stability
VDE	—	vector difference equations

CHAPTER 1: RANDOM MEDIA

This chapter is devoted to the description of random environment, in the form of a Markov chain, a Markov renewal or a semi-Markov process, and its ergodic and martingale properties. We introduce the notions of diffusion processes, semigroups and their generators, and consider the merging property of random environment.

Most of the presentations are very brief, and so we list a few references where the materials of this chapter are based on: Markov chains [15, 23, 26, 48, 59, 65, 66, 72]; Markov processes [24]; semi-Markov processes [49, 53, 54, 77, 80, 84]; ergodicity [22]; semigroups [24, 26, 94].

1.1. Markov Chains

Let us consider a certain random process that retains *no memory* of the past. This means that only the current state of the process can influence where it goes next. Such a process is called a *Markov process.* In this section, we are concerned exclusively with the case where the process can assume only a finite or countable set of states, and such a process is usually called a *Markov chain.* We shall consider chains both in *discrete* time $n \in \mathbf{Z}^+ \equiv \{0, 1, 2, ...\}$ and *continuous time* $t \in \mathbf{R}^+ \equiv [0, +\infty)$. The letters n, m, k will always denote integers, whereas t and s will refer to real numbers. Thus we write $(y_n)_{n\in\mathbf{Z}^+}$ for a discrete-time process and $(y_t)_{t\in\mathbf{R}^+}$ or $y(t)$ for a continuous-time process.

Let us start with a **discrete-time Markov chain.** We noew describe a discrete Markov chains by the following example where the *state space* Y contains three elements (states): $Y = \{1, 2, 3\}$.

In the process, the state is changed from state 1 to state 2 with probability 1; from state 3 either to 1 or to 2 with probability 2/3 or 1/3, respectively, and from state 2 to state 3 with probability 1/4, otherwise stay at 2. Therefore, we obtain a stochastic matrix

$$\mathbf{P} = \begin{pmatrix} p_{11} & p_{12} & p_{13} \\ p_{21} & p_{22} & p_{23} \\ p_{31} & p_{32} & p_{33} \end{pmatrix} = \begin{pmatrix} 0 & 1 & 0 \\ 0 & 3/4 & 1/4 \\ 2/3 & 1/3 & 0 \end{pmatrix}.$$

Now, we let Y be a countable set. Each $i \in Y$ is called a *state* and Y is called the *state space*. We say that a set of real numbers $p = (p_i; i \in Y)$ is a *measure* on Y if $0 \le p_i < +\infty$ for all $i \in Y$. If, in addition, the *total mass* $\sum_{i\in Y} p_i = 1$, then we call p a *distribution*. In what follows, we fix a *probability space* $(\Omega, \mathcal{F}, \mathcal{P})$. A *random variable* y with values in Y is a function $y : \Omega \to Y$. For a random variable y, if we define

$$p_i = \mathcal{P}\{y = i\} \equiv \mathcal{P}\{\omega : y(\omega) = i\},$$

then p defines a distribution, called the distribution of y. We therefore can think of y as modelling a random state which takes the value i with probability p_i.

We say that a matrix $\mathbf{P} = (p_{ij}; i, j \in Y)$ is *stochastic* if every row $(p_{ij}; j \in Y)$ is a distribution, namely, $p_{ij} \geq 0$ and $\sum_{j \in Y} p_{ij} = 1$.

A *semi-stochastic matrix* $\mathbf{P} = (p_{ij}; i, j \in Z_+)$ is one such that if every row $(p_{ij}; j \in Z_+)$ satisfies $0 \leq P_{ij} < +\infty$ and $\sum_{j \in Y} p_{ij} \leq 1$.

We say that $(y_n)_{n \in \mathbf{Z}^+}$ is a Markov chain with *initial distribution* p and *transition matrix* $\mathbf{P}$, if

(a) y_0 has distribution p;

(b) for $n \geq 0$, conditional on $y_n = i$, y_{n+1} has distribution $(p_{ij}; j \in Y)$ independent of $y_0, y_1, ..., y_{n-1}$, namely,

$$\mathcal{P}\{\omega : y_{n+1} = j | y_n = i\} = p_{ij}.$$

More explicitly, we have, for $n \geq 0$ and $i, j, i_0, i_1, ..., i_{n-1} \in Y$, that

(a)
$$\mathcal{P}\{y_0 = i\} = p_i;$$

(b)
$$\mathcal{P}\{y_{n+1} = j | y_0 = i_0, y_1 = i_1, ..., y_{n-1} = i_{n-1}, y_n = i\} = p_{ij}.$$

A Markov chain with a finite phase space is called a *finite Markov chain.* In general, the phase space is not required to be countable. We now describe a general case. Let $(Y, \mathcal{Y})$ be a measurable space, here $\mathcal{Y}$ is a σ-algebra of measurable sets on Y, which can be interpreted as a collaction of observable subsets of states in the random environment. We assume that $\mathcal{Y}$ contains all one-point sets.

A function $P : Y \times \mathcal{Y} \to [0, +\infty)$ is called *a stochastic kernel* in the measurable phase space $(Y, \mathcal{Y})$ if it satisfies the following conditions:

(i) for fixed $y \in Y$, the function $P(y, A)$ is a probability distribution on A and $P(y, Y) = 1$;

(ii) for fixed $A \in \mathcal{Y}$, the function $P(y, A)$ is $\mathcal{Y}$-measurable with respect to $y \in Y$.

If (i) and (ii) are satisfied, except that $P(y, Y) = 1$ being replaced by $P(y, Y) \leq 1$ for $y \in Y$, then the kernel is said to be *semi-stochastic.*

In applications, the stochastic kernel $P(y, A)$ determines the probability of transitions of the random environment under consideration from the state y into the set of states A.

In a discrete phase space $Y = \{1, 2, \ldots, \}$, a stochastic kernel is given by a stochastic matrix $P = (p_{ij}, i, j \epsilon Y)$ with non-negative elements as follows: $P(i, A) = \sum_{j \in A} p_{ij}$ for any $A \in \mathcal{Y}$.

For a Markov chain $(y_n)_{n \in Z^+}$, we can define the initial distribution $P(A) = \mathcal{P}\{y_0 \in A\}$ and the following stochastic kernels $P_n(y, A) = \mathcal{P}\{y_{n+1} \in A | y_n = y\}$, called the *probabilities of the (n-th step) one-step transitions of* $(y_n)_{n \in Z^+}$.

A Markov chain $(y_n)_{n\in Z^+}$ is *homogeneous* if the above probabilities of one-step transitions do not depend on the transition time n.

It is known that $(y_n)_{n\in Z^+}$ is a Markov chain if and only if the following *Markovian property* holds: the joint distributions of the states of this chain are determined only by the initial distribution and the probabilities of one-step transitions as follows

$$\mathcal{P}\{y_n\epsilon A_n, y_{n-1}\epsilon A_{n-1}, \ldots, y_1\epsilon A_1, y_0\epsilon A_0\}$$
$$= \int_{A_0} P(dy_0)\cdot\int_{A_1} P(y_0, dy_1)\cdot\int_{A_2} P(y_1, dy_2)$$
$$\ldots\int_{A_{n-1}} P(y_{n-2}, dy_{n-1})P(y_{n-1}, A_n).$$

According to the Markovian property, the *probability of the n-step transition* $P_n(y, A), n \geq 1$, satisfies the Chapman-Kolmogorov equation

$$P_{n+m}(y, A) = \int_Y P_n(y, dz)P_m(z, A),$$

which describes the Markovian property of the chain $(y_n)_{n\in Z^+}$ analytically: given a state of the chain at a fixed time, the probability law of the future evolution of the environment does not depend on the state of the system in the past.

The main focus of this book is on biological systems in Markov and semi-Markov environment. There are many examples of Markov chains in biological systems itself.

We list a few examples here, but leave the details to Chapters 4. *The first example* of Markov chains in mathematical biology is *Frameshift mutation: a two state Markov chain* (see Section 4.1.1.A) $(y_n)_{n\in Z^+}$, where y_n is equal to all types of a random sample from the nth generation: there are two possible states w (wild) and m (mutant) with probabilities $p_{w,n}$ and $p_{m,n}$, respectively, of the nth generation. If μ is equal to mutation probability for a given gene, and ν the reverse mutation probability, then the stochastic matrix is :

$$\mathbf{P} = \begin{pmatrix} 1-\mu & \mu \\ \nu & 1-\nu \end{pmatrix}.$$

The second example of Markov chains in mathematical biology is the *hypergeometric Markov chain* appearing in Plasmid incompatibility (see Section 4.1.1. B). Here,

$$\mathbf{P} = (p_{ij})_{ij=1}^N = (\frac{C_{2i}^j\, C_{2N-2i}^{n-j}}{C_{2N}^N})_{ij=1}^N,$$

where $C_n^k := \frac{n!}{(n-k)!k!}$ and p_{ij} gives the probability of one daughter having jP^1 and $(N-j)P^{11}-$ plasmids.

We will also consider the *Polya Markov chain*, which is a special case of the probability system known as *Polya's urn* (See Section 4.1.1.E) and the chain associated with the Fisher-Wright model (see Section 4.1.2.B).

Let us consider a simple example from *genetic selection.* Suppose a certain virus can exist in N different straints, and in each generation iteither stays the same or with probability α mutates to another strain (which is chosen at random). What is the probability that the strain in the nth generation is the same as that in the $0th$? We now model this process as an N- state chain, with $N \times N$ transition matrix P given by

$$p_{ii} = 1 - \alpha, \qquad p_{ij} = \alpha/(N-1)$$

for $i \neq j$. Then the question reduces to computing p_{11}^n. At any time a transition is made from the initial state to another with probability α, and a transition from another state to the initial state with probability $\alpha/(N-1)$. Thus we have a two-state Markov chain.

We now turn to *continuous-time Markov chains.* Again, let Y be a countable space. A *continuous-time* random process $(y(t))_{t \in R^+}$ with values in Y is a family of random variables $y(t) : \Omega \to Y$. We shall abuse the notation and using both $y(t)$ and y_t if no confusion arises. There are some problems connected with the process to estimate, for example, $\mathcal{P}(y(t) = i)$. These problems are not presented for discrete-time cases since for a countable disjoint union $\mathcal{P}(\bigcup_n A_n) = \sum_n \mathcal{P}(A_n)$, but there is no such analogue for an uncountable union $\bigcup_{t \geq 0} A_t$. To avoid these problems, we shall restrict our attention to a process $(y(t))_{t \in R^+}$ which is *right-continuous* in the following sense: for all $\omega \in \Omega$ and $t \geq 0$ there exists $\epsilon > 0$ such that $y(s) = y(t)$ for $t \leq s < t + \epsilon$.

Every path $R^+ \ni t \to y(t, \omega)$ of a right-continuous process must remain constant for a while, so there are three cases:

(i) the path makes infinitely many jumps, but only finitely many in any finite interval $[0, t]$;

(ii) the path makes finitely many jumps and then becomes stuck in some state forever;

(iii) the process makes infinitely many jumps in a finite interval.

We are interested, throughtout this book, in the first case. Such processes are called *regular processes.*

We need to introduce the notion of a Q-matrix in order to better describe continuous-time Markov chains. A *Q-matrix* on Y is a matrix $Q = (q_{ij}; i, j \in Y)$ satisfying the following conditions:

(i) $0 \leq -q_{ii} < \infty$ for all i;

(ii) $q_{ij} \geq 0$ for all $i \neq j$;

(iii) $\sum_{j \in Y} q_{ij} = 0$ for all i.

For example, the following matrix is Q-matrix:

$$Q = \begin{pmatrix} -2 & 1 & 1 \\ 1 & -1 & 0 \\ 2 & 1 & -3 \end{pmatrix}.$$

Intuitively, in a Q-matrix, the entry q_{ij} is the rate of changing from state i to state j, and $-q_{ii}$ is the rate of leaving state i.

Now we illustrate how to obtain a continuous-time Markov chain from a discrete-time Markov chain. We first assume the discrete parameter space $\{1, 2, ...\}$ is embedded in the continuous parameter space $[0, \infty)$. A natural way to interpolate the discrete sequence $\{p^n; n \in Z^+\}$ for $p \in (0, \infty)$ is by the function $(\exp\{tq\}; t \geq 0)$, where $q = \log p$. Consider a finite space Y and a matrix $P = (p_{ij}; i, j \in Y)$. Suppose that we can find a matrix Q with $\exp\{Q\} = P$. Then $\exp\{nQ\} = (\exp\{Q\})^n = P^n$. In this way, $(\exp\{tQ\}; t \geq 0)$ fills in the gaps in the discrete sequence. On the other hand, if Q is a given matrix on a finite space Y, and if $P(t) = \exp\{tQ\}$, then $(P(t))_{t\in R^+}$ has the following properties:

(I1)

$$P(s) \cdot P(t) = P(t + s), \forall t, s \geq 0 \qquad \text{(Chapman-Kolmogorov equation);}$$

(I2)

$$\lim_{t\to 0} P(t) = \mathbf{I},$$

the identity matrix.

Conversely, if the properties (I1)-(I2) hold, then $P(t) = [P(t/n)]^n$ for all positive integer n, and hence

$$P(t) = \exp\{Qt\},$$

for all $t \geq 0$, where $Q = P'(0)$ is the derivative of $P(t)$ at $t = 0$. It follows that if $P(t)$ is also a stochastic matrix for every $t \geq 0$, then

$$Q = (q_{ij})$$

satisfies:

$$\begin{cases} q_{ii} & := & \lim_{t\to 0}(p_{ii}(t) - 1)/t \leq 0 \qquad \text{for all } i \in Y, \\ q_{ij} & := & \lim_{t\to 0} p_{ij}(t)/t \geq 0 \qquad \text{for all } i \neq j, \\ \sum_j q_{ij} & = & 0 \qquad \text{for all } i \in Y. \end{cases}$$

Moreover, $P(t)$ satisfies the *backward* equation

$$\frac{dP(t)}{dt} = QP(t), P(0) = I,$$

and *forward* equation

$$\frac{dP(t)}{dt} = P(t)Q, P(0) = I.$$

Furthermore, for $k = 0, 1, 2, ...,$ we have

$$\frac{d^k P(t)}{dt}|_{t=0} = Q^k.$$

It is also easy to show that if Q is a Q-matrix, then for every $t \geq 0$, $P(t)$ is a stochastic matrix.

Therefore, $P(t) = \exp\{tQ\}$ is a stochastic matrix for every $t \geq 0$ if and only if Q is a Q-matrix.

We say $(y(t))_{t\in R^+}$ is a continuous-time Markov chain in a finite phase space Y if there exists a family of stochastic matrices $P(t) = (p_{ij}(t); i, j \in Y)$ satisfying $(I1)$ and $(I2)$ and such that

$$\mathcal{P}\{y(t_{n+1}) = i_{n+1}/y(t_0) = i_0, ..., y(t_n) = i_n) = p_{i_n i_{n+1}}(t_{n+1} - t_n)\}$$

for all $n \in Z^+$, all $0 \leq t_0 \leq ... \leq t_{n+1}$ and all $i_1, ..., i_{n+1} \in Y$.

In particular, $p_{ij}(t) = \mathcal{P}\{y(t) = j|y(0) = i\}$.

Since the process $(y(t))_{t\in R^+}$ is assumed to be right continuous, the system starts from some state y_1, and stays at this state for a duration of θ_1, then moves to a new state y_2 and stays there again for a duration θ_2, and then jumps to a new state y_3 where it stays for a duration θ_3 and this process continues. The resulted discrete-time process $(y_n)_{n\in Z^+}$ is a Markov chain with the transition probability

$$\mathcal{P}(y_{n+1} = j|\theta_1, ..., \theta_n; y_n = i) = q_{ij}/(-q_{ii}),$$

and $(\theta_n)_{n\in Z^+}$ is also a Markov chain with

$$\mathcal{P}(\theta_{n+1} > t|\theta_1, ..., \theta_n; y_{n+1} = j) = \exp(q_j t),$$

where $q_j := q_{jj}$. The transition probability of the two-component chain $((y_n, \theta_n))_{n\in Z_+}$ is given by, for $i \neq j$,

$$\mathcal{P}(y_{n+1} = j, \theta_{n+1} > t|y_n = i) = (q_{ij}/(-q_{ii}))\exp(q_i t).$$

Therefore, a continuous-time Markov chain $(y(t))_{t\in R^+}$ is characterized by an initial distribution p (the distribution of y_0) and a Q-matrix Q such that $P(t) = \exp(Qt)$ (the *generator matrix* of $(y(t))_{t\in R^+}$, which determines how the process involves from its initial state). More specifically, if the chain starts at i, then it stays there for an exponential time of parameter q_i and then jumps to a new state, choosing state j with probabaility $q_{ij}/(-q_{ii})$.

Consider a continuous Markov chain for which the system stays in state 0 for a random time with exponential distribution of parameter $\lambda \in (0, +\infty)$, then jumps to 1. Thus the distribution function of the waiting time T is given by $G_T(t) = 1 - \exp\{-\lambda t\}$, $t \in R^+$. This is the so-called *exponential distribution of parameter* λ. This distribution plays a fundamental role in continuous-time Markov chains because of the the *memoryless property:* $\mathcal{P}(T > t + s|T > s) = \mathcal{P}(T > t)$ for all $s, t \geq 0$. A random variable T has an exponential distribution if and only if it has this memoryless property, see [15, 65, 66, 72].

In the second example, when the system reaches the state 1, it does not stop there but rather, after another independent exponential time of parameter λ jumps to state 2, and so on. The resulting process is called the *Poisson process of rate* λ.

The associated matrix Q is given by:

$$\begin{pmatrix} -\lambda & \lambda & \dots & 0 \\ 0 & -\lambda & \lambda & \dots \\ 0 & 0 & -\lambda & \lambda \\ 0 & \dots & \ddots & \dots \end{pmatrix}.$$

Here, the probability $p_{ij}(t)$ has the following form:

$$p_{ij}(t) = \exp(-\lambda t)((\lambda t)^{j-i}/(j-i)!).$$

Note that if $i = 0$, we obtain the *Poisson probabilities* of parameter λt.

A *birth process* is a generalization of the Poisson process in which λ is allowed to depend on the current state of the process. Thus a birth process is characterized by a sequence of *birth rates* $0 \le q_j < \infty$ for $j = 0, 1, 2, \dots$.

The corresponding matrix Q is given by

$$\begin{pmatrix} -q_0 & q_0 & \dots & 0 \\ 0 & -q_1 & q_1 & \dots \\ 0 & 0 & -q_2 & q_2 \\ 0 & \dots & \ddots & \dots \end{pmatrix}.$$

Finally, a *birth and death process*

by $q_{ii} = -(\lambda_i + \mu_i)$, $q_{ij} = \lambda_j$ if $j = i+1$, $q_{ij} = \mu_j$ if $j = i-1$, and $q_{ij} = 0$ for $j \neq j-1, j+1$,where λ_i are birth rates and μ_i are death rates.

1.2. Ergodicity and Reducibility of Markov Chains

The ergodic properties of homogeneous Markov chains are quite important for the averaging, merging and other limit theorems, and for the stability of populations of biological systems in random environment to be studied in later chapters.

We shall consider a homogeneous Markov chain $(y_n)_{n \epsilon Z^+}$ defined in a measurable phase space of states $(Y, \mathcal{Y})$, and we assume a stochastic kernel $P(y, A), y \epsilon Y, A \epsilon \mathcal{Y}$, is defined by the chain, as explained in Section 1.1.

Let us start with the simple case where $Y = \{1, 2, \dots, \dots\}$, and let $P =$ with $(p_{ij}; i, j = 1, 2, \dots)(p_{ij} \ge 0, \sum_{j=1}^{\infty} p_{ij} = 1,)$ be the transition probability matrix of the chain.

For our example from *genetic selection* (see Section 1.1), using our previous analysis with $\beta = \alpha/(N-1)$, we find that

$$p_{11}^n = 1/N + (1 - 1/N)(1 - \alpha N/(N-1))^n.$$

We write $j \to k$ to denote the fact that the state k *can be reached* from the state j (or j *leads* to k). More precisely, $j \to k$ if there is an integer $n \geq 1$ such that $p_{jk}^{(n)} > 0$, where $p_{jk}^{(n)}$ denotes the jkth element of matrix P^n (the conditional probability of obtaining the kth state from jth state at the nth step). We say j *communicates with* k if $j \to k$. We write $k \leftrightarrow j$, if $k \to j$ and $j \to k$. It is easy to see, that $\leftrightarrow$ is an equivalence relation on Y, and thus particions Y into *communicating classes.*

We call a class of states C *closed* if $p_{jk} = 0$ whenever $j \in C$ and $k \notin C$, namely, no one-step transition is possible from any state belonging to C to any state outside C. Therefore, $p_{jk}^{(n)} = 0$ for all n, if $j \in C$ and $k \notin C$, and so once an environment reaches a state belonging to C it can never subsequently be in any state outside C.

In the special case where C consists of the single state j (that is, $p_{jj} = 1$), j is called an *absorbing state.* A Markov chain in which there are two or more closed sets is said to be *reducible.* The chain is called *irreducible* if there exists no closed set other than the set of all states.

Let $f_j^{(n)}$ denote the probability that the environment starts from the initial state j and returns to state j for the first time after n transitions. The sum $f_j := \sum_{n=1}^{\infty} f_j^{(n)}$ is the probability that the environment ever returns to state j. The state j is *recurrent* (or *persistent*) if $f_j = \sum_{n=1}^{+\infty} f_j^{(n)} = 1$, and *transient* if $f_j = \sum_{n=1}^{+\infty} f_j^{(n)} < 1$. Therefore, if j is transient, then there is a positive probability then the environment will never return to j.

Let $\tau_j := \sum_{n=1}^{\infty} n f_j^{(n)}$ be the *mean recurrence time* for j. If there exists no n such that $f_j^{(n)} = 0$, we let $\tau_j = +\infty$. When τ_j is finite, j is said to be *non-null*; when τ_j is infinite, j is said to be *null.*

To illustrate the above notions, consider the following Markov chain: the states are particioned into three *communicating classes,* (0), $(1, 2, 3)$ and $(4, 5, 6)$. Two of these classes are *closed,* meaning that one cannot escape from either $(1, 2, 3)$ or $(4, 5, 6)$. These closed classes are *recurrent.* The class (0) is *transient,* and the state (0) is null.

We now discuss the *periodicity.* We say that a recurrent state j has period $\mathbf{T}$ if a return to j is impossible except possibly after $\mathbf{T}$, $2\mathbf{T}$, $3\,\mathbf{T}\dots$ transitions and $\mathbf{T}$ is the largest such integer satisfying this property. When $\mathbf{T}$=1, j is said to be *aperiodic.*

A recurrent non-null state which is aperiodic is said to be *ergodic.*

The following results hold for a discrete-time Markov chain:

(P1) j is a transient state if $\sum_{n=1}^{\infty} p_{jj}^{(n)}$ is convergent;

(P2) j is a recurrent state if $\sum_{n=1}^{\infty} p_{jj}^{(n)}$ is divergent;

(P3) $p_{jj}^{(n)} \to \frac{1}{\tau_j}$, if j is ergodic;

(P4) $p_{jj}^{(n)} \to 0$ if j is null;

(P5) $p_{jj}^{(n\mathbf{T})} \to \frac{\mathbf{T}}{\tau_j}$ if j is recurrent, non-null, and with period $\mathbf{T}$;

(P6) In an irreducible Markov chain, all states belong to the same class and they are all transient, they are either all recurrent null or all recurrent non-null, and have the same periodicity.

For a recurrent Markov chain with a countably generated σ-algebra $\mathcal{Y}$ on the phase space Y, there exists a unique (up to a constant factor) distribution $(p_i; i = 1, 2, \ldots)$ such that

$$p_i = \sum_{j=1}^{\infty} p_{ij} p_j. \tag{1.1}$$

Distribution $(p_j)_{j\geq 1}$ defines the *stationary distributions*.

Many long-time properties of a Markov chain are connected with the notion of an *invariant distribution* or *invariant measure.* We say a distribution p is *invariant* if $p\mathbf{P} = p$. The terms *equilibrium* and *stationary* are also used for invariant measures.

The most general two-state chain has the transition matrix

$$\mathbf{P} = \begin{pmatrix} 1-\alpha & \alpha \\ \beta & 1-\beta \end{pmatrix},$$

with $0 < \alpha, \beta \leq 1$.

From the relation $\mathbf{P}^{n+1} = \mathbf{P}^n\mathbf{P}$, we get

$$p_{11}^{(n+1)} = p_{12}^{(n)}\beta + p_{11}^{(n)}(1-\alpha).$$

As $p_{11}^{(n)} + p_{12}^{(n)} = 1$, we can eliminate $p_{12}^{(n)}$ and get a recurrence relation for $p_{11}^{(n)}$ as follows

$$p_{11}^{(n)} = (1-\alpha-\beta)p_{11}^{(n)} + \beta, \qquad p_{11}^{(0)} = 1.$$

This has a unique solution given by

$$p_{11}^{(n)} = \beta/(\alpha+\beta) + (\alpha/(\alpha+\beta))(1-\alpha-\beta)^{(n)}, \qquad \text{if } \alpha+\beta > 0,$$

or

$$p_{11}^{(n)} = 1 \qquad \text{if } \alpha+\beta = 0.$$

In particular,

$$\mathbf{P}^n \to \begin{pmatrix} \beta/(\alpha+\beta) & \alpha/(\alpha+\beta) \\ \beta/(\alpha+\beta) & \alpha/(\alpha+\beta) \end{pmatrix} \qquad \text{as } n \to +\infty,$$

so the distribution $p = (\beta/(\alpha+\beta), \alpha/(\alpha+\beta))$ is invariant, namely, $p\mathbf{P} = p$.

In a general case with a measurable phase space of states $(Y, \mathcal{Y})$, for a recurrent Markov chain $(y_n)_{n\in Z^+}$ in Y with a countably generated σ-algebra $\mathcal{Y}$, there exists a *unique invariant measure* $p(B)$ such that

$$p(B) = \int_Y p(dy)P(y, B), \forall B\epsilon\mathcal{Y}. \tag{1.2}$$

If the invariant measure is finite, then we can assume that it is normalized, i.e., $p(Y) = 1$. It follows from (1.2) that $\mathcal{P}\{y_n\epsilon B\} = p(B)$ for all $n \geq 1$ and $\mathcal{P}\{y_0\epsilon B\} =$

$p(B)$. Hence, the measure $p(B)$ defines the *stationary distribution* of a Markov chain in $(Y, \mathcal{Y})$. Using (1.2), we can get an equivalent definition of an *ergodic Markov chain*: it is an aperiodic Markov chain with a stationary distribution p, which is defined by relation (1.2) and satisfies the condition $p(Y) = 1$.

For an ergodic Markov chain, we have

$$\lim_{n\to+\infty} P^n(y, A) := \lim_{n\to+\infty} \mathcal{P}\{y_n \in A | y_0 = y\} = p(A), \forall A \in \mathcal{Y}, \forall y \in Y. \qquad (1.3)$$

The class structure of a continuous-time Markov chain $(y_t)_{t\in R^+}$ with the associated Q-matrix Q and $P(t) := \exp(tQ)$ is the same as the case of the above described discrete-time Markov chains. We say that i *leads* to j and write $i \to j$ if $\mathcal{P}_i(y(t) = j) := \mathcal{P}\{y(t) = j | y(0) = i\} > 0$ for some $t \geq 0$. The notions of *communication*, *communicating class*, *clossed class*, *absorbing state*, and *irreducibility* are analogous to those for discrete-time Markov chains. We say a state i is *recurrent* if $\mathcal{P}_i([t \geq 0 : y(t) = i]$ is unbounded$) = 1$. A state i is recurrent if $q_i = 0$ or $\mathcal{P}(T_i < \infty) = 1$, where $T_i(\omega) \equiv T_i := \inf(t \geq \tau_1(\omega) : y(t) = i)$ is the *first passage time* of $y(t)$ to state i, and $\tau_1 := \inf(t \geq 0 : y(t) \neq y(0))$. Therefore, a state i is *transient* if $\mathcal{P}_i([t \geq 0 : y(t) = i]$ is unbounded$) = 0$.

The following dichotomy holds: a) if $q_i = 0$, then i is recurrent and $\int_0^\infty p_{ii}(t)dt = \infty$; b) if $q_i > 0$, then i is transient and $\int_0^\infty p_{ii}(t)dt < \infty$.

The notions of invariant distributions and measures play an important role in the study of continuous-time Markov chains $(y_t)_{t\in R^+}$. We say that p is *invariant* if $pQ = 0$.

It is known that if Q is irreducible and recurrent, then Q has an invariant measure p which is unique up to scalar multiplies. Moreover, for all states i, j we have $p_{ij}(t) \to p_j$ as $t \to \infty$. Finally, if Q is irreducible and recurrent, and if p is a measure, then for each fixed $s > 0$ $pQ = 0$ if and only if $pP(s) = p$.

Ergodic Theorem for continuous-time Markov chains asserts that

$$\mathcal{P}((1/t)\int_0^t \mathbf{1}_{(y(s)=i)}ds \to 1/(m_i q_i)) = 1 \qquad \text{as } t \to +\infty,$$

where $m_i := E_i(T_i)$ is the expected return time to state i, and $\mathbf{1}_A$ is the characteristic function of a set A.. Moreover, if $q_i = 0$, then for any bounded function $f : Y \to R$ we have

$$\mathcal{P}((1/t)\int_0^t f(y(s))ds \to \hat{f}) = 1 \qquad \text{as } t \to \infty,$$

where $\hat{f} := \sum_{i\in Y} p_i f_i$ and where $(p_i : i \in Y)$ is the unique invariant distribution.

Let $\mathbf{B}(Y)$ be *a normed space* of Y-measureable bounded functions $f : Y \to R$ with norm $||f|| := \sup_{y\in Y} |f(y)|$. An operator of transition probabilities $\mathbf{P}$ in the space $\mathbf{B}(Y)$ is defined by the stochastic kernel $P(y, A)$ as

$$\mathbf{P}f(y) := \int_Y P(y, dz) f(z). \qquad (1.4)$$

A stationary projector Π in $\mathcal{B}(Y)$ is defined by the stationary distribution $p(A)$ of the ergodic irreducible Markov chain as

$$\Pi f(y) := \int_Y p(dz) f(z) =: \hat{f} \cdot \mathbf{1}(y). \qquad (1.5)$$

Here,

$$\mathbf{1}(y) \equiv 1 \quad \text{for all } y \epsilon Y, \quad \text{and } \hat{f} := \int_Y p(dz) f(z). \tag{1.6}$$

The operator Π defined by (1.5) possesses the projection property, i.e., $\Pi^2 = \Pi$.

An ergodic Markov chain with the operator of transition probabilities $\mathbf{P}$ and the stationary projector Π is called *uniformly ergodic* if

$$\lim_{n \to +\infty} \sup_{||f|| \leq 1} ||(\mathbf{P}^n - \Pi) f|| = 0 \qquad \forall f \in \mathbf{B}(Y). \tag{1.7}$$

For a uniformly ergodic Markov chain, the operator $Q := \mathbf{P} - \mathbf{I}$ is invertibly reducible, namely,

$$\mathbf{B}(Y) = N(Q) \oplus R(Q), \dim N(Q) = 1, \tag{1.8}$$

where $N(Q)$ is the null-space of Q, which consists of all constant functions, and $R(Q)$ is the range of the operator Q, which is closed, dim $N(Q)$ is the dimension of the space $N(Q)$. The stationary projector (1.6) is a projector onto the null-space $N(Q)$.

For a uniformly ergodic Markov chain the following operator, called *the potential*

$$\mathbf{R}_0 = (\mathbf{P} - \mathbf{I} + \Pi)^{-1} - \Pi = \sum_{n=1}^{+\infty} (\mathbf{P}^n - \Pi) \tag{1.9}$$

is well-defined. The boundness of the linear operator $\mathbf{R}_0$ follows from the uniform convergence of the series

$$\sum_{n=1}^{+\infty} ||(\mathbf{P}^n - \Pi) f|| < +\infty \qquad \forall f \epsilon \mathbf{B}(Y)$$

with $||f|| \leq 1$.

We note, that if $(y_n)_{n \in Z^+}$ is a stationary ergodic Markov chain with ergodic distribution $p(dy)$, then for every $f \in \mathbf{B}(Y)$ we have (see, [15] and [23]) that

$$\mathcal{P}\{\lim_{n \to +\infty} \frac{1}{n} \sum_{k=1}^{n} f(y_k) = \int_Y f(y) p(dy)\} = 1. \tag{1.10}$$

For *a reducible Markov chain* $(y_n)_{n \in Z^+}$ in a phase space $(Y, \mathcal{Y})$, we can define a stochastic kernel $P(y, A)$ associated with the *decomposition* of the phase space

$$Y = \bigcup_{v \epsilon V} Y_v \tag{1.11}$$

into disjoint classes Y_v of closed sets of states in the following manner:

$$P(y, Y_v) = \mathbf{1}_v(y) = \begin{cases} 1, & y \in Y_v, \\ 0, & y \notin Y_v \end{cases} \tag{1.12}$$

If $V = \{1, 2, ..., N\}$, for example, then such decomposition defines a merging function $v(y)$ by the relation

$$v(y) = v,$$

if $y \in Y_v$, $\quad v = 1, ..., N$.

It means that for every class Y_v with $y \in Y_v$ corresponds one merged state $v \in V$.

We conclude this section with some remarks about reducible ergodic Markov chains whose closed sets of states $Y_v, v \epsilon V$, are ergodic, so that the stationary distributions $p_v(A), A \epsilon \mathcal{Y}_v (\mathcal{Y}_v$ is a σ-algebra in $Y_v)$

$$p_v(A) = \int_Y p_v(dz) P(z, A), A \epsilon \mathcal{Y}_v, v \epsilon V, \tag{1.13}$$

exist with the normalization property $p_v(Y_v) = 1$. Essentially, a reducible ergodic Markov chain with the decomposition of the phase space (1.11) consists of different irreducible ergodic Markov chains defined by the stochastic kernels

$$P_v(y, B) = P(y, B), \qquad y \in Y_v, \qquad B \in \mathcal{Y}_v. \tag{1.14}$$

For a *reducible Markov chain*, defined on the phase space $(Y, \mathcal{Y})$ with the decomposition (1.11) the definition of uniform ergodicity remains valid, where the stationary projector Π is defined by the relation

$$\Pi f(y) := (\hat{f}_v)_{v \in V},$$

$$\hat{f}_v := \int_{Y_v} p_v(dy) f(y), \qquad \forall y \in Y_v, \qquad \forall v \in V \tag{1.15}$$

Another definition of a Markov chain can be formulated as follows. Let $(\mathcal{F}_u), \mathcal{F}_0 \subseteq \mathcal{F}_1 \subseteq \ldots \subseteq \mathcal{F}$, be a non-decreasing family of σ-algebras and $(\Omega, \mathcal{F}, \mathcal{P})$ is a probability space. A stochastic sequence $(y_n)_{n \in Z^+}$ is called a *Markov chain with respect to the measure* $\mathcal{P}$, if for any $n \geq m \geq 0$ and any $A \epsilon \mathcal{Y}$

$$\mathcal{P}\{y_n \epsilon A | \mathcal{F}_m\} = \mathcal{P}\{y_n \epsilon A | y_m\}. \tag{1.16}$$

In the particular case, where $\mathcal{F}_n = \mathcal{F}_n^Y := \sigma\{\omega : y_0, \ldots y_n\}$ and the stochastic sequence $(y_n)_{n \in Z^+}$ satisfies (1.16), the sequence $(y_n)_{n \in Z^+}$ is the Markov chain defined in the previous sections..

1.3. Markov Renewal Processes

Let us start with an ordinary *renewal process.* This is a sequence of independent identically distributed non-negative random variables $(\theta_n)_{n \in Z^+}$ with a common *distribution function* $F(x)$, where $F(x) := \mathcal{P}\{\omega : \theta_n(\omega) \leq x\}$. The renewal process counts events, and the random variables θ_n can be interpreted as *lifetimes (operating periods, holding times, renewal periods)* of a certain system in a random environment. In particular, θ_n can be regarded as the duration between the $(n-1)th$ and the nth event. From this renewal process we can construct another renewal process $(\tau_n)_{n \in Z^+}$ by $\tau_n := \sum_{k=0}^{n} \theta_k$. The random variables τ_n are called *renewal times* (or *jump times*).

Let $\nu(t) := \sup\{n : \tau_n \leq t\} := \sum_{n=0}^{\infty} \mathbf{1}_{[0,t]}(\tau_n)$. This is called the *counting process*. In the particular case where $F(t) = 1-\exp\{-\lambda t\}, t \geq 0$, the counting process $\nu(t)$ is the homogeneous Poisson process, namly, $E[\nu(t)] = \lambda t$. From the above definitions it follows that $\tau_n \to +\infty$ a.s. as $n \to +\infty$ and $\nu(t) \to +\infty$ a.s. $t \to +\infty$. Also, since $\theta_{\nu(t)} < \infty$ (a.s.), we have that $\theta_{\nu(t)}/\nu(t) \to 0$ a.s. as $t \to +\infty$.

The so-called *renewal function* $m(t) := E[\nu(t)]$ has a number of important properties. First of all, the *Elementary Renewal Theorem* asserts that

$$m(t)/t \to 1/\mu \qquad \text{as } t \to +\infty,$$

where $\mu := E(\theta_1)$ and we regard $1/\infty$ as 0.

Let $\mu = E(\theta_1) < \infty$. Then the *Strong Law of Large Numbers* asserts that

$$\nu(t)/t \to 1/\mu, a.s. \qquad \text{as } t \to +\infty$$

Furthermore, if $0 < \sigma^2 := Var(\theta_1) < \infty$, then the *Central Limit Theorem* asserts that

$$\frac{\nu(t) - t/\mu}{\sqrt{t\sigma^2/\mu^3}} \to N(0,1) \qquad \text{as } t \to +\infty,$$

where $N(0,1)$ is a *normally distributed random variable* with mean value 0 and variance 1, namely, the *density function* $f(x) := F'(x)$ of this variable is equal to $(1/\sqrt{2\pi})\exp\{-x^2/2\}$.

To introduce Markov renewal processes, we need to introduce the notion of a semi-Markov kernel. A function $Q(y, A, t), y\epsilon Y, A\epsilon \mathcal{Y}, t \geq 0$, is called *a semi-Markov kernal* in the measurable space $(Y, \mathcal{Y})$ if it satisfies the following conditions:

(i) $Q(y, A, t)$ is measureable with respect to (y, t);
(ii) for fixed $t > 0, Q(y, A, t)$ is a semi-stochastic kernel in (y, A) and $Q(y, A, t) \leq 1$;
(iii) for fixed $(y, A), Q(y, A, t)$ is a nondecreasing function continuous from the right with respect to $t \geq 0$ and $Q(y, A, 0) = 0$;
(iv) $Q(y, A, +\infty) =: P(y, A)$ is a stochastic kernel;
(v)) for fixed $y \in Y$ and $A \in \mathcal{Y}$, the function $Q(y, Y, t) =: G_y(t)$ is a distribution function with respect to $t \geq 0$.

In a discrete phase space $Y = \{1, 2, 3, \ldots\}$, a semi-Markov kernel is defined by a *semi-Markov matrix* $Q(t) := [Q_{ij}(t); i, j\epsilon Y]$, where $Q_{ij}(t)$ are nondecreasing functions in $t, \sum_{j\epsilon Y} Q_{ij}(t) = G_j(t)$ are distribution functions of t and $[Q_{ij}(+\infty) = p_{ij}; i, j\epsilon Y] =: P$ is a *stochastic matrix.*

A homogeneous two-dimensional Markov chain $(y_n; \theta_n)_{n\in Z^+}$ with values on $Y \times R_+$ is called *a Markov renewal process* (MRP) if its transition probabilities are given by the semi-Markov kernel

$$Q(y, A, t) = \mathcal{P}\{y_{n+1}\epsilon A, \theta_{n+1} \leq t | y_n = y\}, \forall y\epsilon Y, A\epsilon \mathcal{Y}, t\epsilon R^+. \tag{1.17}$$

It follows that the transition probabilities are independent of the second component. This distinquishes a Markov renewal processes from an arbitrary Markov chain with a non-negative second component. The first component $(y_n)_{n\in Z^+}$ of the Markov renewal process forms a Markov chain which is called an *imbedded Markov chain (IMC)*.

The transition probabilities of the IMC can be obtained by setting $t = +\infty$ in (1.17), resulting

$$P(y, A) := Q(y, A, +\infty) = \mathcal{P}\{y_{n+1} \in A | y_n = y\}. \tag{1.18}$$

In a MRP, the non-negative random variables $(\theta_n; n \geq 1)$ define the intervals between the Markov renewal times $\tau_0 := 0$ and $\tau_n := \sum_{k=1}^{n} \theta_k$, for $n \geq 1$. The distribution functions of the renewal times depend on the states of the imbedded Markov chain

$$G_y(t) := \mathcal{P}\{\theta_{n+1} < t | y_n = y\} = Q(y, Y, t). \tag{1.19}$$

For a right-continuous process $(y(t))_{t \in R^+}$, the jump times τ_n and the holding times θ_n are obtained by

$$\tau_0 = 0, \tau_{n+1} := \inf\{t \geq \tau_n : y(t) \neq y(\tau_n)\},$$

for $n \geq 0$ and $\theta_n := \tau_n - \tau_{n-1}$ if $\tau_n < \infty$, or $\theta := 0$ if otherwise.

The discrete-time process $(y_n)_{n \in Z^+}$ given by $y_n := y(\tau_n)$ is called the *imbedded Markov chain* or *jump chain.*

We can now state the *ergodic theorem* for a Markov renewal process (see, [49,54,80] for more details). Let $\alpha(y)$ be a measurable and bounded function of $y \in Y$. Then

$$t^{-1} \sum_{k=1}^{\nu(t)} \alpha(y_k) \to \hat{\alpha} \qquad \text{as } t \to +\infty,$$

where

$$\hat{\alpha} := \int_Y p(dy) \alpha(y)/m$$

and $p(y)$ is a stationary distribution of the Markov chain $(y_n)_{n \in Z^+}$.

1.4. Semi-Markov Processes

The MRP $(y_n, \theta_n)_{n \in Z^+}$ considered in Section 1.3 may serve as a convenient constructive tool to define a *semi-Markov process.*

A semi-Markov process(SMP) $(y(t))_{t \in R^+}$ is defined by the following relations:

$$y(t) := y_{\nu(t)}, \tag{1.20}$$

where

$$\nu(t) := \sup \{n : \tau_n \leq t\}. \tag{1.21}$$

The process $\nu(t)$ in (1.21) is called a *counting process.* It determines the number of renewal times on the segment $[0, t]$. Since the counting process $\nu(t)$ assumes constant values on the intervals $[\tau_n, \tau_{n+1})$ and is continuous from the right, that is,

$$\nu(t) = n, \tau_n \leq t < \tau_{n+1},$$

we conclude that the SMP $(y(t))_{t\in R^+}$ also assumes constant values on the same intervals and is continuous from the right. Namely,

$$y(t) = y_n, \tau_n \le t < \tau_{n+1}. \tag{1.22}$$

Moreover,

$$y(\tau_n) = y_n, n \ge 0. \tag{1.23}$$

Relation (1.23) illustrates the concept of the *imbedded Markov chain* $(y_n)_{n\in Z^+}$.

For the SMP $(y(t))_{t\in R^+}$ in (1.20), the *renewal times (periods)* $\theta_n := \tau_{n+1} - \tau_n$ can be naturally interpreted as the *occupation times (life-times)* in the states $(y_n)_{n\in Z^+}$.

In what follows, we consider only a *regular SMP.*This is a SMP that has finite numbers of renewals on a finite period of time with probability 1. That is,

$$\mathcal{P}\{\nu(t) < +\infty\} = 1 \text{ for all } t > 0. \tag{1.24}$$

The renewal times $(\tau_n)_{n\in Z^+}$ and the IMC $(y_n)_{n\in Z^+}$ form a two-dimensional Markov chain $(y_n, \tau_n)_{n\in Z^+}$ homogeneous with respect to the second component (see [49, 80]), with transition probabilities $P(y, A), y \in Y, A \in \mathcal{Y}$. The two-dimensional Markov chain $(y_n; \tau_n)_{n\in Z^+}$ is also called a *MRP generating the semi-Markov process* $(y(t))_{t\in R_+}$ in (1.20). The process $\gamma(t) := t - \tau_{\nu(t)}$ is sometimes called an *age process* and the two-component process $(y_n, \theta_n)_{n\in Z_+}$ is also called a generating *semi-Markov chain,* although this is actually a Markov chain and the joint distribution of (y_{n+1}, θ_n) depends only on y_n.

Just as in the right-continuous Markov process, the epochs of jumps are regeneration points erasing the influence of the past. The only difference is that sojourn time at a point $y \in Y$ has an arbitrary distribution $G_y(t), y \in Y, t \in R^+$, which depends on the terminal state $y \in Y$.

Let $\alpha(y)$ be a measurable and bounded function on Y. The *ergodic theorem* for a semi-Markov process $(y(t))_{t\in R^+}$ (see, for example [14,15,18]) then states

$$t^{-1}\int_0^t \alpha(y(s))ds \to \tilde{\alpha} \qquad t \to +\infty,$$

where

$$\tilde{\alpha} := \int_Y p(dy)m(y)\alpha(y)/m,$$

$$m := \int_Y p(dy)m(y),$$

$$m(y) := \int_0^{+\infty} tG_y(dt),$$

and $p(dy)$ is the stationary distribution of the Markov chain $(y_n)_{n\ge 1}$, $G_y(t)$ is defined in (1.19).

A Markov renewal process $(y_n; \tau_n)_{n\in Z^+}$, its counting process $\nu(t) := \max\{n : \tau_n \le t\}$, and the associated semi-Markov process $y(t) = y_{\nu(t)}$ together generate several so-called auxiliary processes described below which play an important role in the theory of stochastic evolutionary systems.

The aforementioned auxiliary processes are defined as follows: *A point process* $\tau(t)$ is defined by:

$$\tau(t) := \tau_{\nu(t)}, t \geq 0;$$

an occupation time process $\theta(t)$ is defined by:

$$\theta(t) := \theta_{\nu(t)}, t \geq 0;$$

a running occupation time (defect process, age process) $\gamma(t)$ is given by:

$$\gamma(t) := t - \tau(t), t \geq 0;$$

and *a residual occupation time (excess process)* is given by

$$\gamma^+(t) := \tau_{\nu(t)+1} - t = \tau(t) + \theta(t+1) - t.$$

To describe some properties of these processes, let us first fix a Markov chain $(y_n)_{n \in Z^+}$ and write $\mathcal{F}_n^Y$ for the collection of all sets depending only on $y_0, y_1, y_2, ..., y_n$. The sequence $\mathcal{F}_n^Y$ is called the *filtration* of $(y_n)_{n \in Z^+}$ and we think of $\mathcal{F}_n^Y$ as representing the state of knowledge, or history, of the chain up to time n. We denote this by $\mathcal{F}_n^Y := \sigma\{y_k; 0 \leq k \leq n\}$ and call it the *σ-algebra generated by* chains $y_0, y_1, y_2, ..., y_n$. Let us also fix a continuous-time process $(y(t))_{t \in R^+}$ and write $\mathcal{F}_t^Y$ for the collection of all sets depending only on $y(s)$ for all $0 \leq s \leq t$. The sequence $\mathcal{F}_t^Y$ is called the filtration of $y(t)$ and we can think of $\mathcal{F}_t^Y$ as representing the state of knowledge, or history, of the process up to time t. We denote this by $\mathcal{F}_t^Y := \sigma\{y_s; 0 \leq s \leq t\}$ and call it the *σ-algebra generated by* the process $(y_s; 0 \leq s \leq t)$.

We can then define the following flows of σ-algebras:

$$\mathcal{F}_n^Y := \sigma\{y_0, y_1, \ldots, y_n\} := \sigma\{y_k; 0 \leq k \leq n\},$$

$$\mathcal{H}_t := \sigma\{\nu(t); \tau_n, \ldots, \tau_{\nu(t)}, y_0, y_1, \ldots, y_{\nu(t)}\};$$

$$\mathcal{F}_t^\theta := \sigma\{y(s); \theta(s); 0 \leq s \leq t\};$$

$$\mathcal{F}_t^\gamma := \sigma\{y(s); \gamma(s); 0 \leq s \leq t\};$$

$$\mathcal{F}_t^{\gamma+} := \sigma\{y(s); \gamma^+(s); 0 \leq s \leq t\}.$$

The most important property of the above auxiliary processes is that they complement a semi-Markov process $(y(t))_{t \in R^+}$ to a Markov process with respect to the corresponding flow. That is, each two-component process, $(y(t), \theta(t))$, $(y(t), \gamma(t))$ or $(y(t), \gamma^+(t))$ is a Markov process on the phase space $Y \times R^+$ with respect to the σ-algebra $\mathcal{F}_t^\theta, \mathcal{F}_t^\gamma$ or $\mathcal{F}_t^{\gamma+}$, respectively. See, for example, [49,80].

1.5. Jump Markov Processes

A regular homogeneous jump Markov process $(y(t))_{t\in R^+}$ is defined in terms of the MRP $(y_n;\theta_n)_{n\in Z^+}$ with the semi-Markov kernel

$$Q(y, A, t) = P(y, A)(1 - e^{-\lambda(y)t}) \tag{1.25}$$

as follows

$$y(t) = y_{\nu(t)}, \nu(t) := \max\{n : \tau_n \le t\}. \tag{1.26}$$

Thus, a regular jump Markov process is a SMP with exponentially distributed occupation times with the parameters $\lambda(y) \ge 0, y\epsilon Y$, depending on the states of the IMC $(y_n)_{n\in Z^+}$.

To define a jump Markov process constructively, it we need to have two functions: a stochastic kernel $P(y, A), y\epsilon Y, A\epsilon\mathcal{Y}$, determining the probabilities of jumps of the process (transition probabilities of the imbedded Markov chain), and a non-negative function $\lambda(y) \ge 0, y\epsilon Y$, which fixes the parameters of the exponential distributions of occupation times $\theta_n, n \ge 0$.

In the discrete phase space of states $Y = \{1, 2, 3, \ldots\}$, a jump Markov process is defined by a semi-Markov matrix

$$Q(t) = [Q_{ij}(t); ij\epsilon Y],$$

where

$$Q_{ij}(t) = p_{ij}(1 - e^{-\lambda(i)t}), i, j\epsilon Y, \lambda_i \ge 0. \tag{1.27}$$

We now define *A general Markov process*, which we will use for martingale characterization (see Section 1.9.1) and definition of semigroupsof operators (see Section 1.8). Let $(y(t))_{t\in R^+}$ be a stochastic process defined on a probability space $(\Omega, \mathcal{F}, \mathcal{F}_t, P)$, where $\mathcal{F}_t$ is a nondecreasing family of σ-algebras: $\mathcal{F}_s \subseteq \mathcal{F}_t \subseteq \mathcal{F}$ for all $0 \le s \le t < +\infty$. The *Markov property* of the process $(y(t))_{t\in R_+}$, with respect to $\mathcal{F}_t$ is defined by

$$E[f(y(t))/\mathcal{F}_s] = E[f(y(t))/y(s)], s \le t, \tag{1.28}$$

for any $f\epsilon\mathbf{B}(Y)$. A Markov process is a stochastic process $(y(t))_{t\in R^+}$ satisfying the Markov property. The Markov process $(y(t))_{t\in R^+}$ is uniquely determined by the *transition probabilities*

$$P(s, y; t, A) := \mathcal{P}\{y(t)\epsilon A | y(s) = y\}, s \le t, \tag{1.29}$$

and the initial distribution $p(A) := \mathcal{P}\{y(0)\epsilon A\}$.

The Markov property (1.28) yields the *Kolmogorov-Chapman equation* for the transition probability

$$P(s, y; t, A) = \int_Y P(s, y; u, dz)P(u, z, t, A), s \le u \le t, y\epsilon Y, A\epsilon\mathcal{Y}. \tag{1.30}$$

In the particular case of *time-homogeneous* Markov processes, the transition probabilities are determined by the family of *stochastic kernels*

$$P_t(y, A) := \mathcal{P}\{y(s + t)\epsilon A/y(s) = y\}, s \le t, y\epsilon Y, A\epsilon\mathcal{Y}. \tag{1.31}$$

The corresponding *Kolmogorov-Chapman equation* becomes

$$P_{t+s}(y, A) = \int_Y P_t(y, dz) P_s(z, A). \tag{1.32}$$

1.6. Wiener Processes and Diffusion Processes

A simple example of a *Brownian motion* is a symmetric random walk, choosing the directions with the same probability, in an Euclidean space which takes infinitesimal jumps with infinite frequency. It is named after a botanist who observed such a motion when looking at pollen grains under a microscope. The mathematical object, now called Brownian motion, was actually discovered by N. Wiener, and is thus called the *Wiener process.*

To give the definition of a Wiener process, we start with the definition of *Gaussian distribution with mean* 0 *and variance* t for a real-valued random variable. That is, the random variableb has density function

$$\phi_t(x) = (2\pi t)^{-1/2} \exp(-x^2/2t).$$

A real-valued process $(y(t))_{t \in R_+}$ is said to be *continuous* if

$$\mathcal{P}(\omega : t \to y(t, \omega) \quad \text{is continuous}) = 1.$$

A continuous real-valued process $(w(t))_{t \in R_+}$ is called a *Wiener process* if $w_0 = 0$ and if for all $0 = t_0 < t_1 < ... < t_n$ the increments

$$w_{t_1} - w_{t_0}, ..., w_{t_n} - w_{t_{n-1}}$$

are independent Gaussian random variables of mean 0 and variance $t_1 - t_0, ..., t_n - t_{n-1}$. We note that a Wiener process is a Markov process [18].

One important example of Markov process is the *diffusion process.*

By a *diffusion process* we mean a continuous Markov process $(y(t)_{t \in R^+}$ in $R(Y = R)$ with transition probabilities $P(s, y; t, A), s \leq 5, y \epsilon R, A \epsilon \mathcal{R}$, satisfying the following conditions:

(i)

$$\lim_{\Delta t \to 0} \frac{1}{\Delta t} \int_{|z-y|>\varepsilon} P(s, y, s + \Delta t, dz) = 0,$$

(ii)

$$\lim_{\Delta t \to 0} \frac{1}{\Delta t} \int_{|z-y|\leq\varepsilon} (z - y) P(s, y, s + \Delta t, dz) = a(s, y); \tag{1.33}$$

(iii)

$$\lim_{\Delta t \to 0} \frac{1}{\Delta t} \int_{|z-y|\leq\varepsilon}^{2} (z - y)^2 P(s, y, s + \Delta t, dz) = \sigma^2(s, y), \forall \epsilon > 0$$

The value $a(s, y)$ characterizes a mean trend in the evolution of the random process $(y(t))_{t \in R^+}$ for small interval of time from s to $s + \Delta t$ provided that $y(s) = y$, and is called the *drift coefficient*. The value $\sigma(s,y)$ characterizes a mean square deviation of the process $(y(t))_{t\in R^+}$ from its mean value and is called the *diffusion coefficient*. We have

$$y(s+\Delta t) \approx y(s) + a(s,y(s))\Delta t + \sigma(s,y(s))\Delta w(s), \tag{1.34}$$

where $\Delta w(s)$ is a random variable such that

$$E\Delta w(s) \sim 0 \text{ and } E(\Delta w(s))^2 \sim \Delta t. \tag{1.35}$$

Usually, diffusion processes are represented in differential form as a *stochastic differential equation*

$$dy(t) = a(t,y(t))dt + \sigma(t,y)dw(t)$$

where $(w(t))_{t\in R^+}$ is a standard Wiener process satisfying (1.35).

In such a way, a standard Wiener process is also a diffusion process with drift $a(t,y) \equiv 0$ and diffusion $\sigma(t,y) \equiv 1$.

1.7. Martingales

Let $(\Omega, \mathcal{F}, \mathcal{F}_t, \mathcal{P})$ be a probability space with a non-decreasing flow of σ-algebras

$$\mathcal{F}_s \subseteq \mathcal{F}_t \subseteq \mathcal{F}, \forall s \leq t.$$

An *adapted* (e.g., $\mathcal{F}_t$-measurable) *integrable* (e.g., $E|m(t)| < +\infty$ for all $t \in R^+$) collection $(m(t), \mathcal{F}_t, \mathcal{P})$ is called a *martingale*, if for any $s < t$ and $t \epsilon R^+$, we have

$$E[m(t)|\mathcal{F}_s] = m(s) \qquad \text{a.s.}, \tag{1.36}$$

where $E[\cdot|\mathcal{F}_s]$ is a conditional expectation with respect to σ-algebra $\mathcal{F}_s$. This collection is called *submartingale (supermartingale)* if the relation (1.36) is fulfilled with sign $\geq$ ($\leq$). Submartingales and supermartingales are called *semimartingales*. Later on (in subsections 1.9.2 and 1.9.3), we will give martingale characterizations of Markov and semi-Markov processes, respectively.

Similarly, we can define discrete time martingales $(m_n)_{n\in Z^+}$ (see [18,68]). An *adapted* (e.g., $\mathcal{F}_n$-measurable) *integrable* (e.g., $E|m_n| < +\infty$) process $(m_n)_{n\in Z^+}$ is called a *martingale* if

$$E(m_{n+1}|\mathcal{F}_n) = m_n,$$

where $\mathcal{F}_n$ is the filtration : $\mathcal{F}_n \subseteq \mathcal{F}_m \subseteq \mathcal{F}$, where $n \leq m$. This collection is called *submartingale (supermartingale)* if this relation is fulfilled with sign $\geq$ ($\leq$).

If $(\xi_n)_{n\in Z^+}$ are the sequence of independent random variables such that $E\xi_n = 0$ for all $n \geq 0$ and $\mathcal{F}_n := \sigma(\xi_k; 0 \leq k \leq n)$, then the process $m_n := \sum_{k=0}^n \xi_k$ is martingale with respect to the $\mathcal{F}_n$.

If $E\xi_n = 1$ for all $n \geq 0$, then process $m_n := \prod_{k=1}^n \xi_k$ is also martingale.

In Section 1.9.1 we will give martingale characterization of discrete-time Markov chains. There is a short list of martingales:

(i) A standard Wiener process $(w(t))_{t\in R^+}$ is a martingale with respect to the natural σ-algebra

$$\mathcal{F}_t^w := \sigma\{w(s); 0 \leq s \leq t\}.$$

This is because $Ew(t) = 0$ and because it is a process with independent increments.

(ii) Process $w^2(t) - t$ is also a martingale with respect to $\mathcal{F}_t^w$.
(iii) Process $|w(t)|$ is a submartingale with respect to $\mathcal{F}_t^w$, hence, process $(-|w(t)|)$ is a supermartingale.

We will need *Kolmogorov-Doob inequalities* for semimartingales (see [18,68]). Let $(m_n)_{n\in Z^+}$ be a nonnegative submartingale. Then for every $\lambda > 0$ and every $n \geq 0$, we have

$$\mathcal{P}\{\max_{0\leq k\leq n} |m_k| \geq \lambda\} \leq \frac{Em_n}{\lambda}.$$

If $(m_n)_{n\in Z^+}$ is a supermartingale, then for every $\lambda > 0$ and every $n \geq 0$, we have

$$\mathcal{P}\{\max_{0\leq k\leq n} |m_k| \geq \lambda\} \leq \frac{Em_0}{\lambda}.$$

The same inequalities are true for continuous time martingales.

1.8. Semigroups of Operators and Their Generators.

Let $(B, \mathcal{B}, \|\cdot\|)$ be a real separable Banach space B with σ-algebra of Borel sets $\mathcal{B}$ and the norm $\|\cdot\|$.

A one-parameter family $(\Gamma(t))_{t\in R^+}$ of bounded linear operators on B is called a *semigroup of operators*, if

(i)

$$\Gamma(0) = I,$$

is the identity operator;

(ii)

$$\Gamma(t+s) = \Gamma(t)\cdot\Gamma(s), \tag{1.37}$$

for all $s, t \geq 0$.

The semigroup $(\Gamma(t))_{t\in R^+}$, is said to be a *contraction* semigroup of $\| \Gamma(t) \| \leq 1, \forall t \geq 0$; and *strongly continuous* semigroup if

$$\lim_{t\to 0} \| (\Gamma(t) - I)f \| = 0, \forall f \in B.$$

The generator (or infinitesimal operator) of a semigroup $(\Gamma(t))_{t\in R^+}$ is the linear operator A defined by

$$Af := \lim_{t\to 0} t^{-1}[(\Gamma(t) - I)f], \tag{1.38}$$

with the *domain* $D(A)$ gievn by

$$D(A) := \{f \in B : \lim_{t\to 0} t^{-1}[(\Gamma(t) - I)f] \quad \text{exists}\}.$$

It is known (see, for example, [24, 94]), that the generator A of a strongly continuous semigroup $(\Gamma(t))_{t\in R^+}$ on B is a densely defined closed operator, i.e., $\overline{D(A)} = B$. Moreover, the set $\bigcap_{n=1}^{+\infty} D(A^n)$ is dense in B.

Also, for a strongly continuous semigroup $(\Gamma(t))_{t\in R^+}$ on **B** with the generator A, we have

$$\begin{cases} \frac{d\Gamma(t)}{dt} = \Gamma(t) \cdot A = A \cdot \Gamma(t), \\ \Gamma(0) = I. \end{cases} \tag{1.39}$$

We now give a few examples of semigroups of operators and generators.

(E1). *Exponential semigroup.* Let A be a bounded linear operator on B. Define the following one-parameter family

$$\Gamma(t) = e^{tA} := \sum_{k=0}^{+\infty} \frac{t^k}{k!} A^k, t \geq 0.$$

It is easy to verify that $(\Gamma(t))_{t \in R^+}$ defined above is a strongly continuous semigroup with generator A. Here $D(A) = B$.

(E2). *Uniform motion on the real line.* Let $B = C(R)$ be a Banach space of bounded continuous functions on R equipped with sup-norm, and let $\Gamma(t)f(x) := f(x + vt)$, where $v > 0$ is a constant velocity, $f \epsilon C(R)$. $(\Gamma(t))_{t \in R^+}$ is a semigroup with $Af(x) = v \cdot f'(x)$ and $D(A) = C^1(R)$ is the space of differentiable functions on R with continuous derivatives.

(E3). *Motion with velocity depending on the state.* Let $q(t, z)$ solves the Cauchy problem:

$$\frac{dq(t,z)}{dt} = v(q(t,z)), q(0,z) = z.$$

Then

$$\Gamma(t)f(z) := f(q(t,z)), f \epsilon C(R),$$

gives a strongly continuous contraction semigroup and

$$\Gamma f(z) = v(z) f'(z), \quad \forall f \epsilon C'(R).$$

The semigroup property follows from the equality

$$q(t+s, z) = q(s, q(t,z)), \forall z \ \epsilon, \forall s, t \ \epsilon R^+.$$

(E4). *Continuous-time Markov chain and its infinitesimal matrix.* Here, $\Gamma(t) = \mathbf{P}(t)$ is the infinite dimentional matrix with

$$\mathbf{P}(t) = (p_{ij}(t); i, j = 1, 2, \ldots), t \geq 0,$$

and

$$A = Q = (q_{ij}; i, j = 1, 2, \ldots). \tag{1.40}$$

(E5). *Bellman-Harris branching process.* Let $\xi(t)$ be a Bellman-Harris process (a homogeneous Markov branching process) with generating function

$$\Phi(t, u) := Eu^{\xi(t)}, |u| \leq 1,$$

and $b(a) := a[p(u) - u]$, where $a > 0$ is the intensity of the exponential distribution of the lifetimes of particles of $\xi(t)$ and

$$p(u) := \sum_{k=0}^{\infty} p_k u^k$$

is the generating function of the number of direct descendants of one particle. Then

$$\begin{cases} \frac{d\Phi(t,u)}{dt} = b(\Phi(t,u)) \\ \Phi(0,u) = u. \end{cases} \tag{1.41}$$

Define $\Gamma(t)f(u) := f(\Phi(t,u)), f \epsilon C(R) = B$. Then we obtain a semigroup $(\Gamma(t))_{t\in R_+}$ with the generator given by

$$\Gamma f(u) = b(u)\frac{df(u)}{du}, \quad f \in C^1(R) = D(\Gamma).$$

(E6). *Diffusion processes.* Let $y(t)$ be a diffusion process with drift $a(t,y) \equiv a(y)$ and diffusion $\sigma(t,y) \equiv \sigma(y)$. As these are independent of t, we obtain the so-called *homogeneous diffusion process* with transition probabilities $P(t,y,A), t \in R^+, y \in R, A \in \mathcal{R}$. The associated semigroup and its generator are

$$\Gamma(t)f(z) := \int_Y f(y)P(t,z,dy), f(y)\epsilon C(R), \tag{1.43}$$

and

$$Af(z) = a(y)\frac{df(z)}{dz} + 1/2\, \sigma^2(y)\frac{d^2 f(z)}{dz^2}, f\epsilon C^2(R). \tag{1.44}$$

In the special case of a Wiener process, we will have

$$P(t,y,A) = \frac{1}{\sqrt{2\pi t}}\int_A \exp^{-\frac{(z-y)^2}{2t}} dz \tag{1.45}$$

and

$$Af(z) = 1/2\frac{d^2 f(z)}{dz^2}.$$

(E7). *Jump Markov process.* For a regular homogeneous jump Markov process, the semigroup defined by

$$\Gamma(t)f(y) = \int_Y P(t,y,dz)f(z) = \int_Y P(y,dz)f(z)\cdot(1 - e^{-\lambda(y)t}), \tag{1.46}$$

where $f \in C'(R)$, is a strongly continuous contraction semigroup, with the generator

$$Af(y) = \lambda(y)\cdot\int_Y [P(y,dz)f(z) - f(y)], \forall f(y)\epsilon C(R) := D(A). \tag{1.47}$$

(E8). *Semi-Markov process.* Let $y(t) := y_{\nu(t)}$ be the semi-Markov process introdused in Section 1.4. As we mentioned earlier, each of the auxiliary processes $\theta(t), \gamma(t)$ and $\gamma^+(t)$ compliments $y(t)$ to a Markov process. In particular, for $\gamma(t) := t - \tau_{\nu(t)}$, $(y(t), \gamma(t))$ is a Markov process on $Y \times R^+$ with the generator

$$Af(y,t) = \frac{df(y,t)}{dt} + \frac{g_y(t)}{\bar{G}_y(t)}[Pf(y,0) - f(y,t)], \tag{1.48}$$

where

$$g_y(t) := \frac{dG_y(t)}{dt}, \bar{G}_y(t) := 1 - G_y(t), \qquad f \in C(Y) \times C^1(R^+). \tag{1.49}$$

We should mention that a semi-Markov process $y(t)$ does not generate a semigroup due to the arbitrary distribution function for the sojourn times, rather than the exponential one in the case of a Markov process.

1.9. Martingale Characterization of Markov and Semi-Markov Processes

We now describe martingale properties of Markov chains, Markov processes and semi-Markov processes.

1.9.1. MARTINGALE CHARACTERIZATION OF MARKOV CHAINS

Let $(y_n)_{n \in Z^+}$ be a homogeneous Markov chain on a measurable phase space $(Y, \mathcal{Y})$ with stochastic kernel $P(y, A), y \epsilon Y, A \epsilon \mathcal{Y}$. Let P be the operator on the space $\mathbf{B}(Y)$

$$Pf(y) := \int_Y P(y, dz) f(z) = E[f(y_n)|y_{n-1} = y] := E_y[f(y_n)], \tag{1.50}$$

generated by $P(y, A)$, and $\mathcal{F}_n^Y := \sigma\{y_k; 0 \leq k \leq n\}$ be a natural filtration generated by $(y_n)_{n \in Z^+}$. The Markov property can be described by

$$E[f(y_n)|\mathcal{F}_{n-1}^Y] = Pf(y_{n-1}). \tag{1.51}$$

This is because (1.50) and the Markov property of the chain $(y_n)_{n \in Z_+}$

$$E[f(y_n)|\mathcal{F}_{n-1}^Y] = E[f(y_n)|y_{n-1} = y\} = Pf(y_{n-1}). \tag{1.52}$$

We note that

$$Pf(y_n) - f(y) = \sum_{k=0}^{n} [P - I] f(y_n), y_0 = y. \tag{1.53}$$

Hence, from (1.52) and (1.53) it follows that

$$E(f(y_n) - f(y) - \sum_{k=0}^{n} [P - I] f(y_n) | y_{n-1} = y) = 0,$$

or

$$E(f(y_n) - f(y) - \sum_{k=0}^{n} [P - I] f(y_n) | \mathcal{F}_{n-1}^Y) = 0.$$

Consequently,

$$M_n := f(y_n) - f(y) - \sum_{k=0}^{n}[P - I]f(y_k) \tag{1.54}$$

is an $\mathcal{F}_n^Y$-martingale.

The *quadratic variation*

$$< m_n > := \sum_{k=1}^{n} E[(m_k - m_{k-1})^2 | \mathcal{F}_{k-1}^Y]$$

of the martingale m_n in (54) is given by

$$< m_n > = \sum_{k=0}^{n}[Pf^2(y_k) - (Pf(y_k))^2]. \tag{1.55}$$

1.9.2. MARTINGALE CHARACTERIZATION OF MARKOV PROCESSES

Let $(y(t))_{t \in R^+}$ be a homogeneous Markov process on a measurable phase space $(Y, \mathcal{Y})$ with transition probabilities $P(t, y, A), t \in R^+$, $y \in Y$, $A \in \mathcal{Y}$. The transition probabilities $P(t, y, A)$ generate the contraction semigroup $\Gamma(t)$ on the Banach space $\mathbf{B}(Y)$ by the following formula

$$\Gamma(t)f(y) := \int_Y P(t, y, dz)f(z) = E[f(y(t))|y(0) = y]. \tag{1.56}$$

Let Q be the infinitesimal operator of the Markov process $(y(t))_{t \in R_+}$. Then

$$\Gamma(t)f(y) - f(y) = \int_0^t Q\Gamma(s)f(y(s))ds = \int_0^t \Gamma(s)Qf(y(s))ds. \tag{1.57}$$

From this and the Markov property it follows that

$$E[f(y(t)) - f(y) - \int_0^t Qf(y(s))ds | \mathcal{F}_s^Y] = 0, \tag{1.58}$$

where

$$\mathcal{F}_s^Y := \sigma\{(y(u)); 0 \le u \le s\}.$$

Therefore,

$$m(t) := f(y(t)) - f(y) - \int_0^t Qf(y(s))ds \tag{1.59}$$

is $\mathcal{F}_t^Y$-martingale. The quadratic variation of $m(t)$ is

$$< m(t) > := \int_0^t [Qf^2(y(s)) - 2f(y(s))Qf(y(s))]ds. \tag{1.60}$$

1.9.3. MARTINGALE CHARACTERIZATION OF SEMI-MARKOV PROCESSES

Let $y(t) := y_{\nu(t)}$ be a semi-Markov process, constructed by Markov renewal process $(y_n, \theta_n)_{n \in Z^+}$ on $Y \times R^+$ and and $\gamma(t) := t - \tau_{\nu(t)}$ be a defect process. Then the process $(y(t), \gamma(t))_{t \in R^+}$ on $Y \times R^+$ is a homogeneous Markov process with the generator

$$\hat{Q}f(t, y) = \frac{d}{dt}f(t, y) + \frac{g_y(t)}{\bar{G}_y(t)}[Pf(0, y) - f(t, y)], \qquad f \epsilon C^1(R^+) \times C(Y). \tag{1.61}$$

Let $\hat{\Gamma}(t)$ be the semigroup generated by transition probabilities $\hat{P}(t,(s,y),\cdot)$ of the process $(y(t),\gamma(t))_{t\in R^+}$. Then

$$\hat{\Gamma}(t)f(0,y) - f(0,y) = \int_0^t \hat{Q}\hat{\Gamma}(s)f(\gamma(s),y(s))ds. \tag{1.62}$$

This, together with the Markov property, implies

$$E[f(\gamma(t),y(t)) - f(0,y) - \int_0^t \hat{Q}f(\gamma(s),y(s))ds|\hat{\mathcal{F}}_s] = 0, \tag{1.63}$$

where

$$\hat{\mathcal{F}}_s := \sigma\{\gamma(u),y(u); 0 \le u \le s\}.$$

Consequently,

$$\hat{m}(t) := f(\gamma(t),y(t)) - f(0,y) - \int_0^t \hat{Q}f(\gamma(u),y(u))du \tag{1.64}$$

is an $\hat{\mathcal{F}}_t$- martingale.

The quadratic variation $<\hat{m}(t)>$ of the martingale $\hat{m}(t)$ in (1.64) is given by

$$<\hat{m}(t)>= \int_0^t [\hat{Q}f^2(\gamma(u),y(u)] - 2f(\gamma(u),y(u))\hat{Q}f(\gamma(u),y(u))du. \tag{1.65}$$

1.9.4. MERGING OF SEMI-MARKOV PROCESSES

Consider a Markov renewal process $(y_n^\varepsilon,\theta_n)_{n\in Z^+}$ in the phase space $(Y,\mathcal{Y})$ with a semi-Markov kernel

$$Q_\varepsilon(y,dz,t) := P_\varepsilon(y,dz)G_y(t), \tag{1.66}$$

where the stochastic kernel $P_\varepsilon(y,dz)$, which defines the transition probabilities of the perturbed Markov chain, $(y_n^\varepsilon)_{n\in Z^+}$ is given by

$$P_\varepsilon(y,B) = P(y,B) - \varepsilon P_1(y,B), y\epsilon Y, B\epsilon\mathcal{Y}, \tag{1.67}$$

where $P_\varepsilon(y,B)$ is a transition probability of basic non-perturbed Markov chain $(y_n)_{n\in Z^+}$ and $P_1(y,B)$ is a some probability measure.

We assume that the stochastic kernel $P(y,B)$ is associated with the following given decomposition of phase space $(Y,\mathcal{Y})$

$$Y = \bigcup_{v\epsilon V} Y_v, Y_v \bigcap Y_{v'} = \phi, v \neq v' \tag{1.68}$$

as disjoint classes of closed sets of states by

$$P(y,Y_v) = \mathbf{1}_v := \begin{cases} 1, & y\varepsilon Y_v \\ 0, & y \not\varepsilon Y_v. \end{cases} \tag{1.69}$$

In each class $Y_v, v\epsilon V$, the basic non-perturbed Markov chain is uniformly ergodic by $v\epsilon V$ with stationary distribution $p_v(A), v\epsilon V, A\epsilon\mathcal{Y}$ given by

$$p_v(A) = \int_{y_v} p_v(dy)P(y,A), A \subset Y_v, p_v(Y_v) = 1.$$

The decomposition (1.68) defines a merging function

$$v(y) = v \text{ if } y\epsilon Y_v, v\epsilon V.$$

Here $(V, \mathcal{V})$ is a measurable merged phase space.

Let

$$m_v = \int_{Y_v} p_v(dy)m(y), \hat{P}_1(v,H) := \int_{Y_v} p_v(dy)P_1(y,Y_H), v\epsilon V, H\epsilon\mathcal{V}$$

and

$$Y_H := \bigcup_{v\epsilon H} Y_v \in \mathcal{Y}, H \subset V. \tag{1.70}$$

It is known (see [80]) that the kernel

$$\hat{Q}(v,H) := \hat{P}_1(v,H)/m_v, v \not\epsilon H, \tag{1.71}$$

with function $q(v)$ such that

$$0 < g(v) := -\hat{P}_1(v)/m_v = \int_{Y_v} p_v(dy)P_1(y,Y_v)/m_v, v \not\epsilon H, \tag{1.72}$$

defines a jump Markov process $\hat{y}(t)$ in the phase space $(V, \mathcal{V})$ with the stochastic kernel

$$\hat{Q}(v,H,t) := \hat{P}(v,H)\cdot(1-e^{-q(v)\cdot t}), \tag{1.73}$$
$$\hat{P}(v,H) := \hat{Q}(v,H)/g(v),$$

where $\hat{Q}(V,H)$ and $q(v)$ are defined in (1.71) and (1.72), respectively.

Semi-Markov process $v(y^{\varepsilon}_{\nu(t/\varepsilon)})$ converges weakly under conditions (1.66)-(1.73) as $\varepsilon \to 0$ to the jump Markov process $\hat{y}(t)$ with stochastic semi-Markov kernel $\hat{Q}(v,H,t)$.

We can now state a *merging theorem* for additive and integral functionals of perturbed Markov chains and semi-Markov processes, respectively. Let $\alpha(y)$ be a measurable and bounded function on Y. Then for additive functionals of the Markov chain y^{ϵ}_n, we have

$$\epsilon \sum_{k=1}^{\nu(t/\epsilon)} \alpha(y^{\epsilon}_k) \to_{\epsilon\to 0} \int_0^t \hat{\alpha}(\hat{y}(s))ds,$$

where $\hat{\alpha}(v) := \int_{Y_v} p(dy)\alpha(y)/m_v$; and for the integrale functional of the semi-Markov process $y^{\epsilon}(t)$ we have

$$\epsilon \int_0^{t/\epsilon} \alpha(y^{\epsilon}(s))ds \to_{\epsilon\to 0} \int_0^t \tilde{\alpha}(\hat{y}(s))ds,$$

where $\tilde{\alpha}(v) := \int_{Y_v} p(dy)m(y)\alpha(y)/m_v$.

We note that if we have one ergodic class Y, then these merging theorems reduce to the usual ergodic theorems for additiv functionals of Markov renewal process and for integral functionals of semi-Markov processes.

1.10. General Representation and Measurability of Biological Systems in Random Media.

In this section we give a general representation of all the models, considered in sections 3.1-3.6, in random environment and state their measurability.

1.10.1. GENERAL REPRESENTATION OF THE BIOLOGICAL SYSTEMS

Let $g(x,y)$ be a bounded and measurable function on $X \times Y$, where X is a linear measurable space, for example, $X = R^d, d \geq 1$, and Y is a phase statae space of Markov process $(y_n; n \geq 0)$.

Then all the population biological systems in random environment may be represented in the following forms:

1) *discrete-time models in Markov random environment*

$$x_{n+1} = g(x_n, y_{n+1}), x_0 = x \in X; \tag{1.74}$$

2) *continuous-time models in Markov renewal random environment*

$$x_{\nu(t)+1} = g(x_{\nu(t)}, y_{\nu(t)+1}), x_0 = x \in X; \tag{1.75}$$

3) *continuous-time models in semi-Markov random environment*

$$\frac{dx(t)}{dt} = g(x(t), y(t)), x(0) = x \in X; \tag{1.76}$$

where $y(t) := y_{\nu(t)}$.

Indeed, for example, for logistic growth model in sec. 3.6. function $g(x,y)$ has the following form:

$$g(x,y) = r(y)x \cdot (1 - x/K(y)), x_0 = x \in R = X,$$

where $r(y)$ and $K(y)$ are bounded and measurable functions on Y. The representations 1)-3) we will use in our study of population biological systems in RM.

1.10.2. MEASURABILITY OF THE GENERAL SYSTEMS

Let $(\Omega, \mathcal{F}, \mathcal{F}_n, P)$ be a probability space with $(\mathcal{F}_n)_{n \epsilon N}$ denoting sub-σ-algebras increasing to $\mathcal{F}$, and $(y_n; \theta_n)_{n \in Z^+}$ be a Markov renewal process (MRP) on $Y \times R^+$ (see Chapter 1).

We note, that (y_n, θ_n) is $\mathcal{F}_n/\mathcal{Y} \times \mathcal{R}^+$-measurable for each $n \epsilon N$, where $\mathcal{Y}$ and $\mathcal{R}^+$ are σ-algebras generated by Borel sets in Y and R^+, respectively.

Semi-Markov process $y(t)$ associated with MRP $(y_n, \theta_n)_{n \in Z^+}$ is defined by

$$y(t) = y_n(w) \quad for \quad \tau_n(w) \leq t < \tau_{n+1}(w),$$

where

$$\tau_n = \sum_{k=0}^{n} \theta_k.$$

Let for $n\epsilon N$

$$\mathcal{F}_n^Y := \sigma\{y_0, y_1, \dots, y_n\} \tag{1.77}$$

be a σ-algebra generated by $(y_1, \dots, y_n)$, and let

$$\mathcal{G}_t := \sigma\{\nu(t); \tau_1, \dots, \tau_{\nu(t)}; y_0, y_1, \dots y_{\nu(t)+1}\}, \tag{1.78}$$

be a σ-algebra generated by $\nu(s), \tau_1, \dots, \tau_{\nu(s)}$, and $y_0, y_1, \dots, y_{\nu(t)+1}$ on the interval [0,t], and let

$$\mathcal{H}_t := \sigma\{y(s); \gamma(s) := s - \tau_{\nu(s)}; 0 \le s \le t\} =$$
$$= \sigma\{\nu(t); \tau_1, \dots, \tau_{\nu(t)}; y_0, \dots y_{\nu(t)}\}, \tag{1.79}$$

be a σ-algebra generated by $\nu(s)$, $\tau_1, \tau_2, \dots, \tau_{\nu(s)}; y_0, \dots, y_{\nu(s)}$ on the interval $[0, t]$. We note that $\mathcal{F}_n^Y \subset \mathcal{F}_n, \forall n \in Z^+$, and $\mathcal{H}_t \subset \mathcal{G}_t, \forall t\epsilon R_+$.

In this way, process x_{n+1} in (1.74) is $\mathcal{F}_{n+1}/\mathcal{X}$ -measurable, where $\mathcal{X}$ is a σ-algebra in X, and $\mathcal{F}_{n+1}$ is defined in (1.77). The process $x_{\nu(t)+1}$ in (1.75) is $\mathcal{G}_t/\mathcal{X}$-measurable, where $\mathcal{G}_t$ is defined in (1.78).

The process $x(t)$ in (1.76) is $\mathcal{H}_t/\mathcal{X}$-measurable, where $\mathcal{H}_t$ is defined in (1.79).

CHAPTER 2: LIMIT THEOREMS FOR DIFFERENCE EQUATIONS IN RANDOM MEDIA

Our focus is related to the problem of small perturbations. For the sake of illustration, we first consider the *dynamical system in series scheme*

$$dz_t^\epsilon/dt = \epsilon g(z_t^\epsilon, y(t)), \qquad z_0^\epsilon = z,$$

where y(t) is a certain stochastic process, $\epsilon > 0$ is a small parameter and g is a given function. If function g does not increase too fast, then the solution of z_t^ϵ converges to $z_t^0 \equiv z$ as $\epsilon \to 0$, uniformly on every finite time interval $[0, T]$. However, the behavior of z_t^ϵ on time intervals of order ϵ^{-1} or of higher orders (for example, ϵ^{-2}) is usually of great interest and, on these intervals significant changes occur.

For the study of this system on intervals of the form $[0, T\epsilon^{-1}]$, it is convenient to introduce new coordinates in order to work on time intervals independent of ϵ. For example, if we set $X_t^\epsilon = z_{t/\epsilon}^\epsilon$, then the equation for X_t^ϵ become

$$dX_t^\epsilon/dt = g(X_t^\epsilon, y(t/\epsilon)), \qquad X_0^\epsilon = z,$$

and the study of this system on a finite time interval $[0, T]$ is equivalent to the study of *original system* on time intervals of order ϵ^{-1}. We further assume that *averaging in time condition*

$$\lim_{t \to +\infty} \frac{1}{T} \int_0^T g(x, y(s))ds = \hat{g}(x)$$

is fulfilled for all $x \in R$, and assume that g is Lipschitz. This condition is satisfied if $y(t)$ is a periodic or ergodic process, for example. In the latter case, for example, $\hat{g}(x) = \int_Y \pi(dy)g(x, y)$, where $\pi(dy)$ is stationary distribution of $y(t)$. We note that

$$X_\Delta^\epsilon - x = \Delta\left(\frac{\epsilon}{\Delta} \int_0^{\Delta/\epsilon} g(x, y(s))ds) + \gamma_\epsilon(\Delta)\right),$$

the coefficient of Δ converges to $\hat{g}(x)$ as $\epsilon/\Delta \to 0$, and there term $\gamma_\epsilon(\Delta)$ satisfies the inequality $|\gamma_\epsilon(\Delta)| < K\Delta^2$, if there exists $K > 0$ such that $|g(x_1, y) - g(x_2, y)| \leq K|x_1 - x_2|$ for all $x_2, x_2 \in R$ and $y \in Y$. Consequently, the displacement of the trajectory X_t^ϵ over a small time interval differs from the displacement of the trajectiry $\hat{X}_t$ of the *averaged differential equation*

$$d\hat{x}_t/dt = \hat{g}(\hat{x}_t), \qquad \hat{x}_0 = x,$$

only by an infinitely small quantity of Δ if $\Delta \to 0$ and $\frac{\epsilon}{\Delta} \to 0$. If convergence in *averaging in time condition* is uniform in x, then X_t^ϵ converges to the solution of *the averaged differential equation* uniformly on every finite time interval as $\epsilon \to 0$. The assertion that X_t^ϵ is close to $\hat{X}_t$ is called the the *averaging principle.* In this sense, solutions of the *original differential equation* may be approximated by $\hat{x}(\epsilon t)$.

If $\hat{g}(x) = 0$ then a solution of the *original differential equation* during time $1/\epsilon$ does not leave a small neighbourhood of its initial state. In this case, we shall study the solution of *the original equation* with a change of time t/ϵ^2 as follows

$$dz_{t/\epsilon^2}^\epsilon/dt = \frac{1}{\epsilon}g(z_{t/\epsilon^2}^\epsilon, y(t/\epsilon^2)), \qquad z_0^\epsilon = z.$$

In this case, the limiting function $\hat{z}_t$ satisfies the *stochastic differential equation* (SDE)

$$d\hat{z}_t = \hat{a}(\hat{z}_t)dt + \hat{b}(\hat{z}_t)dw_t$$

with a drift coefficient $\hat{a}(z)$ and diffusion coefficient $\hat{b}(z)$ depending on the function g and its derivative, here w_t is a Wiener process. $\hat{z}_t$ in this case is a *diffusion process.* The assertion that $z_{t/\epsilon^2}^\epsilon$ is close to $\hat{z}_t$ is called the *diffusion approximation principle.*

Even if $\hat{g}(x) \neq 0$ we can obtain *diffusion approximation.* In the deterministic case ($y(t)$, for example, is a periodic deterministic function) the relation $(X_t^\epsilon - \hat{x}_t)/\epsilon$ is bounded for enough small $\epsilon > 0$. But in the stochastic case this is no longer true. We should therefore study the normalized difference

$$\eta_t^\epsilon := (X_t^\epsilon - \hat{x}_t)/\sqrt{\epsilon},$$

where X_t^ϵ and $\hat{x}_t$ are the solutions of *the original eqaution* and *the averaged equation*, respectively. η_t^ϵ also converges to some *diffusion process* $\hat{\eta}_t$ as $\epsilon \to 0$.. The assertion that η_t^ϵ is close to $\hat{\eta}_t$ is called the *normal deviations principle.*

If we have a stochastic process $y^\epsilon(t)$ such that its phase space Y has more than one ergodic class (or if Y is splitted into several ergodic classes), then process $y^\epsilon(t)$ converges to the *merged process* $\hat{y}(t)$ with some merged phase space V which has the same number of states as Y has the ergodic classes. In this case, the process X_t^ϵ

$$dX_t^\epsilon/dt = g(X_t^\epsilon, y^\epsilon(t/\epsilon)), \qquad X_0^\epsilon = z,$$

converges to the process $\tilde{x}_t$ such that

$$\tilde{x}_t/dt = g(\tilde{x}_t, \hat{y}(t)), \qquad \tilde{x}_0 = z.$$

The assertion that x_t^ϵ is close to $\tilde{x}_t$ is called the *merging principle.* The *averaging principle,* the *diffusion approximation principle*, the *normal deviations principle* and the *merging principle* will be the central subjects of this Chapter.

Let $(Y, \mathcal{Y})$ be a measurable space, X a linear space and $g : X \times Y \to X$ be a function which determines the dynamics in a random environment discribed by a semi-Markov process $y(t)$. Let ϵ be a small positive parameter. In this chapter, we consider a dynamical system described by the following iteration:

$$X_{\nu(t/\epsilon^i)+1}^\epsilon = X_{\nu(t/\epsilon^i)}^\epsilon + \varepsilon g(X_{\nu(t/\epsilon^i)}^\epsilon, y_{\nu(t/\epsilon^i)+1}),$$

for $t \in R^+$, where $X_0^\epsilon = X_0 = x$ is given, $\nu(t)$ is a counting process, $i = 1$ or 2. Major issues to be addressed in this chapter include:

(A) *Averaging* ($i = 1$) *and diffusion approximation* ($i = 2$) of solutions of the equation as $\varepsilon \to 0$ developed in [81];

(B) *Normal deviations* of the process $X^\varepsilon_{\nu(t/\varepsilon)}$ ($i = 1$) from averaged one $\tilde{x}^\epsilon_{\nu(t/\epsilon)}$, developed in [82], namely, the limit

$$z^\varepsilon(t) := [X^\varepsilon_{\nu(t/\varepsilon)} - \tilde{x}^\varepsilon_{\nu(t/\varepsilon)}]/\sqrt{\varepsilon} \text{ as } \varepsilon \to 0,$$

where $\tilde{x}^\varepsilon_n$ is defined by the averaged difference equation

$$\tilde{x}^\varepsilon_{n+1} - \tilde{x}^\varepsilon_n = \varepsilon \tilde{g}(\tilde{x}^{\varepsilon)}_n,$$

with

$$\tilde{g}(x) := \int_Y p(dy) g(x,y)/m,$$

where $(p(A), A \in \mathcal{Y})$ is a stationary distribution of $(y_n)_{n \in Z^+}$, and m is the mean sojourn time;

(C) *Merging of solutions* of the following equation

$$X^\varepsilon_{\nu(t/\varepsilon)+1} - X^\varepsilon_{\nu(t/\varepsilon)} = \varepsilon g(X^\epsilon_{\nu(t/\varepsilon)}, y^\epsilon_{\nu(t/\varepsilon)+1}),$$

where $(y^\varepsilon_n)_{n \in Z^+}$ is a perturbed Markov chain in the splitted phase space $Y = \bigcup_{v \in V} Y_v$, of distinct classes Y_v developed in [82];

(D) *Stability properties* for difference equations in random media, described by a semi-Markov process: we consider a family of difference equations labelled by a parameter $\epsilon \to 0$ such that when $\epsilon \to 0$ the limiting equation becomes an averaged or diffusion equations. We assume that the limiting equation has some stability property, and we show that the corresponding stability property holds for the original difference equation in random media in series scheme when $\epsilon > 0$ is sufficiently small. These results are recently obtained in [83].

To obtain the afovementioned averaging, diffusion approximation, normal deviation , and merging results for difference equations in random environment we need various limit theorems for random evolutions in series scheme.

2.1. Limit Theorems for Random Evolutions

In this section, we consider the general theory of random evolutions. Definitions and classifications of random evolutions will be given. Martingale methods and their applications to the limit theorems (averaging, merging, diffusion approximation, and normal deviations) for random evolutions will be considered.

2.1.1. DEFINITIONS AND CLASSIFICATIONS OF RANDOM EVOLUTIONS

Let $(\Omega, \mathcal{F}, \mathcal{F}_t, \mathcal{P})$ be a probability space, $t \in R^+$, $(Y, \mathcal{Y})$ be a measurable phase space, and $(\mathbf{B}, \mathcal{B}, \|\cdot\|)$ be a separable Banach space.

We consider a Markov renewal process $(y_n, \theta_n)_{n \in Z_+}, y_n \in Y, \theta_n \in R^+$, with the stochastic kernel

$$\begin{cases} Q(y,A,t) & := & P(y,A)G_y(t), \\ P(y,A) & := & \mathcal{P}\{y_{n+1}\in A|y_n=y\}, \\ G_y(t) & := & \mathcal{P}\{\theta_{n+1}<t|y_n=y\}, \end{cases} \tag{2.1}$$

for all $y \in Y, A \in \mathcal{Y}, t \in R^+$. Recall that the process $y_t := y_{\nu(t)}$ is a semi-Markov process, where $\nu(t) := \max\{n : \tau_n \leq t\}, \tau_n := \sum_{k=0}^{n} \theta_k, y_n = y_{\tau_n}, \mathcal{P}\{\nu(t) < +\infty, \forall t \in R^+\} = 1$. Recall also that if $G_y(t) = 1 - e^{-\lambda(y)t}$, where $\lambda(x)$ is a measurable and bounded function on X, then y_t is called a jump Markov process.

Let $(\Gamma(y))_{y\in Y}$ be a family of operators on a dense subspace $\mathbf{B}_0 \in \mathbf{B}$, which is the common domain for $\Gamma(y)$, independent of y. $\Gamma(y)$ are noncommuting and unbounded in general, but we assume that the map $\Gamma(y)f : Y \to \mathbf{B}$ is strongly $\mathcal{Y}/\mathcal{B}$- measurable for every given $f \in \mathbf{B_0}$. Also, let $(D(y))_{y\in Y}$ be a family of bounded linear operators on $\mathbf{B}$ such that map $D(y)f : Y \to \mathbf{B}$ is $\mathcal{Y}/\mathcal{B}$-measurable, for every $f \in \mathbf{B}$. We also assume that the operators $(\Gamma(y))_{y\in Y}$ generate semigroups $\Gamma_y(t)$ on $\mathbf{B}$.

A random evolution (RE) is defined as the solution of the following stochastic operator integral equation in the separable Banach space $\mathbf{B}$

$$V(t)f = f + \int_0^t \Gamma(y_s)V(s)fds + \sum_{k=1}^{\nu(t)}[D(y_k) - I]V(\tau_k-)f, \tag{2.2}$$

where I is an identity operator on $\mathbf{B}, \tau_k- := \tau_k - 0, f \in \mathbf{B}$. In the literature, the random evolution $V(t)$ is also called a *discontinuous* RE [44,72].

If y_t given above is a Markov (or semi-Markov) process, then $V(t)$ in (2.2) is called a *Markov or (semi-Markov) RE.*

If $D(y) \equiv I$ for every $y \in Y$, then $V(t)$ in (2.2) is called *a continuous RE.*

If $\Gamma(y) \equiv 0$ for every $y \in Y$, then $V(t)$ in (2.2) is called *a jump RE.*

A RE $V_n := V(\tau_n)$ is called *a discrete-time RE* or, shortly, *discrete RE.*

Intuitively, operators $\{\Gamma(y)\}_{y\in Y}$ describe a continuous component $V^c(t)$ of the RE $V(t)$ in (2.2), and operators $\{D(y)\}_{y\in Y}$ describe a jumb component $V^d(t)$ of the RE $V^d(t)$ in (2.2). In such a way, a RE is described by two objects:

(i) an operator dynamical system $\{V(t)\}_{t\in R^+}$;
(ii) a random process $(y_t)_{t\in R^+}$.

Under the above conditions, the solution $V(t)$ of (2.2) is unique and has the representation:

$$V(t) = \Gamma_{y_t}(t - \tau_{\nu(t)}) \prod_{k=1}^{\nu(t)} D(y_k)\Gamma y_{k-1}(\theta_k), \tag{2.3}$$

where $\{\Gamma_y(t)\}_{t\in R^+}$ are the semigroups generated by the operators $\{\Gamma(y)\}_{y\in Y}$.

This can be proved by a constructive method described in [45,71].

We now provide several examples of random evolutions. First of all, note that if

$$\Gamma(y) := v(y)\frac{d}{dz},$$

$$D(y) \equiv I,$$
$$\mathbf{B} = \mathbf{C}^1(R),$$

then (2.2) is a transport equation which describes the motion of a particle with random velocity $v(y_t)$. Consequently, various interpretations of the operators $\Gamma(y)$ and $D(y)$ yield random evolutions in many applications.

(E1). Impulse traffic process. Let $\mathbf{B} = \mathbf{C}(R)$ and assume operators $\Gamma(y)$ and $D(y)$ are defined by

$$\Gamma(y)f(z) := v(z,y)\frac{d}{dz}f(z),$$
$$D(y)f(z) := f(z + a(y)), \tag{2.4}$$

where functions $v(z, y)$ and $a(y)$ are continuous and bounded on $R \times Y$ and Y respectively, $z \in R$, $y \in Y$ and $f \in \mathbf{C}^1(R) := \mathbf{B}_0$. Then equation (2.2) takes the form

$$f(z_t) = f(z) + \int_0^t v(z_s, y_s)\frac{d}{dz}f(z_s)ds + \sum_{k=1}^{\nu(t)}[f(z_{\tau_k -} + a(y_k)) - f(z_{\tau_k -})], \tag{2.5}$$

and the RE $V(t)$ is defined by the relation

$$V(t)f(z) = f(z_t),$$

$$z_0 = z.$$

Equation (2.5) is a functional equation for the **impulse traffic process** z_t, which satisfies the equation

$$z_t = z + \int_0^t v(z_s, y_s)ds + \sum_{k=1}^{\nu(t)} a(y_k). \tag{2.6}$$

We note that the impulse traffic process z_t in (6) is a realization of a discontinuous RE .

(E2). Summation on a Markov chain. Let $v(z, x) \equiv 0$ with $z \in R$ and $x \in X$ in (2.6). Then the process

$$z_t = z + \sum_{k=1}^{\nu(t)} a(y_k) \tag{2.7}$$

is a summation on a Markov chain $(y_n)_{n \in Z^+}$ and it is a realization of a jump RE. Let $z_n := z_{\tau_n}$ in (2.7). Then the discrete process

$$z_n = z + \sum_{k=1}^{n} a(y_k)$$

is a realization of a discrete RE.

(E3). Diffusion process in random media. Let $\mathbf{B} = \mathbf{C}(R)$, $\mathbf{B}_0 = C^2(R)$, $P_x(t, z, A)$ be a Markov continuous distribution function with respect to the diffusion process $\xi(t)$, that is the solution of the stochastic differential equation in R with semi-Markov switchings:

$$d\xi(t) = \mu(\xi(t), y_t)dt + \sigma(\xi(t), y_t)dw_t,$$

$$\xi(0) = z, \tag{2.8}$$

where y_t is a semi-Markov process independent of a standard Wiener process w_t, coefficients $\mu(z, y)$ and $\sigma(z, y)$ are bounded and continuous functions on $R \times Y$. Let us define the following contraction semigroups of operators on $\mathbf{B}$:

$$\Gamma y(t)f(z) := \int_R P_y(t, z, dz)f(z), f(z) \in \mathbf{B}, y \in Y. \tag{2.9}$$

Their infinitesimal operators $\Gamma(y)$ are given by

$$\Gamma(y)f(z) = \mu(z, y)\frac{d}{dz}f(z) + 2^{-1}\sigma^2(z, y)\frac{d^2}{dz^2}f(z),$$

$$f(z) \in \mathbf{B}_0.$$

The process $\xi(t)$ is a continuous one, that is why the operators $\mathcal{D}(y) \equiv I, \forall y \in Y$, are identify operators. Then the equation (2.2) takes the form

$$f(\xi(t)) = f(z) + \int_0^t [\mu(\xi(s), y_s)\frac{d}{dz} + 2^{-1}\sigma^2(\xi(s), y_s)\frac{d^2}{dz^2}]f(\xi(s))ds, \tag{2.10}$$

and RE $V(t)$ is defined by the relation

$$V(t)f(z) = E[f(\xi(t))/y_s; 0 \leq s \leq t; \xi(0) = x].$$

Equation (2.10) is a functional one for diffusion process $\xi(t)$ in (2.8) in semi-Markov random media y_t. We note that diffusion process $\xi(t)$ in (2.8) is a realization of continuous RE.

(E4). Biological systems in random media. Let $\mathbf{B}$ be the same space as in **E1**. Let us define the operators $\Gamma(y)$ and $D(y)$ in the following way: $\Gamma(y) := I$ and $D(y)f(x) := f(x + g(x, y))$, where $g(x, y)$ is bounded and continuous function. Then equation (2.2) takes the form

$$f(X_{\nu(t)}) = f(x) + \sum_{k=1}^{\nu(t)}[f(X_{\tau_k-} + g(X_k, y_{k+1})) - f(X_{\tau_k-})],$$

and RE $V(t)$ is defined by the relation for $t = \tau_{\nu(t)}$:

$$V(\tau_{\nu(t)})f(x) = f(X_{\nu(t)}),$$

$$X_0 = x.$$

The equation for $f(X_t)$ is a functional one for many biological systems in random media, which satisfiy the equation

$$X_{\nu(t)+1} = X_{\nu(t)} + g(X_{\nu(t)}, y_{\nu(t)+1}).$$

For example, for logistic growth model, $g(x, y) := r(y)x(1-x/K(y))$ for two positive functions r and K.

2.1.2. MARTINGALE CHARACTERIZATION OF RANDOM EVOLUTIONS

The main approach to the study of REs are *martingale characterization.*

Here the main idea is that a process

$$M_n f := V_n f - f - \sum_{k=0}^{n-1} E[(V_{k+1} - V_k)f|\mathcal{F}_k], V_0 = I, \tag{2.11}$$

is an $\mathcal{F}_n$-martingale in $\mathbf{B}$, where

$$\mathcal{F}_n := \sigma\{y_k, \tau_k; 0 \le k \le n\},$$

$$V_n := V(\tau_n),$$

E is an expectation with respect to probability $\mathcal{P}$. Representation of the martingale M_n (see(2.4)) in the form of the martingale-difference

$$M_n f = \sum_{k=0}^{n-1}[V_{k+1}f - E(V_{k+1}f/\mathcal{F}_k)] \tag{2.12}$$

gives us the possibility of calculating the *weak characteristics (or weak quadratic variation)* of the martingale M_n :

$$< l(M_n f) >:= \sum_{k-0}^{n-1} E[l^2((V_{k+1} - V_k)f)|\mathcal{F}_k], \tag{2.13}$$

where $l \in \mathbf{B}^*$, and $\mathbf{B}^*$ is a dual space to $\mathbf{B}$, dividing points of $\mathbf{B}$.

The martingale method leading to various limit theorems for the sequence of REs based on the solution of the following problems:

(i) *weak compactness* of the family of measures generated by the sequences of REs;
(ii) any limiting point of this family of measures is the *solution of a martingale problem*;
(iii) the solution of the martingale problem is *unique.*

The solutions (i)-(ii) quarantee the existence of a weakly converging subsequence, and the solution of (iii) gives the uniqueness of the weak limit. It follows from (i)-(iii) that sequence of RE converges weakly to the unique solution of the martingale problem. The weak convergence of RE in a series scheme will be obtained from the criterion of weakly compactness of the processes with values in separable Banach space. The limit RE will be obtained from the solution of some martingale problem in the form of some integral operator equations in Banach space $\mathbf{B}$. We also use the representation

$$V_{k+1} - V_k = [\Gamma_{y_k}(\theta_{k+1})D(y_{k+1}) - I]V_k,$$

$$V_k := V(\tau_k), \tag{2.14}$$

and the following expression for semigroups of operators $\Gamma_y(t)$:

$$\Gamma_y(t)f = f + \sum_{k=1}^{n-1} \frac{t^k}{k!}\Gamma^k(y)f + ((n-1)!)^{-1}\int_0^t (t-s)^n \Gamma_y(s)\Gamma^n(y) f ds,$$

$$\forall y \in Y, \forall f \in \bigcap_{y\in Y} \text{Dom } (\Gamma^n(y)). \tag{2.15}$$

Taking into account (2.11)-(2.15), we obtain the limit theorems for RE. In the previous subsection we considered the evolution equation associated with random evolutions by using the jump structure of the semi-Markov process or jump Markov process.

We now formulate a random evolution in terms of a *martingale problem.* It has been shown by Stroock and Varadhan in their book "Multidimensional Diffusion Processes", Springer, (1969) that the entire theory of multi-dimentional diffusion processes (and many other continuous parameterized Markov processes) can be so formulated.

For a given evolution equation

$$\frac{df}{dt} = Gf. \tag{2.16}$$

the *martingale problem* is to find for a RE $V(t)$ and a Markov process $(y(t))_{t\in R^+}$ with infinitesimal operator Q, and a RE $V(t)$ so that for every smooth functions f,

$$V(t)f(y(t)) - \int_0^t V(s)Gf(y(s))ds \tag{2.17}$$

is a martingale. The solution of this martingale formulation provides a solution to (2.16) in the following sense: the operator

$$f \to T(t)f := E_y[V(t)f(y(t))]$$

defines a semigroup of operators on the Banach space **B**, whose infinitesimal generator can be computed by taking the expectation:

$$E_y[V(t)f(y(t))] - f(y) = E_y[\int_0^t V(s)Gf(y(s))ds],$$

and then

$$\lim_{t\to 0}((E_y[V(t)f(y(t))] - f(y))/t) = \lim_{t\to 0}(\int_0^t E_y[V(s)Gf(y(s))]ds/t) = Gf(y).$$

The quadratic variation $< m(t) >$ for martingale $m(t)$ in (45) has the following form:

$$< m(t) >= \int_0^t [Gf^2(y(s)) - 2f(y(s))Gf(y(s))]ds. \tag{2.18}$$

We note that operators Q and G are connected in the following way: $G = Q + \Gamma(y)$, where $\Gamma(y)$ are defined in Section 2.1.1.

The following result gives a solution to the martingale problem for Markov chain y_n with infinitesimal operator $\mathbf{P} - I$. Let us consider a homogeneous Markov chain $(y_n)_{n\in Z_+}$ on a measurable phase space $(Y, \mathcal{Y})$ with stocastic kernel $P(y, A), y \in Y, A \in \mathcal{Y}$, respected to the operator $\mathbf{P}$ on the Banach space $\mathbf{C}(Y)$:

$$\mathbf{P}f(y) = \int_Y P(y, dz)f(z) = E[f(y_n)|y_{n-1} = y].$$

Since

$$\mathbf{P}f(y_n) - f(y) = \sum_{k=0}^{n}[\mathbf{P} - I]f(y_n), y_0 = y,$$

and

$$E[f(y_n) - f(y) - \sum_{k=0}^{n}[\mathbf{P} - I]f(y_k)|y_{n-1} = y] = 0,$$

the process

$$m_n := f(y_n) - f(y) - \sum_{k=0}^{n}[\mathbf{P} - I]f(y_k) \tag{2.19}$$

is $\mathcal{F}_k^Y$-martingale, where $\mathcal{F}_n^Y := \sigma\{y_k; 0 \le k \le n\}$.

The inverse is also true: if we have martingale in (2.19), then the process y_n is a Markov chain with infinitesimal operator $\mathbf{P} - I$ (See [44]). We note that the quadratic variation $< m_n >$ for martingale m_n in (2.19) has the following form:

$$< m_n >= \sum_{k=0}^{n}[\mathbf{P}f^2(y_k) - (\mathbf{P}f(y_k))^2]. \tag{2.20}$$

Examples. 1. If $V(t) = I$ in (2.17), then we have the solution of the martingale problem for Markov process $y(t)$ with infinitesimal operator Q : process

$$f(y(t)) - f(y) - \int_0^t Qf(y(s))ds$$

is a martingale.

2. Let $y(t) = w(t)$ be a Wiener process and $V(t) = I$. Then the solution of martingale the problem for Wiener process has the following form: process

$$f(w(t)) - f(0) - \frac{1}{2}\int_0^t \frac{d^2 f(w(t))}{dx^2} ds$$

is a martingale.

Remark. We note that a measurable process $V(t)$ is a solution of the martingale problem for operator A if and only if for all $l, l_k \in \mathbf{B}^*$, we have

$$E[l(V(t_{k+1})f - V(t_k)f - \int_{t_k}^{t_{k+1}} V(s)Gfds)] \cdot \prod_{k=1}^{n} l_k(V(t_k)f) = 0, \tag{2.21}$$

where $0 \le t_1 < t_2 < \cdots < t_{n+1}, \forall f \in \mathrm{Dom}(A)$, and $k = 1, ..., n$. Consequently, the statement that a measurable process is a solution of a martingale problem is a statement about its finite-dimensional distributions.

Remark. It is known that {convergence of finite-dimensional distributions of the process}+
{tightness of sequence of the processes}={weak compactness of the processes}.

In connection with the latter Remark, (2.21) and the previous statement, we conclude that: {the solution of martingale problem for the sequence of the processes} +{tightness of sequence of the processes}={weak compactness of the processes}.

2.1.3 LIMIT THEOREMS FOR RANDOM EVOLUTIONS

The martingale method to obtain various limit theorems (averaging and diffusion approximation, for example) for the sequence of SMRE is based on the solution of the following problems:

(i) weak compactness of the family of measures generated by the sequence of SMRE;
(ii) verification of the fact that every limiting point of the above family of measures is a solution of the martingale problem;
(iii) the uniqueness solution of the above martingale problem.

In particular, the conditions (i)-(ii) guarantee the existence of a weakly convergent subsequence, and condition (iii) gives the uniqueness of a weak limit. From (i)-(iii) it follows that consequence of SMRE converges weakly to the unique solution of the martingale problem.

Weak Convergence of Random Evolutions: We now propose a *weak convergence of SMRE* in series scheme, that we obtain from the criterion of weak compactness of the process with values in a separable Banach space [45,71].

The main idea is that the process

$$M_n f := V_n f - f - \sum_{k=0}^{n-1} E[(V_{k+1} - V_k) f | \mathcal{F}_k], V_0 f = f, \tag{2.22}$$

is an $\mathcal{F}_n$ – *martingale* in $\mathbf{B}$, where

$$\mathcal{F}_n := \sigma\{y_k; \tau_k; 0 \le k \le n\}, V_n := V(\tau_n),$$

E is an expectation by probability $\mathcal{P}$ on a probability space $(\Omega, \mathcal{F}, \mathcal{P})$. This can be achived as follows. Representation of the martingale M_n in the form of *martingale differences*

$$M_n = \sum_{k=0}^{n-1} [V_{k+1} - E(V_{k+1} | \mathcal{F}_k)] \tag{2.23}$$

gives us the possibility to calculate the *weak quadratic variation:*

$$< l(M_n f) > := \sum_{k=0}^{n-1} E[l^2((V_{k+1} - V_k) f) / \mathcal{F}_k], \tag{2.24}$$

where $l \in \mathbf{B}^*$, and $\mathbf{B}^*$ is the dual space to $\mathbf{B}$, dividing points of $\mathbf{B}$.

From (2.23) it follows that

$$V_{k+1} - V_k = [\Gamma_{y_k}(\theta_{k+1})D(y_{k+1}) - I]V_k. \tag{2.25}$$

We also need the following expression for semigroup of operators $\Gamma_y(t)$:

$$\Gamma_y(t)f = f + \sum_{k=1}^{n-1} \frac{t^k}{k!}\Gamma^k(y)f + \frac{1}{(n-1)!}\int_0^t (t-s)^n \Gamma_y(s)\Gamma^n(y) f ds,$$

$$\forall y \in Y, \forall f \in \bigcap_y \text{Dom }(\Gamma^n(y)). \tag{2.26}$$

Then the aforementioned result follows from (2.22)-(2.25).

In what follows, we assume that the following conditions are satisfied:

(A) there exists Hilbert spaces $\mathbf{H}$ and $\mathbf{H}^*$ which compactly imbedded in Banach spaces $\mathbf{B}$ and $\mathbf{B}^*$ respectively, that is $\mathbf{H} \subset \mathbf{B}$ and $\mathbf{H}^* \subset \mathbf{B}^*$, where $\mathbf{B}^*$ is a dual space to $\mathbf{B}$;

(B) operators $\Gamma(y)$ and $(\Gamma(y))^*$ are dissipative on the Hilbert spaces $\mathbf{H}$ and $\mathbf{H}^*$, respectively;

(C) operators $D(y)$ and $D^*(y)$ are contractive on the Hilbert spaces $\mathbf{H}$ and $\mathbf{H}^*$, respectively;

(D) $(y_n)_{n\in Z_+}$ is a uniformly ergodic Markov chain with stationary distribution $p(A), A \in \mathcal{Y}$;

(E) $m_i(y) := \int_0^\infty t^i G_y(dt)$ are uniformly integrable, $\forall i = 1, 2, 3$, where

$$G_y(t) := \mathcal{P}\{\omega : \theta_{n+1} \le t | y_n = y\}; \tag{2.27}$$

(F)

$$\int_Y p(dy) \parallel \Gamma(y)f \parallel^k < +\infty;$$

$$\int_Y p(dy) \parallel PD_j(y)f \parallel^k < +\infty;$$

$$\int_Y p(dy) \parallel \Gamma(y)f \parallel^{k-1} \cdot \parallel PD_j(y)f \parallel^{k-1} < +\infty;$$

$$\forall k = 1, 2, 3, 4, j = 1, 2, f \in B, \tag{2.28}$$

where $\mathbf{P}$ is the operator generated by the transition probabilities $P(y, A)$ of the Markov chain $(y_n)_{n\in Z_+}$:

$$P(y, A) := \mathcal{P}\{\omega : y_{n+1} \in A | y_n = y\}, \tag{2.29}$$

and

$$\{D_j(y)\}_{y\in Y}, \quad j = 1, 2\}$$

is a family of closed operators defined by the jumps operators $\{D^\epsilon(y); y \in Y\}$, which define a jump part of the semi-Markov RE in series scheme.

In particular, if $\mathbf{B} := \mathbf{C}_0(R)$, then $\mathbf{H} := \mathbf{W}^{l,2}(R)$ is a Sobolev space, and $\mathbf{W}^{l,2}(R) \subset \mathbf{C}_0(R)$ is a compact imbedding. For the spaces $\mathbf{B} := \mathbf{L}_2(R)$ and $\mathbf{H} := \mathbf{W}^{l,2}(R)$, the same result holds.

It follows from the conditions (A)-(B) that operators $\Gamma(y)$ and $(\Gamma(y))^*$ generate strongly continuous contractive semigroups of operators $\Gamma_y(t)$ and $\Gamma_y^*(t), \forall y \in Y$, in $\mathbf{H}$ and $\mathbf{H}^*$ respectively. From the conditions (A)-(C) it follows that the SMRE $V(t)$ in (2.2) is a contractive operator in $\mathbf{H}, \forall t \in R_+$, and $\| V(t)f \|_{\mathbf{H}}$ is a semimartingale $\forall f \in \mathbf{H}$. In such a way, conditions (A)-(C) imply the following result:

SMRE $V(t)f$ is a tight process in $\mathbf{B}$, namely, $\forall \Delta > 0$ there exists a compact set K_Δ such that

$$\mathcal{P}\{V(t)f \in K_\Delta; 0 \le t \le T\} \ge 1 - \Delta. \tag{2.30}$$

This result follows from the Kolmogorov-Doob inequality for the semi-martingale $\| V(t)f \|_{\mathbf{H}}$. See [44] for related references.

Inequality (2.30) provides a main tool towards limit theorems and rates of convergence for the sequence of SMRE in series scheme.

Averaging of Random Evolutions. Let us consider the following SMRE in series scheme:

$$V_\varepsilon(t)f = f + \int_0^t \Gamma(y(s/\varepsilon))V_\varepsilon(s)f ds + \sum_{k=1}^{\nu(t/\varepsilon)} [D^\varepsilon(y_k) - I]V_\varepsilon(\varepsilon\tau_k-)f, \tag{2.31}$$

where

$$D^\varepsilon(y) = I + \varepsilon D_1(y) + o(\varepsilon), \tag{2.32}$$

and

$$\{D_1(y)\}_{y \in Y}$$

is a family of closed linear operators, $\| o(\varepsilon)f \| /\varepsilon \to 0$ as $\varepsilon \to 0$,

$$f \in \mathbf{B}_0 := \bigcap_{y \in Y} \text{Dom } (\Gamma^2(y)) \cap \text{Dom}(D_1^2(y)). \tag{2.33}$$

Another form for $V_\varepsilon(t)$ in (2.31) is

$$V_\varepsilon(t) = \Gamma_{y(t/\varepsilon)}(t - \varepsilon\tau_{\nu(t/\varepsilon)}) \prod_{k=1}^{\nu(t/\varepsilon)} D^\varepsilon(y_k)\Gamma_{y_{k-1}}(\varepsilon\theta_k). \tag{2.34}$$

Under conditions (A)-(C) we know that the sequence of SMRE $V_\varepsilon(t)f$ is tight (see (2.30)) a.s..

Under conditions (D) and (E) with $i = 2$, and condition (F) with $k = 2$ and $j = 1$, the sequence of SMRE $V_\varepsilon(t)f$ is weakly compact in $\mathbf{D_B}[0, +\infty)$ with limit points in $\mathbf{C_B}[0, +\infty), f \in \mathbf{B}_0$.

We now consider the following process in $\mathbf{D_B}[0, +\infty)$:

$$M^\varepsilon_{\nu(t/\varepsilon)} f^\varepsilon := V^\varepsilon_{\nu(t/\varepsilon)} f^\varepsilon - f^\varepsilon - \sum_{k=0}^{\nu(t/\varepsilon)-1} E_p[V^\varepsilon_{k+1} f^\varepsilon_{k+1} - V^\varepsilon_k f^\varepsilon_k / \mathcal{F}_k], \tag{2.35}$$

where $V^\varepsilon_n := V_\varepsilon(\varepsilon\tau_n)$ (see (2.34)),

$$f^\varepsilon := f + \varepsilon f_1(y(t/\varepsilon)),$$

$$f_k^\varepsilon := f^\varepsilon(y_k),$$

function $f_1(x)$ is defined from the equation

$$(\mathbf{P}-I)f_1(y) = [(\hat{\Gamma}+\hat{D}) - (m(y)\Gamma(y) + \mathbf{P}D_1(y))]f,$$

$$\hat{\Gamma} := \int_Y p(dy)m(y)\Gamma(y),$$

$$\hat{D} := \int_Y p(dy)D_1(y),$$

$$m(y) := m_1(y) \tag{2.36}$$

(see (E)), $f \in \mathbf{B}_0$.

The process $M^\varepsilon_{\nu(t/\varepsilon)}f^\varepsilon$ is an $\mathcal{F}^\varepsilon_t$- martingale with respect to the σ-algebra $\mathcal{F}^\varepsilon_t := \sigma\{y(s/\varepsilon); 0 \le s \le t\}$.

The martingale $M^\varepsilon_{\nu(t/\varepsilon)}f^\varepsilon$ in (2.35) has the asymptotic representation:

$$M^\varepsilon_{\nu(t/\varepsilon)}f^\varepsilon = V^\varepsilon_{\nu(t/\varepsilon)}f^\varepsilon f - f - \varepsilon \sum_{k=0}^{\nu(t/\varepsilon)} (\hat{\Gamma}+\hat{D})V_k^\varepsilon f + o_f(\varepsilon), \tag{2.37}$$

where $\hat{\Gamma}, \hat{D}, f, f^\varepsilon$ are defined in (2.35)-(2.36) and

$$\| o_f(\varepsilon) \| /\varepsilon \to \text{const as } \varepsilon \to 0, \forall f \in \mathbf{B}_0.$$

We have used (2.25)-(2.26) as $n = 2$, and representation (2.32)and (2.33) in (2.37).

The families $l(M^\varepsilon_{\nu(t/\varepsilon)}f^\varepsilon)$ and

$$(\sum_{k=0}^{\nu(t/\varepsilon)} E_p[(V^\varepsilon_{k+1}f^\varepsilon_{k+1} - V_k^\varepsilon f_k^\varepsilon)/\mathcal{F}_k])$$

are weakly compact for all $l \in \mathbf{B}_0^*$ is a dense subset from $\mathbf{B}^*$.

Let $V_0(t)$ be a limit process for $V_\varepsilon(t)$ as $\varepsilon \to 0$. Since (see (2.34))

$$[V_\varepsilon(t) - V^\varepsilon_{\nu(t/\varepsilon)}] = [\Gamma_{y(t/\varepsilon)}(t - \varepsilon\tau_{\nu(t/\varepsilon)}) - I] \cdot V^\varepsilon_{\nu(t/\varepsilon)} \tag{2.38}$$

and the right hand side in (2.38) tends to zero as $\varepsilon \to 0$, the limits for $V_\varepsilon(t)$ and $V^\varepsilon_{\nu(t/\varepsilon)}$ are the same, namely, $V_0(t)$, p-a.s.

The sum $\varepsilon \sum_{k=0}^{\nu(t/\varepsilon)} (\hat{\Gamma}+\hat{D})V_k^\varepsilon f$ converges strongly as $\varepsilon \to 0$ to the integral

$$m^{-1}\int_0^t (\hat{\Gamma}+\hat{D})V_0(s)f ds.$$

The quadratic variation of the martingale $l(M^\varepsilon_{\nu(t/\varepsilon)}f^\varepsilon)$ tends to zero, and hence

$$M^\varepsilon_{\nu(t/\varepsilon)}f^\varepsilon \to 0 \text{ as } \varepsilon \to 0, \forall f \in \mathbf{B}_0, \forall l \in \mathbf{B}_0^*.$$

Passing to the limit in (2.37) as $\varepsilon \to 0$ and taking into account all the previous discussins, we obtain that the limit process $V_0(t)$ satisfies the equation

$$0 = V_0(t)f - f - m^{-1}\int_0^t (\hat{\Gamma}+\hat{D})V_0(s)f ds, \tag{2.39}$$

where

$$m := \int_X p(dx)m(x), f \in \mathbf{B}_0, t \in [0,T].$$

Diffusion Approximation of Random Evolutions. Let us consider SMRE $V_{\varepsilon(t/\varepsilon)}$, where $V_\varepsilon(t)$ is defined in (2.31) or (2.34), with the operators

$$D^\varepsilon(y) := I + \varepsilon D_1(y) + \varepsilon^2 D_2(y) + 0(\varepsilon^2), \tag{2.40}$$

$\{D_i(y); y \in Y, i = 1,2\}$ are closed linear operators and $\| o(\varepsilon^2) f \| / \varepsilon^2 \to 0, \varepsilon \to 0$

$$\forall f \in \mathbf{B}_0 := \bigcap_{x,y\in X} \text{Dom }(\Gamma^4(y)) \bigcap \text{Dom }(D_2(y)),$$

$$\begin{aligned} &\text{Dom }(D_2(y)) \subseteq \text{Dom}(D_1(y)); D_1(y) \subseteq \text{Dom }(D_1(y)),\\ &\forall y \in Y, \Gamma^i(y) \subset \text{Dom }(D_2(y)), \quad i = 1,2,3. \end{aligned} \tag{2.41}$$

In this way

$$V_\varepsilon(t/\varepsilon) = \Gamma_{y(t/\varepsilon^2)}(t/\varepsilon - \varepsilon\tau_{\nu(t/\varepsilon^2)}) \prod_{k=1}^{\nu(t/\varepsilon^2)} D^\varepsilon(y_k)\Gamma_{y_{k-1}}(\varepsilon\theta_k), \tag{2.42}$$

where $D^\varepsilon(y)$ are defined in (2.40).

Under conditions (A)-(C), the sequence of SMRE $V_\varepsilon(t/\varepsilon)f$ is tight (see (2.30)) a.s.

Under conditions (D) and (E) with $i = 3$, and condition (F) with $k = 4$, the sequence of SMRE $V_\varepsilon(t/\varepsilon)f, f \in \mathbf{B}_0$ is weakly compact in $\mathbf{D_B}[0,+\infty)$ with limit points in $\mathbf{C_B}[0,+\infty)$, $f \in \mathbf{B}_0$.

Assume now the *balance condition*

$$\int_Y p(dy)[m(y)\Gamma(y) + D_1(y)]f = 0, \forall f \in \mathbf{B}_0 \tag{2.43}$$

is satisfied. Let us consider the following process in $\mathbf{D_B}[0+\infty)$:

$$M^\varepsilon_{\nu(t/\varepsilon^2)} f^\varepsilon := V^\varepsilon_{\nu(t/\varepsilon^2)} f^\varepsilon - f^\varepsilon - \sum_{k=0}^{\nu(t/\varepsilon^2)} E_p[V^\varepsilon_{k+1} f^\varepsilon_{k+1} - V^\varepsilon_k f^\varepsilon_k / \mathcal{F}_k], \tag{2.44}$$

where $f^\varepsilon := f + \varepsilon f_1(y(t/\varepsilon^2)) + \varepsilon^2 f_2(y(t/\varepsilon^2))$, and functions f_1 and f_2 are defined from the following equations:

$$\begin{aligned} (\mathbf{P} - I) f_1(y) &= -[m(y)\Gamma(y) + D_1(y)]f,\\ (\mathbf{P} - I) f_2(y) &= [\hat{L} - L(x)]f,\\ \hat{L} &:= \int_Y p(dy) L(y),\\ L(y) := (m(y)\Gamma(y) &+ D_1(y))(R_0)(m(y)\Gamma(y) + D_1(y))\\ +m_2(y)\Gamma^2(y)/2 &+ m(y)D_1(y)\Gamma(y) + D_2(y), \end{aligned} \tag{2.45}$$

where $\mathbf{R}_0$ is a potential operator of $(y_n)_{n\in Z_+}$.

The balance condition (2.43) and condition $\prod(\hat{L} - L(y)) = 0$ give the solvability of the equations in (2.45).

The process $M_{\nu(t/\varepsilon^2)}f^\varepsilon$ is an $\mathcal{F}_t^\varepsilon$-martingale with respect to the σ-algebra $\mathcal{F}_t^\varepsilon := \sigma\{y(s/\varepsilon^2); 0 \le s \le t\}$. This martingale has the following asymptotic representation:

$$M_{\nu(t/\varepsilon^2)}f^\varepsilon = V_{\nu(t/\varepsilon^2)}f^\varepsilon - f - \varepsilon^2 \sum_{k=0}^{\nu(t/\varepsilon^2)} \hat{L}V_k^\varepsilon f + O_f(\varepsilon), \tag{2.46}$$

where $\hat{L}$ is defined in (2.45) and

$$\| O_f(\varepsilon) \| / \varepsilon \to const,$$

as $\varepsilon \to 0$, for all $f \in \mathbf{B}_0$. Here we have used (2.25), (2.26) as $n = 3$, and representation (2.40) and (2.45) in (2.46).

The families $l(M_{\nu(t/\varepsilon^2)}f^\varepsilon)$ and $l(\sum_{k=0}^{\nu(t/\varepsilon^2)} E_p[(V_{k+1}^\varepsilon f_{k+1}^\varepsilon - V_k^\varepsilon f_k^\varepsilon)/\mathcal{F}_k])$ are weakly compact for all $l \in \mathbf{B}_0^*, f \in \mathbf{B}_0$.

From (34) we obtain that the limits for $V_\varepsilon(t/\varepsilon)$ and $V_{\nu(t/\varepsilon^2)}^\varepsilon$ are the some, namely, $V^0(t)$.

The sum $\varepsilon^2 \sum_{k=0}^{\nu(t/\varepsilon^2)} \hat{L}V_k^\varepsilon f$ converges strongly as $\varepsilon \to 0$ to the integral

$$m^{-1} \int_0^t \hat{L}V^0(s) f ds.$$

Let $M^0(t)f$ be a limit martingale for $M_{\nu(t/\varepsilon^2)}^\varepsilon f^\varepsilon$ as $\varepsilon \to 0$. Then, from (2.44)-(2.46) and previous discussions, we have as $\varepsilon \to 0$:

$$M^0(t)f = V^0(t)f - f - m^{-1} \cdot \int_0^t \hat{L}V^0(s) f ds. \tag{2.47}$$

The quadratic variation of the martingale $M^0(t)f$ has the form

$$< l(M^0(t)f) >= \int_0^t \int_Y l^2(\sigma(y)\Gamma(y)V^0(s)f)p(dy)ds, \tag{2.48}$$

where

$$\sigma^2(y) := [m_2(y) - m^2(y)]/m.$$

The solution of martingale problem for $M^0(t)$ (namely, to find the representation of $M^0(t)$ with quadratic variation (2.48)) is expressed by the integral over Wiener orthogonal martingale measure $W(dy, ds)$ with quadratic variation $p(dy)ds$:

$$M^0(t)f = \int_0^t \int_Y \sigma(y)\Gamma(y)V^0(s)fW(dy, ds). \tag{2.49}$$

In this way, the limit process $V^0(t)$ satisfies the following equation (see (2.47) and (2.49)):

$$V^0(t)f = f + m^{-1} \cdot \int_0^t \hat{L} \cdot V^0(s) f ds + \int_0^t \int_Y \sigma(y)\Gamma(y)V^0(s)fW(dy, ds). \tag{2.50}$$

If the operator $\hat{L}$ generates a semigroup $U(t)$ then the process $V^0(t)f$ in (2.50) satisfied the following equation:

$$V^0(t)f = U(t)f + \int_0^t \int_Y \sigma(y)U(t-s)\Gamma(y)V^0(s)fW(dy, ds). \tag{2.51}$$

The *uniqueness* of the limit evolution $V_0(t)f$ in the *averaging* scheme follows from the equation (2.39) and the fact that if the operator $\hat{\Gamma} + \hat{D}$ generates a semigroup, then $V_0(t)f = \exp\{(\hat{\Gamma} + \hat{D}) \cdot t\}f$ and this representation is unique.

The *uniqueness* of the limit evolution $V^0(t)f$ in *diffusion approximation* scheme follows from the uniqueness of the solution of martingale problem for $V^0(t)f$ (see (2.47)-(2.49)). The latter is proved by *dual SMRE* in series scheme by the constructing the limit equation in diffusion approximation and by using a dual identity. See [44,72].

Averaging of Random Evolutions in Reducible Phase Space and Merged Random Evolutions. We now assume that the following conditions hold true:

(a) *decomposition* of phase space X (*reducible* phase space):

$$Y = \bigcup_{v \in V} Y_v, Y_v \bigcap Y_{\tilde{v}} = \emptyset, v \neq \tilde{v} : \tag{2.52}$$

where $(V, \mathcal{V})$ is a measurable phase space (*merged* phase space);

(b) Markov renewal process $(y_n^\varepsilon; \theta_n)_{n \in Z_+}$ on $(Y, \mathcal{Y})$ has the *semi-Markov kernel*

$$Q_\varepsilon(y, A, t) := P_\varepsilon(y, A)G_y(t), \tag{2.53}$$

where $P_\varepsilon(y, A) = P(y, A) - \varepsilon^1 P_1(y, A), y \in Y, A \in \mathcal{Y}, P(y, A)$ are the transition probabilities of the *supporting non-perturbed* Markov chain $(y_n)_{n \in Z_+}$, P_1 is a probability measure;

(c) the stochastic kernel $P(y, A)$ is adapted to the decomposition (2.52) in the following form:

$$P(y, Y_v) = \begin{cases} 1, & x \in Y_v \\ 0, & x \notin Y_v, \quad v \in V; \end{cases}$$

(d) the Markov chain $(y_n)_{n \in Z_+}$ is uniformly ergodic with stationary distributions

$$p_v(B) = \int_{Y_v} P(y, B)p_v(dy), \forall v \in V, \forall B \in \mathcal{Y}; \tag{2.54}$$

(e) there is a family $\{p_v^\varepsilon(A); v \in V, A \in \mathcal{Y}, \varepsilon > 0\}$ of stationary distributions of perturbed Markov chain $(y_n^\varepsilon)_{n \in Z_+}$;

(f)

$$b(v) := \int_{Y_v} p_v(dx)P_1(x, Y_v) > 0, \qquad \forall v \in V,$$

$$b(v, \Delta) := -\int_{Y_v} p_v(dx)P_1(x, Y_\Delta) > 0, \qquad \forall v \notin \Delta, \qquad \Delta \in V; \tag{2.55}$$

(g) the operators $\Gamma(v) := \int_{Y_v} p_v(dy)m(y)\Gamma(y)$ and

$$\hat{D}(v) := \int_{Y_v} p_v(dx) \int_{Y_v} P(x, dy)D_1(y) \tag{2.56}$$

are closed, $\forall v \in V$, with a common domain $\mathbf{B}_0$, and operators $\hat{\Gamma}(v) + \hat{D}(v)$ generate semigroup of operators, $\forall v \in V$.

Decomposition (2.52) in (a) defines the *merging* function

$$v(y) = v, \qquad \forall y \in Y_v, \qquad v \in V. \tag{2.57}$$

We note that σ-algebras $\mathcal{Y}$ and $\mathcal{V}$ are coordinated such that

$$Y_\Delta = \bigcup_{v \in \Delta} Y_v, \qquad \forall v \in V, \qquad \Delta \in \mathcal{V}. \tag{2.58}$$

We set $\pi_v f(v) := \int_{Y_v} p_v(dy) f(y)$ and $y^\varepsilon(t) := y^\varepsilon_{\nu(t/\varepsilon)}$.

SMRE in reducible phase space X is defined by the solution of the equation:

$$V_\varepsilon(t) = I + \int_0^t \Gamma(y^\varepsilon(s/\varepsilon)) V_\varepsilon(s) ds + \sum_{k=0}^{\nu(t/\varepsilon)} [D^\varepsilon(y^\varepsilon_k) - I] V_\varepsilon(\varepsilon\tau_k^-), \tag{2.59}$$

where $D^\varepsilon(y)$ are defined in (2.32).

Consider the martingale

$$\begin{aligned} M^\varepsilon_{\nu(t/\varepsilon)} f^\varepsilon(y^\varepsilon(t/\varepsilon)) \ &:= V^\varepsilon_{\nu(t/\varepsilon)} f^\varepsilon(y^\varepsilon(t/\varepsilon) - f^\varepsilon(y) \\ &- \textstyle\sum_{k=0}^{\nu(t/\varepsilon)-1} E^\varepsilon_{p_u} [V^\varepsilon_{k+1} f^\varepsilon_{k+1} - V^\varepsilon_k f^\varepsilon_k / \mathcal{F}^\varepsilon_k], \end{aligned} \tag{2.60}$$

where

$$\mathcal{F}^\varepsilon_n := \sigma\{y^\varepsilon_k, \theta_k; 0 \le k \le n\},$$

$$f^\varepsilon(y) := \hat{f}(v(y)) + \varepsilon f^1(y), \hat{f}(v) := \int_{Y_v} p_v(dy) f(y), \tag{2.61}$$

$$\begin{aligned} (\mathbf{P} - I) f_1(y) \ &= [-(m(y)\Gamma(y) + D_1(y)) + \hat{\Gamma}(v) \\ &+ \hat{D}(v) + (\Pi_v - I) P_1] \hat{f}(v), \end{aligned} \tag{2.62}$$

$$f^\varepsilon_k := f^\varepsilon(y^\varepsilon_k), V^\varepsilon_n := V_\varepsilon(\varepsilon\tau_n),$$

and $V_\varepsilon(t)$ is defined in (2.59), P_1 is an operator generated by $P_1(y, A)$ (see (2.53)).

The following representation holds (see [44,45]):

$$\Pi^\varepsilon_u = \Pi_u - \varepsilon^r \Pi_u P_1 R_0 + \varepsilon^{2r} \Pi^\varepsilon_u (P_1 R_0)^2, r = 1, 2, \tag{2.63}$$

where $\Pi^\varepsilon_v, \Pi_v, P_1$ are the operators generated by p^ε_v, p_v and $P_1(y, A)$ respectively, $y \in Y, A \in \mathcal{Y}, v \in V$.

It follows from (63) that for every continuous and bounded function $f(x)$

$$E^\varepsilon_{p_v} f(x) \to E_{p_v} f(x), \qquad \forall_v \in V$$

as $\epsilon \to 0$. Here we used calculations similar to those in the above discussion on averaging RE replacing E_{p_u} by $E_{p^\varepsilon_u}$.

Under conditions (A)-(C) the sequence of SMRE $V_\varepsilon(t)f$ in (2.17) with $f \in \mathbf{B}_0$ is tight (see (2.30)) p_u – a.s., $\forall v \in V$.

Under conditions (D) and (E) with $i = 2$, and condition (F) with $k = 2$ and $j = 1$, the sequence of SMRE $V_\varepsilon(t)f$ is weakly compact p_v – a.s., $\forall v \in V$, in $\mathbf{D_B}[0, +\infty)$ with limit points in $\mathbf{C_B}[0, +\infty)$.

We note that $u(y^\varepsilon(t/\varepsilon)) \to \hat{y}(t)$ as $\varepsilon \to 0$, where $\hat{y}(t)$ is a *merged* jump Markov process in $(V, \mathcal{V})$ with *infinitesimal operator* $\Lambda(\hat{\mathbf{P}} - I)$,

$$\Lambda \hat{f}(v) := [b(v)/m(v)]\hat{f}(v),$$

$$\hat{\mathbf{P}}\hat{f}(v) := \int_V [b(v, dv')/b(v)]\hat{f}(v'),$$

$$m(v) := \int_{Y_v} p_v(dx)m(x), \tag{2.64}$$

$b(v)$ and $b(v, \triangle)$ are defined in (2.55). We also note that

$$\Pi_v P_1 = \Lambda(\hat{\mathbf{P}} - I), \tag{2.65}$$

where Π_v is defined in (2.58), P_1-in (2.53), Λ and $\hat{\mathbf{P}}$-in (2.64).

Using (2.25), (2.26) with $n = 2$, and (2.61)-(2.62), (2.63) with $r = 1$ and (2.65), we obtain the following representation:

$$M^\varepsilon_{\nu(t/\varepsilon}f^\varepsilon(y^\varepsilon(t/\varepsilon)))$$

$$\begin{aligned} &= V^\varepsilon_{\nu(t/\epsilon)}\hat{f}(u(y^\epsilon(t/\epsilon))) - \hat{f}(u(x)) \\ &-\varepsilon \textstyle\sum_{k=0}^{\nu(t/\varepsilon)}[m(u)\hat{\Gamma}(u) + m(u)\hat{D}(u) \\ &+m(u)\Lambda(\hat{P} - I)V^\varepsilon_k \hat{f}(u(x^\varepsilon_k)) + 0_f(\varepsilon), \end{aligned} \tag{2.66}$$

where $\| O_f(\varepsilon) \| /\varepsilon \to const$ as $\varepsilon \to 0, \forall f \in \mathbf{B}_0$. Since the third term in (66) tends to the integral

$$\int_0^t [\Lambda(\hat{\mathbf{P}} - I) + \hat{\Gamma}(y(s)) + \hat{D}(\hat{y}(s))]\hat{V}_0(s)\hat{f}(\hat{y}(s))ds,$$

and the quadratic variation of the martingale $l(M^\varepsilon_{\nu(t/\varepsilon)}f^\varepsilon(y^\varepsilon(t/\varepsilon)))$ tends to zero as $\varepsilon \to 0$ (and, hence $M^\varepsilon_{\nu(t/\varepsilon)}f^\varepsilon(y^\varepsilon(t/\varepsilon))) \to 0, \varepsilon \to 0)\forall l \in \mathbf{B}^*_0$, we obtain from (2.51) that the limit evolution $\hat{V}_0(t)$ satisfies equation:

$$\hat{V}_0(t)\hat{f}(\hat{x}(t)) = \hat{f}(u) + \int_0^t [\Lambda(\hat{\mathbf{P}} - I) + \hat{\Gamma}(\hat{x}(s)) + \hat{D}(\hat{x}(s))]\hat{V}_0(s)\hat{f}(\hat{x}(s))ds. \tag{2.67}$$

Normal Deviations of Random Evolutions.

The averaged evolution obtained in averaging and merging schemes can be considered as the first approximation to the initial evolution. The diffusion approximation of the SMRE determine the second approximation to the initial evolution, since the first approximation under balance condition–the averaged evolution-appears to be trivial.

Here we consider the *double approximation* to the SMRE-the averaged and the diffusion approximation- when the balance conditions fails.

We introduce the *deviation process* as the normalized difference between the initial and averaged evolutions. In the limit we obtain the *normal deviations* of the initial SMRE from the averaged one.

Consider the SMRE $V_\varepsilon(t)$ in (2.31) and the averaged evolution $V_0(t)$ in (2.39), and consider the deviation of the initial evolution $V_\varepsilon(t)f$ from the averaged one $V_0(t)f$:

$$W_\varepsilon(t)f := \varepsilon^{-1/2}[V_\varepsilon(t) - V_0(t)]f, \forall f \in \mathbf{B}_0 \tag{2.68}$$

(see (2.33)).

From equaations (2.31) and (2.30) we obtain the following relation for $W_\varepsilon(t)$:

$$\begin{aligned} W_\varepsilon(t)f &= \varepsilon^{-1/2} \int_0^t (\Gamma(y(s/\varepsilon)) - \hat{\Gamma})V_\varepsilon(s)f ds + \int_0^t \hat{\Gamma} W_\varepsilon(s) f ds \\ &+\varepsilon^{-1/2}[V_\varepsilon^d(t) - \int_0^t \hat{D}\cdot V_0(s)ds]f, \forall f \in B_0, \end{aligned} \tag{2.69}$$

where

$$V_\varepsilon^d(t)f := \sum_{k=1}^{\nu(t/\varepsilon)} [D^\varepsilon(y_k) - I]V_\varepsilon(\varepsilon\tau_k^-)f,$$

and $\hat{\Gamma}$ and $\hat{D}$ are defined in (2.36).

If the process $W_\varepsilon(t)f$ has the weak limit $W_0(t)f$ as $\varepsilon \to 0$ then we obtain:

$$\int_0^t \hat{\Gamma} W_\varepsilon(s) f ds \to \int_0^t \hat{\Gamma} W_0(s) f ds, \varepsilon \to 0. \tag{2.70}$$

Since the operator $(\Gamma(y) - \hat{\Gamma})$ satisfies to the balance condition

$$\Pi(\Gamma(y) - \hat{\Gamma})f = 0,$$

the diffusion approximation of the first term in the righthand side of (2.69) gives

$$\varepsilon^{-1/2} \int_0^t l(\Gamma(y(s/\varepsilon)) - \hat{\Gamma})f)ds \to l(\sigma_1 f)w(t), \varepsilon \to 0 \tag{2.71}$$

where

$$l^2(\sigma_1 f) = \int_Y \rho(dx)[m(x)l((\Gamma(x) - \hat{\Gamma})f)\mathbf{R}_0\Gamma(x) - \hat{\Gamma})f) + 2^{-1}\cdot m_2(x)l^2((\Gamma(x) - \hat{\Gamma})f)]/m,$$

$\forall l \in \mathbf{B}_0, w(t)$ is a standard Wiener process.

Since $\Pi(\mathbf{P}D_1(y) - \hat{D})f = 0$, the diffusion approximation of the third term in the right-hand side of (2.69) gives the following limit:

$$\varepsilon^{-1/2} l(V_\varepsilon^d(t)f - \int_0^t \hat{D}V_0(s)f ds) \to l(\sigma_2 f)w(t), \tag{2.72}$$

as $\varepsilon \to 0$, where

$$l^2(\sigma_2 f) := \int_Y \rho(dx) l((D_1(y) - \hat{D})f)\mathbf{R}_0 l((D_1(y) - \hat{D})f).$$

Passing to the limit as $\varepsilon \to 0$ in the representation (2.68) and using (2.69)-(2.72) we obtain the equation for $W_0(t)f$:

$$W_0(t)f = \int_0^t \hat{\Gamma} W_0(s) f ds + \sigma f w(t), \tag{2.73}$$

where the variance opeator σ is determined from the relation:

$$l^2(\sigma f) := l^2(\sigma_1 f) + l^2(\sigma_2 f), \qquad \forall l \in \mathbf{B}_0, \qquad \forall l \in \mathbf{B}_0^*, \tag{2.74}$$

and operators σ_1 and σ_2 are defined in (2.71) and (2.72), respectively.

Double approximation of the SMRE has the form:

$$V_\varepsilon(t)f \approx V_0(t)f + \sqrt{\varepsilon} W_0(t)f$$

for small ε, which perfectly fits the standard form of the Central Limit Theorem with non-zero limiting mean value.

2.2. Averaging of Difference Equations in Random Media

In this section, we return to the method of random evolution, developed in last section by considering two types of difference equations: a) difference equations with Markov random pertubations as a random media and discrete parameter; b) difference equations with semi-Markov random pertubations as a random media and continuous parameter.

2.2.1. AVERAGING IN MARKOV RANDOM MEDIA

We consider a system in a linear phase space X with discrete time $n \in Z^+$ which is perturbed by a Markov chain $(y_n)_{n\epsilon Z^+}$ defined on a measurable space $(Y, \mathcal{Y})$. The system depends on a small parameter $\epsilon > 0$. Let $X_n^\epsilon \in X$ denote the state of the system at time n. We suppose that X_n^ϵ is determined by the recurrence relations:

$$X_{n+1}^\epsilon - X_n^\epsilon = \varepsilon g(X_n^\epsilon, y_{n+1}), \qquad X_0^\epsilon = x_0, \tag{2.75}$$

where $X_0 = x$ is a given initial value and $g : X \times Y \to X$ is a given function.

We shall concentrate on system (2.75) in the following phase spaces $X = R^d$ with $d \geq 1$, and we assume always that function $g(x, y)$ is measurable in y and continuous in $x \in X$. Our goal is to investigate the asymptotic behaviour of the system as $\varepsilon \to 0$ and $n \to \infty$.

We first rewrite equation (2.75) in the form

$$X_{n+1}^{\epsilon} = X_0 + \varepsilon \sum_{k=1}^{n} g(X_k^{\epsilon}, y_{k+1}). \tag{2.76}$$

Consider the following family of operators $D^{\epsilon}(y)$ on $\mathbf{B} = \mathbf{C}_0^1(X)$:

$$D^{\varepsilon}(y)f(x) := f(x + \varepsilon g(x, y)), f(x) \in \mathbf{C}_0^1(X). \tag{2.77}$$

Therefore, $f(X_{n+1}^{\epsilon})$ can be represented as

$$f(X_{n+1}^{\epsilon}) = f(X_0 + \varepsilon \sum_{k=1}^{n} g(X_k^{\epsilon}, y_{k+1})) = \prod_{k=0}^{n} D^{\epsilon}(y_{k+1}) f(x) =: V_n^{\epsilon} f(x),$$
$$X_0 = x. \tag{2.78}$$

We note that operators $D^{\epsilon}(y)$ are linear contractive and has the following asymptotic expansion:

$$D^{\epsilon}(y)f(x) = f(x) + \varepsilon g(x, y)\frac{d}{dx}f(x) + \varepsilon O_{\epsilon}(1)f(x) \tag{2.79}$$

as $\varepsilon \to 0$, where $\| O_{\epsilon}(1)f(x) \| \to 0$ as $\varepsilon \to 0$, and $\| \cdot \|$ is a norm in the space $\mathbf{C}_0^1(X)$. Namely,

$$D^{\epsilon}(y)f(x) = f(x) + \varepsilon D_1(y)f(x) + \varepsilon O_{\epsilon}(1)f(x), \tag{2.80}$$

where

$$D_1(y)f(x) := g(x, y)\frac{d}{dx}f(x). \tag{2.81}$$

We now suppose that the process $(y_n)_{n \in Z^+}$ is a stationary ergodic process with ergodic distribution $p(dy)$, namely, for every function $f : Y \to R$ with $\int_Y |f(y)|p(dy) < +\infty$, we have

$$P\{\lim_{n \to +\infty} \frac{1}{n} \sum_{k=1}^{n} f(y_k) = \int_Y f(y)p(dy)\} = 1. \tag{2.82}$$

Let

$$\hat{D}f(x) := \int_Y D_1(y)f(y)p(dy), \tag{2.83}$$

and consider the following equation:

$$\hat{V}_t f(x) - f(x) - \int_0^t \hat{D}\hat{V}_s f(x)ds = 0, \ \forall\ f \in \mathbf{C}_0^1(X). \tag{2.84}$$

From the discussion on averaging of RE in Section 2.1.3 it follows that if

$$\int_Y p(dy)\|D_1(y)f(x)\|^2 < +\infty \tag{2.85}$$

and if there exists a compact set $K_T^{\Delta} \subset \mathbf{C}^1(X)$ such that

$$\lim_{\varepsilon \to 0} \inf \mathcal{P}\{V_{[t/\epsilon]}^{\epsilon} f \in K_T^{\Delta}; 0 \le t \le T\} \ge 1 - \Delta, \quad \forall\ \Delta > 0, T > 0, \tag{2.86}$$

then the sequence $V_{[t/\epsilon]}^{\epsilon}$ is relatively compact in $\mathbf{D}_B[O, +\infty]$ with limit points in $\mathbf{C}[0, +\infty]$. Moreover, if the operator $\hat{D}$ in (2.83) generates a semigroup, then $V_{[t/\epsilon]}^{\epsilon}$ converges weakly as $\epsilon \to 0$ to the process $\hat{V}_t$ in (2.84), $\forall\ t \in [0, T]$.

The family $V^{\epsilon}_{[t/\epsilon]}$, corresponds to the process (see (2.4)) $X^{\epsilon}_{[t/\epsilon]+1}$, since

$$\begin{aligned} V^{\epsilon}_{[t/\epsilon]+1}f(x) &= \prod_{k=0}^{[t/\epsilon]+1} D^{\epsilon}(y_{k+1})f(x) \\ &= f(X_0 + \varepsilon \sum_{k=o}^{[t/\epsilon]+1} g(X^{\epsilon}_k, y_{k+1})) \\ &= f(X^{\epsilon}_{[t/\epsilon]+1}). \end{aligned} \tag{2.87}$$

We now assume that the following conditions are satisfied:

$$\int_Y |g(x,y)|^2 p(dy) \ < \ +\infty \text{ for all fixed x}\epsilon\text{X}, \tag{2.88}$$

and $g^l_x(x,y)$ is bounded and continuous, where g^l_x is the l-th derivitive with respect to $x, l \geq 1$.

To obtain compactness condition (2.86) for our process $X^{\epsilon}_{[t/\epsilon]}$, we need to construct a compact set in the Banach space $\mathbf{B}$, so that we can construct a Hilbert space $\mathbf{H}$ that is compactly embedded in $\mathbf{B} := \mathbf{C}^1_0(X)$. This can be easily achived since the Sobolev imbedding theorem [70] states that bounded sets in $\mathbf{W}^{l,2}(R^d)$ are compacts in $\mathbf{C}^1_0(R^d)$ provided $2l \geq d$. Therefore, we have natural choice

$$\mathbf{B} = \mathbf{C}^1_0(R^d), \quad \mathbf{H} = \mathbf{W}^{l,2}(R^d),$$

with $2\,\ell \geq d$. We note that if $d = 1$ then it sufficient to take $l = 1$.

It is easy to show that if (2.88) holds, then so do (2.85) and (2.86). It means that family of measures, generated by process $X^{\epsilon}_{[t/\epsilon]}$, is relatively compact and hence, there exists a unique limiting process $\hat{x}_t$ of $X^{\epsilon}_{[t/\epsilon]}$ process $\hat{X}_t$ as $\varepsilon \to 0$, in therms of weak convergence.

From (2.85) and (2.86) it follows that

$$V^{\epsilon}_{[t/\epsilon]}f(x) \to_{\varepsilon\to 0} \hat{V}_t f(x), \tag{2.89}$$

and from (2.87) and (2.89) we obtain

$$f(X^{\epsilon}_{[t/\epsilon]}) = V^{\epsilon}_{[t/\epsilon]}f(x) \to \hat{V}_t f(x) = f(\hat{x}_t)$$

as $\epsilon \to 0$. Namely,

$$f(X^{\epsilon}_{[t/\epsilon]}) \to f(\hat{x}_t) \tag{2.90}$$

as $\epsilon \to 0$. Moreover, from (2.84), (2.87) and (2.89)-(2.90) we obtain

$$f(\hat{x}) - f(x) - \int_o^t \hat{D}f(\hat{x})ds = 0. \tag{2.91}$$

Recall that

$$\hat{D}f(x) = \int_Y p(dy)g(x,y)\frac{d}{dx}f(x) = \hat{g}(x)\frac{d}{dx}f(x), \tag{2.92}$$

we obtain

$$f(\hat{x}) - f(x) - \int_0^t \hat{g}(\hat{x}_s)\frac{d}{dx}f(\hat{x}_s)ds = 0. \tag{2.93}$$

Therefore, $f(\hat{x}_t)$ satisfies

$$\begin{cases} \frac{df(\hat{x}_t)}{dt} &= \hat{g}(\hat{x})\frac{d}{dx}f(\hat{x}_t) \\ f(\hat{x}_0) &= f(x), \end{cases}$$

and $\hat{x}_t$ satisfies

$$\begin{cases} \frac{d\hat{x}_t}{dt} &= \hat{g}(\hat{x}_t) \\ \hat{x}_0 &= x. \end{cases}$$

In summary, we obtain the following

Theorem 1: Under conditions (2.82), (2.88) and (2.89) the process $X^\epsilon_{[t/\epsilon]}$ in (2.75) converges weakly to the process $\hat{x}_t$ given in (2.93) as $\epsilon \to 0$, where $\hat{g}(x) := \int_Y p(dy)g(x,y)$.

We now conclude this subsection with a few remarks.

Remark 1: For the process:

$$\begin{aligned} X^\epsilon_t &= X^\epsilon_{[t/\epsilon]} + (t/\epsilon - [t/\epsilon])(X^\epsilon_{[t/\epsilon]+1} - X^\epsilon_{[t/\epsilon]} \\ &= X^\epsilon_{[t/\epsilon]}) + \epsilon(t/\epsilon - [t/\epsilon])g(X^\epsilon_{t/\epsilon]}, y_{[t/\epsilon]+1}). \end{aligned} \tag{2.94}$$

by Theorem 1, the process X^ϵ_t converges weakly to the process $\hat{x}(t)$ as $\epsilon \to 0$.

Remark 2: By Theorem 1 it follows that $\sup_{n\epsilon\leq t}||X^\epsilon_n - \hat{x}(\epsilon n)|| \to_{\epsilon\to 0} 0$ for any $t > 0$, where X^ϵ_n and $\hat{x}(t)$ are defined in (2.75) and (2.93), respectively.

Remark 3: We also note that Theorem 1, in Hoppensteadt, Salehi and Skorohod [36, p. 466] is a special core of our Theorem 1.

2.2.2 AVERAGING IN SEMI-MARKOV RANDOM MEDIA

Assume that the random medium is described by a Markov renewal process $(y_n; \theta_n)_{n\in Z^+}$ with stochastic kernel

$$Q(y, dz, dt) := \mathcal{P}\{y_{n+1} \in dz, \theta_{n+1} \leq dt/y_0 = y\} = P(y, dz)G_y(dt), \tag{2.95}$$

and let

$$\nu(t) = \max\{\mathrm{n} : \tau_{\mathrm{n}} \leq \mathrm{t}\} \tag{2.96}$$

be a counting process, where

$$\tau_n = \sum_{k=1}^n \theta_k, \quad \theta_0 = 0.$$

We consider the following difference equation in semi-Markov random media:

$$\begin{cases} X^{\epsilon}_{\nu(t/\epsilon)+1} - X^{\epsilon}_{\nu(t/\epsilon)} &= \epsilon\, g(X^{\epsilon}_{\nu(t/\epsilon)}, y_{\nu(t/\epsilon)+1}) \\ X^{\epsilon}_{0} &= X_0 = x \in X, \end{cases} \tag{2.97}$$

where $y_{\nu(t)}$ is a semi-Markov process and $\nu(t)$ is defined in (2.96).

We note that if $t \in [\varepsilon\tau_n, \varepsilon\tau_{n+1})$, where τ_n is defined in (2.97), then $X^{\epsilon}_{\nu(t/\epsilon)}$ satisfies the equation:

$$X^{\varepsilon}_{n+1} - X^{\varepsilon}_{n} = \varepsilon g(X^{\epsilon}_{n}, y_{n+1}).$$

(comparing (2.75)). We now assume the semi-Markov process is regular, namely

$$P\{\nu(t) < +\infty\} = 1, \qquad \forall t \in R_+. \tag{2.98}$$

To investigate the asymptotic behavior of the system (2.25) as $\epsilon \to 0$, we rewrite equation (2.97) as

$$X^{\epsilon}_{\nu(t/\epsilon)+1} = X_0 + \varepsilon \sum_{k=0}^{\nu(t/\epsilon)} g(X^{\epsilon}_{k}, y_{k+1}). \tag{2.99}$$

Therefore, for a given $f \in \mathbf{C}^1_0(X)$, we have

$$\begin{aligned} f(X_{\nu(t/\epsilon)+1}) &= f(X_0 + \varepsilon \textstyle\sum_{k=0}^{\nu(t/\epsilon)} g\,(x^{\epsilon}_{k}, y_{k+1})) = \prod_{k=0}^{\nu(t/\epsilon)} D^{\epsilon}(y_{k+1}) f(x) \\ &=: V^{\epsilon}_{\nu(t/\epsilon)} f(x), \end{aligned} \tag{2.100}$$

where the operators $D^{\epsilon}(y)$ are defined in (2.77), and V^{ϵ}_{u} in (2.78). We note that product in (2.100) is finite since condition (2.98) is satisfied. We note that operators $D^{\epsilon}(y)$ are linear and contractive uniformly with respect to y and admits the representation (2.79) (or equivalently, (2.80)).

We suppose that

$$m_2(y) := \int_0^{\infty} t^2 G_y(dt) \tag{2.101}$$

is uniformly integrable, where $G_y(dt) := \mathcal{P}\{\theta_{n+1} \le dt | y_0 = y\}$. Let

$$\begin{aligned} \tilde{D}f(x) &= \textstyle\int_Y p(dy) D_1(y) f(y)/m = \int_Y p(dy) g(x,y) \frac{d}{dx} f(x)/m \\ &=: \tilde{g}(x) \tfrac{d}{dx} f(x), \end{aligned} \tag{2.102}$$

where

$$m := \int_Y m(y) p(dy), m(y) := \int_0^{\infty} t G_y(dt). \tag{2.103}$$

We consider the following euation

$$\tilde{V}_t f(x) - f(x) - \int_0^t \tilde{D}\tilde{V}_s f(x) ds = 0, \quad \forall f \in \mathbf{C}^1_0(X). \tag{2.104}$$

From the theory of semi-Markov random evolutions developed in section 4.1.3 it follows that if conditions (2.82), (2.85), (2.93) and (2.101) are satisfied and if there exists a compact set $K_T^{\triangle} \subset \mathbf{C}_0^1(X)$ such that

$$\lim_{\varepsilon\to 0}\inf \mathcal{P}\{V_{\nu(t/\epsilon)}f(x)\epsilon K_T^{\triangle}; 0 \le t \le T\} \ge 1-\triangle, \quad \forall\triangle > 0, \forall T > 0, \tag{2.105}$$

then the sequence $V_{\nu(t/\epsilon)}^{\varepsilon}$ is relatively compact in $\mathbf{D_B}[0,+\infty)$ with limit points in $\mathbf{C}[0,+\infty)$. Moreover, if the operator $\tilde{D}$ in (2.102) generates asemigroup, then $V_{\nu(t/\epsilon)}^{\varepsilon}$ converges weakly as $\epsilon \to 0$ to the process $\tilde{V}_t$ in (104), $\forall\ t\epsilon[0,T]$.

From condition (2.88), (2.89), (2.98) and (2.101) it follows that conditions (2.85) and (2.105) are satisfied. Therefore, the family of measures, generated by processes $X_{\nu(t/\epsilon)}^{\epsilon}$, is relatively compact and there exists a unique limiting for $X_{\nu(t/\epsilon)}^{\epsilon}$ process $\tilde{x}_t$ as $\epsilon \to 0$, in the sense of weak convergence.

From (2.100) and (2.104) it follows that

$$f^{\epsilon}(X_{\nu(t/\epsilon)}) = V_{\nu(t/\epsilon)}^{\epsilon}f(x) \to \tilde{V}_t f(x) = f(\tilde{x}_t) \tag{2.106}$$

as $\epsilon \to 0$. Moreover, from (2.102), (2.104), (2.106) and (2.107) we obtain that

$$f(\tilde{x}_t) - f(x) - \int_0^t \tilde{D}f(\tilde{x}_s)ds = 0, \tag{2.107}$$

where $\tilde{D}$ is defined in (2.102). Consequently, taking into account (2.107) we have that $f(\tilde{x}_t)$ satisfies the equation:

$$f(\tilde{x}_t) - f(x) - \int_0^t \tilde{g}(\tilde{x}_s)\frac{d}{dx}f(\tilde{x}_s)ds = 0, \forall f(x) \in \mathbf{C}_0^1(X).$$

This means that $\tilde{x}_t$ satisfies

$$\frac{d\tilde{x}_t}{dt} = \tilde{g}(\tilde{x}_t), \quad \tilde{x}_0 = x_0 = x. \tag{2.108}$$

In summary, we obtain the following

Theorem 2: Under above mentioned conditions (2.82), (2.88), (2.89), (2.98) and (2.101), the process $X_{\nu(t/\epsilon)}^{\epsilon}$ in (2.25) converges weakly to the process $\tilde{x}_t$ in (2.36) as $\epsilon \to 0$, where

$$\tilde{g}(x) = \int_Y p(dy)g(x,y)/m,$$

and m is defined in (2.108).

Remark 4: The following process

$$\begin{aligned} X(t)^{\epsilon} &:= X_{\nu(t/\epsilon)}^{\epsilon} + (t/\epsilon - \tau_{\nu(t/\epsilon)})(X_{\nu(t/\epsilon)+1}^{\epsilon} - X_{\nu(t/\epsilon)}^{\epsilon}) \\ &= X_{\nu(t/\epsilon)}^{\epsilon} + \varepsilon(t/\epsilon - \tau_{\nu(t/\epsilon)})g(X_{\nu(t/\epsilon)}^{\epsilon}, y_{\nu(t/\epsilon)+1}^{\epsilon}), \end{aligned} \tag{2.109}$$

converges weakly to the process $\tilde{x}_t$ in (2.108) as $\epsilon \to 0$. This follows from Theorem 2, representation (2.109) and the following result [45, p. 163]:

for every bounded and continuous function $f(y)$ and every $t \in [0,T]$ we have

$$\lim_{\epsilon \to 0} E_y[(t/\epsilon - \tau_{\nu(t/\epsilon)})f(y_{\nu(t/\epsilon)})] = t\int_Y p(dy)m_2(y)f(y)/2m,$$

where m and $m_2(y)$ are defined in (2.103) and (2.101), respectively.

Remark 5: We also note that the corresponding semi-Markov random evolution $V^\epsilon(t)$ to the process $X^\epsilon(t)$ in (2.109) has the form

$$V^\epsilon(t)f(x) = V^\epsilon_{\nu(t/\epsilon)} + (t/\epsilon - \tau_{\nu(t/\epsilon)})(V^\epsilon_{\nu(t/\epsilon)+1} - V^\epsilon_{\nu(t/\epsilon)}).$$

2.3. Diffusion Approximation of Difference Equations in Random Media

In this section, using the method of random evolutions in diffusion approximation scheme we obtain diffusion approximation of difference equations in Markov and semi-Markov random media under *balance condition.*

2.3.1. DIFFUSION APPROXIMATION IN MARKOV MEDIA

We assume in this subsection that the following *balance condition* holds

$$\hat{g}(x) := \int_Y p(dy)g(x,y) = 0, \forall x \epsilon X. \tag{2.110}$$

Under this condition $\hat{x}_t = \tilde{x}_t = x_0$, where $\hat{x}_t$ and $\tilde{x}_t$ are defined in (2.93) and (2.108), respectively.

We consider the following difference equation

$$X^\epsilon_{n+1} = X^\epsilon_n + \epsilon g(X^\epsilon_n, y_{n+1}), \quad X^\epsilon_0 = X_0 = x \in X, \tag{2.111}$$

where $(y_n)_{n \in Z^+}$ and $g(x,y)$ are defined in Section 2.2.

For the operators $D^\epsilon(y)$ in (2.77), using the representation (2.111), we get that operators $D^\epsilon(y)$ have the expansion:

$$\begin{aligned} D^\epsilon(y)f(x) &= f(x) + \epsilon g(x,y)f_x(x) + \tfrac{\varepsilon}{2}^2(g(x,y)g_x(x,y)(f_x)^2 \\ &+ g(x,y)^2 f_{xx}) + \varepsilon^2 O_\epsilon(1)f(x), \end{aligned} \tag{2.112}$$

where $||O_\epsilon(1)f(x)|| \to 0$, as $\epsilon \to 0$ for $f(x) \in \mathbf{C}^2(X)$. Thus,

$$D^\epsilon(y)f(x) = f(x) + \epsilon D_1(y)f(x) + \epsilon^2 D_2(y)f(x) + \epsilon^3 O_\epsilon(1)f(x), \tag{2.113}$$

where

$$\begin{aligned} D_1(y)f(x) &:= g(x,y)f_x, \\ D_2(y)f(x) &:= \frac{1}{2}g^2 f_{xx}. \end{aligned} \tag{2.114}$$

Define now the following discrete random evolutions

$$V_n^\epsilon f(x) = \prod_{k=0}^{n} D^\epsilon(y_{k+1}) f(x) = f(X_{n+1}^\epsilon), \tag{2.115}$$

where X_{n+1}^ϵ and $D^\epsilon(y)$ are defined in (2.111) and (2.113), respectively. Let

$$\bar{L}f(x) := \int_Y p(dy)[D_1(y)\mathbf{R}_0 D_1(y) + D_2(y)]f(x), \tag{2.116}$$

and consider the following

$$\bar{M}_t f(x) := \bar{V}f(x) - f(x) - \int_o^t \bar{L}\bar{V}_s f ds, \forall f(x) \epsilon B. \tag{2.117}$$

From the theory of random evolutions developed in Section 2.1.3 regarding diffusion approximation of RE we know that if condition (2.82) is satisfied, if the third moment of $G_y(dt)$

$$m_3(y) := \int_0^\infty t^3 G_y(dt) \tag{2.118}$$

is uniformly integrable, if

$$\int_Y p(dy)||D_1(y)f||.||D_2(y)f|| \ < \ +\infty; \quad \int_Y p(dy)||D_1(y)f||^3 \ < \ \infty;$$

and

$$\int_Y p(dy)||D_2(y)f||^2 \ < \ +\infty, \quad \forall f \in Dom(D_2(y)), \tag{2.119}$$

and if there exists a compact set $K_T^\Delta \subset B$ such that

$$\lim_{\epsilon\to 0} \inf \mathcal{P}\{V_{[t/\epsilon^2]}^\epsilon f \in K_T^\Delta; 0 \le t \le T\} \ge 1 - \Delta, \quad \forall \Delta > 0, \quad \forall T > 0, \tag{2.120}$$

then the sequence $V_{[t/\epsilon^2]}^\epsilon$ is relatively compact in $\mathbf{D_B}[0, +\infty]$ with limit points in $\mathbf{C_B}[0, +\infty]$. Moreover, if the operator $\bar{L}$ in (2.115) generates a semigroup, then $V_{[t/\epsilon^2]}^\epsilon$ converges weakly as $\epsilon \to 0$ to the process $\bar{V}_t$ in (2.117) such that $\bar{M}_t f(x)$ is a continuous $\mathcal{F}_t$-martingale, where $\mathcal{F}_t := \sigma\{y(s) := y_{\nu(s)}; 0 \le s \le t\}$.

Now note that the family $V_{[t/\epsilon^2]}^\epsilon$ corresponds to the process (see (2.111) and (2.115)) $X_{[t/\epsilon^2]}^\epsilon$, as

$$V_{[t/\epsilon^2]}^\epsilon f(x) = \prod_{k=o}^{[t/\epsilon^2]} D^\epsilon(y_{k+1}) f(x) = f(x + \epsilon \sum_{k=0}^{[t/\epsilon^2]} g(X_k^\epsilon, y_{k+1})) = f(X_{[t/\epsilon^2]}^\epsilon). \tag{2.121}$$

Also note that if

$$\int_Y p(dy)|g(x,y)|^4 < +\infty$$

and

$$\int_Y p(dy)|g_x'(x,y)|^2 < +\infty, \qquad \text{for all } x \in X, \tag{2.122}$$

then condition (2.119) is satisfied.

In order to obtain the compactness condition (2.120) for the process $X^{\epsilon}_{[t/\epsilon^2]}$, we need again to construct a compact set in the Banach space $\mathbf{B} = \mathbf{C}^2(X)$. More precisely, we need to construct a Hilbert space $\mathbf{H}$ compactly embedded in $\mathbf{B}$.

The Sobolev embedding theorem [70] states that bounded sets in $\mathbf{W}^{l,2}(R^d)$ are compact in $\mathbf{C}^2_0(R^d)$ provided $2l \geq d$. Therefore, for

$$\mathbf{B} = \mathbf{C}^2(R^d), \mathbf{H} = \mathbf{W}^{l,2}(R^d),$$

with $2l \geq d$ under conditions (2.122) and (2.88), assumptions (2.119) and (2.120) are satisfied.

Therefore, the family of measures generated by the process $X^{\epsilon}_{[t/\epsilon^2]}$ is relatively compact and there exists a unique limiting $\bar{x}_t$ as $\epsilon \to 0$ for $X^{\epsilon}_{[t/\epsilon]}$ process in the sense of weak convergence.

From (2.119) and (2.120) it follows that

$$V^{\epsilon}_{[t/\epsilon^2]} f(x) \to \hat{V}_t f(x) \tag{2.123}$$

as $\epsilon \to 0$, and from (2.121) and (2.123) we obtain

$$f(X^{\epsilon}_{[t/\epsilon^2]}) = V^{\epsilon}_{[t/\epsilon^2]} f(x) \to \hat{V}_t f(x) = f(\bar{X}_t),$$

as $\epsilon \to 0$. Furthermore, from (2.117), (2.119) and (2.124) we obtain that

$$f(\bar{X}_t) - f(x) - \int_o^t \bar{L} f(\bar{X}_s) ds \tag{2.125}$$

is a continuous $\mathcal{F}_t$ martingale.

We now calculate the operator $\bar{L}$ in (2.116) with $D_1(y)$ and $D_2(y)$ given in (2.114):

$$\begin{aligned} \bar{L} f(x) &= \textstyle\int_Y p(dy)[g f_x \mathbf{R}_0 g f_x + 1/2 g^2 f_{xx}] \\ &= \textstyle\int_y p(dy)[g \mathbf{R}_0 g_x f_x + g \mathbf{R}_0 g f_{xx} + 1/2 g^2 f_{xx}] \\ &= \alpha(x) f_x + 1/2 \beta^2(x) f_{xx}, \end{aligned} \tag{2.126}$$

where

$$\alpha(x) := \int_Y p(dy)[g \mathbf{R}_0 g_x],$$

$$\beta^2(x) := 2 \int_Y p(dy)[g \mathbf{R}_0 g + \frac{1}{2} g^2], \tag{2.127}$$

and $\mathbf{R}_0$ is a potential of Markov chain $(y_n)_{n \in Z^+}$ [45,71].

From (2.125)-(2.127) it follows that the process $\bar{X}_t$ is a diffusion process with infinitesimal operator $\bar{L}$ in (2.126), with drift coefficient $\alpha(x)$ and diffusion coefficient $\beta(x)$ in (2.127).

Hence, process $\bar{X}_t$ satisfies the following stochastic differential equation:

$$d\bar{X}_t = \alpha(\bar{X}_t) dt + \beta(\bar{X}_t) dw_t, \tag{2.128}$$

where w_t is a standard Wiener process and the coefficients α and β are defined in (2.127).

Theorem 3: Under conditions (2.82), (2.88), (2.110), (2.118) and (122), the process $(X^\epsilon_{[t/\epsilon^2]})$ converges weakly to the process $\bar{X}_t$ given in (2.128) as $\epsilon \to 0$ with coefficients α and β in (2.127).

2.3.2 DIFFUSION APPROXIMATION IN SEMI-MARKOV RANDOM MEDIA

We now switch to a Markov renewal process $(y_n, \theta_n)_{n \in Z^+}$ with stochastic kernel $Q(y, dz, dt)$ given in (2.95), counting process given in (2.96) with regular condition (2.98). We also suppose that the *balance condition* (2.110) is satisfied.

We study the following difference equation

$$\begin{cases} X^\epsilon_{\nu(t/\epsilon^2)+1} - X^\epsilon_{\nu(t/\epsilon^2)} &= \epsilon g(X^\epsilon_{\nu(t/\epsilon^2)}, y^\epsilon_{\nu(t/\epsilon^2)+1}) \\ X^\epsilon_0 &= X_0 = x \in X \end{cases} \tag{2.129}$$

In particular, if $t \in [\epsilon^2 \tau_n, \epsilon^2 \tau_{n+1})$, where τ_n is defined in (2.97), then $X^\epsilon_{\nu(t/\epsilon^2)+1}$ satisfies equation (2.111) already considered above.

Let us rewrite equation (2.129) as

$$X^\epsilon_{\nu(t/\epsilon^2)+1} = X_0 + \epsilon \sum_{k=1}^{\nu(t/\epsilon^2)} g(X^\epsilon_k, y_{k+1}). \tag{2.130}$$

Using (2.115), we can express $f(X^\epsilon_{\nu(t/\epsilon^2)})$ as

$$\begin{aligned} f(X^\epsilon_{\nu(t/\epsilon^2)}) &= f(X_0 + \epsilon \textstyle\sum_{k=1}^{\nu(t/\epsilon^2)} g(X^\epsilon_k, y_{k+1})) \\ &= \textstyle\prod_{k=1}^{\nu(t/\epsilon^2)} D^\epsilon(y_{k+1}) f(x) = V^\epsilon_{\nu(t/\epsilon^2)} f(x), \end{aligned} \tag{2.131}$$

where the operators $D^\epsilon(y)$ are defined in (2.77), and V^ϵ_n in (2.78).

Put

$$\tilde{L} f := \bar{L} f / m, \tag{2.132}$$

where operator $\bar{L}$ is defined in (2.116) and m is defined in (2.103). Then we consider the following equation

$$\tilde{V}(t) f(x) = f(x) + \int_0^t \tilde{L} \tilde{V}(s) f(x) ds + \tilde{M}(t) f(x), \tag{2.133}$$

in a Banach space $\mathbf{B}$ with the process $\tilde{M}(t) f(x)$ being a continuous $\mathcal{F}_t$- martingale.

From the theory of random evolutions developed in Subsection 2.1.3 regarding diffusion approximation of RE, we conclude that under conditions of Subsection 2.2.2 and condition (2.98), the process $V^\epsilon_{\nu(t/\epsilon)^2}$ is relatively compact in $\mathbf{D_B}[0, +\infty]$ with limit points in $\mathbf{C_B}[0, +\infty]$. Moreover, if the operator $\tilde{L}$ in (2.132) generates a semigroup, then $V^{\epsilon}_{\nu(t/\epsilon^2)}$ converges weakly (as $\epsilon \to 0$) to the process $\tilde{V}(t)$ in (133). Therefore, we obtain

$$V^{\epsilon}_{\nu(t/\epsilon^2)} \to \tilde{V}(t)f, \tag{2.134}$$

as $\epsilon \to 0$, and (2.131) yields

$$f(X^{\epsilon}_{\nu(t/\epsilon^2)}) = V^{\epsilon}_{\nu(t/\epsilon)^2}f(x) \to \tilde{V}(t)f(x) := f(\tilde{X}(t)) \tag{2.135}$$

as $\epsilon \to 0$, and

$$f(X^{\epsilon}_{\nu(t/\epsilon^2)}) \to f(\tilde{X}(t)) \tag{2.136}$$

as $\epsilon \to 0$. Furthermore, the family of measures generated by the processes $X^{\epsilon}_{\nu(t/\epsilon^2)}$ is relatively compact and there exists a unique limiting process $\tilde{X}(t)$ as $\epsilon \to 0$ for $X^{\epsilon}_{\nu(t/\epsilon^2)}$ (due to (2.134)-(2.136)), in the sense of weak convergence.

Finally, from (2.134)-(2.136) and (2.133) we obtain that

$$f(\tilde{X}(t)) - f(x) - \int_0^t \tilde{L}f(\tilde{X}(s)ds \tag{2.137}$$

is a continous $\mathcal{F}_t$ - martingale.

Using (2.132) and noting that $\bar{L}$ calculated in (2.126), we get

$$\tilde{L} = \tilde{\alpha}(x)\frac{d}{dx} + 1/2\tilde{\beta}^2(x)\frac{d^2}{dx^2}, \tag{2.138}$$

where

$$\tilde{\alpha}(x) := \frac{\alpha(x)}{m}, \quad \tilde{\beta}^2(x) := \frac{\beta^2(x)}{m}, \tag{2.139}$$

and $\alpha(x), \beta^2(x)$ are defined in (2.127). From (2.136)-(2.138) we obtain that the process $\tilde{X}(t)$ is the diffusion process with infinitessimal operator $\tilde{L}$ given in (2.138) and with drift coefficients $\tilde{\alpha}(x)$ and diffusion coefficient $\tilde{\beta}^2(x)$ in (2.139).

Hence, the process $\tilde{X}(t)$ satisfies the following stochastic differential equation

$$d\tilde{X}(t) = \tilde{\alpha}(\tilde{X}(t))dt + \tilde{\beta}(\tilde{X}(t))dw(t), \tag{2.140}$$

where $w(t)$ is a standard Wiener process.

We summarize the discussion in the following

Theorem 4: Under conditions of Theorem 3 and (2.98), the process $X^{\epsilon}_{\nu(t/\epsilon^2)}$ in (2.129) converges weakly as $\epsilon \to 0$ to the diffusion process $\tilde{X}(t)$ in (2.140) with infinitesimal operator $\tilde{L}$ given in (2.138) and drift coefficient $\tilde{\alpha}(x)$ and diffusion coefficient $\tilde{\beta}(x)$ in (2.139).

2.4. Normal Deviations of Difference Equations in Random Media

Consider

$$X^{\varepsilon}_{n+1} - X^{\varepsilon}_{n} = \epsilon g(X^{\varepsilon}_{n}, y_{n+1}), X^{\varepsilon}_0 = x_0, \tag{2.141}$$

where $x_0 \in R^d$ is given, $g : X \times Y \to X$ is a given function that is measurable in y and continuous in x, $(y_n)_{n \in Z^+}$ is a stationary ergodic Markov chain in a measurable space

$(Y, \mathcal{Y})$ with ergodic distribution $p(A), A \in \mathcal{Y}$. We also assume the function $g(x, y)$ has continuous second derivative with respect to $x \in X$, and $\int_Y ||g(x,y)||^2 p(dy) < +\infty, \forall x \in X$. Finally, we assume that transition probabilities satisfy a *strong mixing condition* described below:

$$\sum_{k=1}^{+\infty} \sup_{y \in Y, A \in \mathcal{Y}} |P_k(y, A) - p(A)| < +\infty.$$

In Section 2.2 we proved that under the above conditions the process $x^{\varepsilon}_{[t/\varepsilon]}$ converges as $\varepsilon > 0$ to the process $\hat{x}_t$, where

$$\frac{d\hat{x}_t}{dt} = \hat{g}(\hat{x}_t), \quad \hat{x}_0 = x, \tag{2.142}$$

and

$$\hat{g}(x) := \int_Y p(dy) g(x, y). \tag{2.143}$$

Moreover, if $\mathcal{P}\{\nu(t) < +\infty)\} = 1, \forall t \epsilon R_+, \nu(t) := \max\{u : \tau_n \le t\}, \tau_n := \sum_{k=1}^{u} \theta_k$, and $\{\theta_n; n \ge 0\}$ is a sojourn time with distribution function $G_y(dt)$ such that $\int_0^\infty t^2 f_y(dt) := m_2(y)$ is uniformly integrable, then the process $X^{\varepsilon}_{\nu(t/\varepsilon)+1}$ converges weakly as $\varepsilon \to 0$ to the process $\hat{x}(t)$ where

$$\frac{d\tilde{x}(t)}{dt} = \tilde{g}(\tilde{x}(y)), \tilde{x}(0) = x_0, \tag{2.144}$$

and

$$\tilde{g}(x) := \hat{g}(x)/m, m := \int_Y p(dy) m(y),$$

$$m(y) := \int_0^\infty t G_y(dt), \tag{2.145}$$

and $\hat{g}(x)$ is defined in (2.143).

Consider the following deterministic sequence $\tilde{X}^{\varepsilon}_n$ given by

$$\hat{X}^{\varepsilon}_{n+1} - \hat{X}^{\varepsilon}_n = \epsilon \hat{g}(\tilde{X}^{\varepsilon}_n), \tag{2.146}$$

where $\hat{g}(x)$ is defined in (2.143) under condition $\hat{g}(x) \ne 0, \forall x \epsilon X$.

The main focus of this section is on the normal deviations of the solution of the perturbed system (2.141) from the solution of the averaged system (2.146). Let

$$Z^{\varepsilon}_n := [X^{\varepsilon}_n - \hat{X}^{\varepsilon}_n]/\sqrt{\varepsilon}, \tag{2.147}$$

where X^{ε}_n is defined in (2.141) and $\hat{X}^{\varepsilon}_n$ is defined in (2.146), we shall show that Z^{ε}_n converges weakly to a diffusion process, under the following condition:

(C) there exists a measurable function $h : Y \to R_+$ such that

$$\int_Y h(y) p(dy) := \hat{h} < +\infty, \tag{2.148}$$

$$||g(x, y) - g(\tilde{x}, y)|| \le h(y) ||x - \tilde{x}||, \tag{2.149}$$

for all $y \in Y$, $x, \tilde{x} \in R^d$. We note that Z_n^ε in (2.147) satisfies the following

$$\begin{aligned} Z_n^\varepsilon &= \sqrt{\varepsilon} \cdot \sum_{k=0}^n [g(X_k^\varepsilon, y_{k+1}) - \hat{g}(\hat{X}_k^\varepsilon)] \\ &= \sqrt{\varepsilon} \cdot \sum_{k=0}^u [g(X_k^\varepsilon, y_{k+1}) - g(\hat{X}_n^\varepsilon, y_{k+1})] \\ &+ \sqrt{\varepsilon} \cdot \sum_{k=0}^u [g(\hat{X}_k^\varepsilon, y_{k+1}) - \hat{g}(\hat{X}_k^\varepsilon)]. \end{aligned} \tag{2.150}$$

From condition (**C**) it follows that

$$||Z_n^\varepsilon|| \leq \epsilon \sum_{k=0}^n h(y_{k+1}) ||Z_k^\varepsilon|| + \sqrt{\varepsilon} \sum_{k=0}^n [g(\hat{X}_k^\varepsilon, y_{k+1}) - \hat{g}(\hat{X}_k^\varepsilon)],$$

and

$$\sup_{k \leq n} ||Z_k^\varepsilon|| \leq \sup_{k \leq n} \sqrt{\varepsilon} \cdot ||S_k^\varepsilon|| \exp\{\varepsilon \cdot \sum_{k=0}^n h(y_{k+1})\}, \tag{2.151}$$

where

$$S_n^\varepsilon := \sum_{k=0}^n [g(\hat{X}_k^\varepsilon, y_{k+1}) - \hat{g}(\hat{X}_k^\varepsilon)]. \tag{2.152}$$

We note that

$$\lim_{\varepsilon \to 0} \varepsilon \sum_{\varepsilon n \leq t_0} h(y_{n+1}) = t_0 \int_Y h(y) p(y), \tag{2.153}$$

due to the ergodicity of $(y_n)_{n \in Z^+}$, and

$$\lim_{\varepsilon \to 0} \varepsilon \sum_{n \leq \nu(t/\varepsilon)} h(y_{n+1}) = t_0 \int_Y h(y) p(dy)/m, \tag{2.154}$$

due to (2.153) and renewal theorem (see Chapter 1), where m is defined in (2.145).

From (2.151)-(2.154) it follows that $\sup_{n \leq \nu(t/\varepsilon)} ||Z_n^\varepsilon||$ is bounded in probability as $\varepsilon \to 0$, if only $\sup_{n \leq \nu(t/\varepsilon)} \sqrt{\varepsilon} ||S_n^\varepsilon||$ converges. Let us consider the expression $\sqrt{\varepsilon} S_n^\varepsilon$. We note that the function $G(x, y) := (g(x, y) - \hat{g}(x))$ satisfies the balance condition

$$\int_Y p(dy) G(x, y) = \int_Y p(dy)(g(x, y) - \hat{g}(x)) = 0. \tag{2.155}$$

From the theory of random evolutions developed in Subsection 2.1.3 regarding normal deviations of RE, it follows that under condition (2.155) the process

$$\sqrt{\varepsilon} \sum_{k=0}^{\nu(t/\varepsilon)} [g(\hat{X}_k^\varepsilon, y_{k+1}) - \hat{g}(\hat{X}_k^\varepsilon]$$

converges weakly as $\varepsilon \to 0$ to the stochastic Ito integral

$$\int_0^t \sigma(\hat{x}_s) dw(s), \tag{2.156}$$

with diffusion coefficient $\sigma(x)$ given by

$$\begin{aligned}\sigma^2(x) \;&:=\; \textstyle\int_Y p(dy)[(g(x,y)-\hat{g}(x))\mathbf{R}_0(g(x,y)-\hat{g}(x)) \\ &+\; (g(x,y)-\hat{g}(x))^2/2]/m.\end{aligned} \tag{2.157}$$

where $w(s)$ is a standard Wiener process and $\mathbf{R}_0$ is a potential of the Markov chain $(y_n)_{n\in Z^+}$.

The second term in the righthand side of (2.150) converges weakly as $\varepsilon \to 0$ to the integral (2.156).

We now consider the first term on the righthand side of (2.150). By Taylor formula, we have

$$\begin{aligned}&g(X_k^\varepsilon, y_{k+1}) - g(\hat{X}_k^\varepsilon, y_{k+1}) \\ &= \sqrt{\varepsilon} g_x(\hat{X}_k^\varepsilon, y_{k+1}) \cdot Z_n^\varepsilon + \tfrac{1}{2}\varepsilon (Z_n^\varepsilon)^2 g_{xx}(\hat{X}_k^\varepsilon + \sqrt{\varepsilon}\theta Z_n^\varepsilon, y_{k+1}),\end{aligned} \tag{2.158}$$

where $0 < \theta < 1$. Therefore, the first term on the righthand side of (2.150) is

$$\varepsilon \sum_{k=0}^{n} g_x(\hat{X}_k^\varepsilon, y_{k+1}) Z_k^\varepsilon + \frac{1}{2}(\varepsilon)^{3/2} \sum_{k=0}^{n} g_{xx}(\hat{X}_k^\varepsilon + \sqrt{\varepsilon}\theta Z_k^\varepsilon, y_{k+1}) Z_n^{\varepsilon^2}. \tag{2.159}$$

The second term in (2.159) converges weakly to zero as $\varepsilon \to 0$, due to the ergodicity $(y_n)_{n\in Z^+}$ and the continuosity of g_{xx}.

For the first term in (2.159) we consider the process

$$Z_t^\varepsilon := \sum_{k=1}^{\infty} Z_k^\varepsilon \mathbf{1}\{\tau_k \le t/\varepsilon < \tau_{k+1}\}. \tag{2.160}$$

This process is tight in $\mathbf{D}[0,T]$ [45,71], since

$$E||Z_{t_1}^\varepsilon - Z_{t_2}^\varepsilon||^4 \le C \cdot |t_1 - t_2|^4,$$

that follows from the strong mixing condition of $(y_n)_{n\in Z^+}$ (see [25]), where C does not depend on n and ϵ. Hence, the sequence Z_t^ε converges weakly in $\mathbf{D}[0,T]$ to some process $\tilde{z}_t$ in $\mathbf{C}[0,T]$. Thus, for the first term in (2.159), we obtain

$$\varepsilon \sum_{k=0}^{\nu(t/\varepsilon)} g_x(\hat{X}_k^\varepsilon, y_{k+1}) Z_k^\varepsilon \to \int_0^t \tilde{g}_x(\tilde{X}_x)\tilde{z}_s ds, \tag{2.161}$$

as $\epsilon \to 0$, where

$$\tilde{g}_x(x) := \int_Y p(dy) g_x(x,y)/m. \tag{2.162}$$

From (2.159) and (2.161) we obtain that the first term in the righthand side of (2.150) converges weakly as $\varepsilon \to 0$ to the limit in (2.161).

Combining (2.155)-(2.157) and (2.158)-(2.161), we obtain the following result.

Theorem 5: Under mentioned above *strong mixing condition* and condition (**C**), the process Z_t^ε in (2.160) converges weakly as $\varepsilon \to 0$ to the process $\tilde{z}_t$, that satisfies the following stochastic differential equation

$$\tilde{z}_t = \int_0^t \tilde{g}_x(\tilde{x}_s)\tilde{z}_s ds + \int_0^t \sigma(\tilde{x}_s)dw(s), \tag{2.163}$$

where $\tilde{g}_x(x)$ is defined in (2.162), and $\sigma(x)$ is defined in (2.156), $w(t)$ is a standard Wiener process.

2.5. Merging of Difference Equations in Random Media

Let $(Y, \mathcal{Y})$ be a measurable space and $g : R^d \times Y \to R^d$ be a function.

Consider a Markov renewal process $(y_n^\varepsilon, \theta_n)_{n\in Z+}$ in the phase space $(Y, \mathcal{Y})$ and semi-Markov kernel

$$Q_\varepsilon(y, dz, t) := P_\varepsilon(y, dz)G_y(t), \tag{2.164}$$

where stochastic kernel $P_\varepsilon(y, dz)$, which defines the transition probabilities of a perturbed Markov chain $(y_n^\varepsilon)_{n\in Z+}$, has the following form

$$P_\varepsilon(y, B) := P(y, B) - \varepsilon P_1(y, B), y \in Y, B \in \mathcal{Y}. \tag{2.165}$$

here $P(y, B)$ is the transition probabilities of a non-perturbed Markov chain $(y_n)_{n\in Z+}$ and $P_1(y, B)$ is a probability measure.

Our basic assumptions consist of the hypothesis that the stochastic kernal $P(y, B)$ is associated with the following given decomposition of the phase space $(Y, \mathcal{Y})$

$$Y = \bigcup_{v\in V} Y_v, Y_v \bigcap Y_{\tilde{v}} = \emptyset, v \neq \tilde{v}, \tag{2.166}$$

via

$$P(y, Y_v) = \mathbf{1}_v(y) := \begin{cases} 1, & y \in Y_v, \\ 0, & y \notin Y_v. \end{cases} \tag{2.167}$$

We assume that in each class $Y_v, v \in V$, the non-perturbed Markov chain is uniformly ergodic with respect to $v \in V$ with stationary distribution $p_v(A), v \in V, A \in \mathcal{Y}$: $p_v(B) = \int_{Y_v} p_v(dy)P(y, B)$, $B \subset Y_v, p_v(Y_v) = 1$. The decomposition (2.166) defines a merging function by

$$v(y) = v \qquad \text{if } y \in Y_v, \quad v \in V.$$

Here $(V, \mathcal{V})$ is a measurable merged phase space introduced in Subsection 1.9.4.

Let

$$m_v := \int_{Y_v} p_v(dy)m(y),$$

$$\hat{P}_1(v, H) := \int_{Y_v} p_v(dy)P_1(y, Y_H), \quad v \in V, \quad H \in V,$$

$$Y_H := \bigcup_{v\in H} Y_v \in Y, \quad H \subset V. \tag{2.168}$$

It is known [44,72] that the kernel

$$\hat{Q}(v,H) := \hat{P}_1(v,H)/m_v, \quad v \notin H, \tag{2.169}$$

with function $q(v)$, such that

$$0 < q(v) := -\hat{P}_1(v)/m_v = \int_{Y_v} p_v(dy)P_1(y,Y_v)/m_v, v \notin H, \tag{2.170}$$

defines a jump Markov process $\hat{y}(t)$ in phase space (V,v) with stochastic kernal

$$\hat{Q}(v,H,t) := \hat{P}(v,H)\cdot(1-e^{-q(v)t}), \hat{P}(v,H) := \hat{Q}(v,H)/q(v), \tag{2.171}$$

where $\hat{Q}(v,H)$ and $q(v)$ are defined in (2.169) and (2.170), respectively.

Namely, the semi-Markov process $y^{\varepsilon}_{\nu(t/\varepsilon)}$ converges weakly under conditions (2.164)-(2.171) as $\varepsilon\to 0$ to the jump Markov process $\hat{y}(t)$.

This Markov process $\hat{y}(t)$ is called a *merged Markov process* and the phase space $(V,\mathcal{V})$ is called a *merged phase space.*

The infinitesimal operator $\hat{Q}$ of the merged Markov process $\hat{y}(t)$ acts as:

$$\hat{Q}\hat{f}(v) := q(v)\cdot\int_V \hat{P}(v,dv')\cdot[\hat{f}(v')-\hat{f}(v)],$$

$$\hat{f}(v) := \int_{Y_v} p_v(dy)f(y). \tag{2.172}$$

The goal of this section is to study the behaviour of the solution of the following difference equation as $\varepsilon\to 0$:

$$X^{\varepsilon}_{\nu(t/\varepsilon)+1)} - X^{\varepsilon}_{\nu(t/\varepsilon)} = \epsilon g(X^{\varepsilon}_{\nu(t/\varepsilon)}, y^{\varepsilon}_{\nu(t/\varepsilon)+1)}). \tag{2.173}$$

For this purpose, we first consider the following family of operators $D^{\varepsilon}(y)$ in $\mathbf{B} := \mathbf{C}^1(R^d)$

$$D^{\varepsilon}(y)f(x) := f(x+\varepsilon g(x,y)). \tag{2.174}$$

We note that the operators $D^{\varepsilon}(y)$ are linear contractive uniformly with respect to y and the following holds:

$$D^{\varepsilon}(y)f(x) = f(x) + \varepsilon g(x,y)\frac{d}{dx}f(x) + O_{\varepsilon}(1)f(x) \quad \text{as } \varepsilon\to 0, \tag{2.175}$$

where $||O_{\varepsilon}(1)f||\to 0$ as $\epsilon\to 0$, and $||\cdot||$ is the norm in $\mathbf{C}^1(R^d)$.

Put

$$D_1(y)f(x) := g(x,y)\frac{d}{dx}f(x), \forall f(x)\epsilon\mathbf{C}^1(R^d), \tag{2.176}$$

and consider the operator

$$\hat{D}(v) := \int_Y p_v(dy)D_1(y)/m_v, \tag{2.177}$$

where $D_1(y)$ and m_v are defined in (2.176) and (2.168), respectively. Let the opeator $\hat{V}(t)$ be a solution of the following

$$\hat{V}(t)f - f - \int_0^t \hat{D}(\hat{y}(s))\hat{V}(s)fds = 0, \forall f\epsilon\mathbf{B}, \tag{2.178}$$

where the operator $\hat{D}(y)$ is defined in (2.177).

From (2.173), we get

$$X^{\varepsilon}_{\nu(t/\varepsilon)+1} = X_0 + \varepsilon \sum_{k=0}^{\nu(t/\varepsilon)} g(X^{\varepsilon}_k, y^{\varepsilon}_{k+1}). \tag{2.179}$$

On the other hand, for $f \in \mathbf{C}^1(R^d)$, we have

$$\begin{aligned} f(X^{\varepsilon}_{\nu(t/\varepsilon)}) &= f(x_0 + \varepsilon \cdot \textstyle\sum_{k=0}^{\nu(t/\varepsilon)} g(X^{\varepsilon}_k, y^{\varepsilon}_{k+1})) \\ &= \textstyle\prod_{k=0}^{\nu(t/\varepsilon)} D^{\varepsilon}(y^{\varepsilon}_{k+1}) f(x) := V^{\varepsilon}_{\nu(t/\varepsilon)} f(x), \end{aligned} \tag{2.180}$$

and the process $V^{\varepsilon}_{\nu(t/\varepsilon)}$ is a semi-Markov random evolution. From the theory of semi-Markov random evolutions developed in Subsection 2.1.3 regarding merging of RE it follows that under the conditions of Theorem 1 and conditions (164)-(171), the family of random evolutions $V^{\varepsilon}_{\nu(t/\varepsilon)}$ converges weakly as $\varepsilon \to 0$ to the *merged Markov random evolution* $\hat{V}(t)$ which is defined by the solution of the equation (2.178). Namely,

$$V^{\varepsilon}_{\nu(t/\varepsilon)} \to \hat{V}(t) \tag{2.181}$$

as $\epsilon \to 0$. We now calculate the operator $\hat{D}(v)$ in (2.177). Using (2.175)-(2.176), we get

$$\begin{aligned} \hat{D}(v) f(x) &= \textstyle\int_{Y_v} p_v(dy) D(y)/m_v = \int_{Y_v} p_v(dy) g(x,y) \frac{d}{dx} f(x)/m_v \\ &=: \hat{g}(x,v) \tfrac{d}{dx} f(x). \end{aligned}$$

Namely,

$$\hat{D}(v) f(x) = \hat{g}(x,v) \frac{d}{dx} f(x) = \int_{Y_v} p_v(dy) g(x,y) \frac{d}{dx} f(x)/m(v). \tag{2.182}$$

From (2.180) and (2.181), we obtain

$$f(X^{\varepsilon}_{\nu(t/\varepsilon)}) = V^{\varepsilon}_{\nu(t/\varepsilon)} f(x) \to \hat{V}(t) f(x) := f(\hat{x}(t)), \tag{2.183}$$

as $\epsilon \to 0$, where $\hat{x}(t)$ is a limiting process for $X^{\varepsilon}_{\nu(t/\varepsilon)}$, and

$$f(\hat{x}(t)) - f(x) - \int_0^t \hat{g}(\hat{x}(s), \hat{y}(s)) \frac{d}{dx} f(\hat{x}(s)) ds = 0, \tag{2.184}$$

which follows directly from (2.178), (2.182) and (2.183).

Therefore, $\hat{x}(t)$ solves the following initial value problem

$$\frac{d\tilde{x}(t)}{dt} = \hat{g}(\hat{x}(t), \hat{y}(t)), \quad \hat{x}(0) = X_0 = x. \tag{2.185}$$

In conclusion, we have obtained the following

Theorem 6: Under the conditions of Theorem 1 and (2.164)-(2.171), random process $X^{\varepsilon}_{\nu(t/\epsilon)}$ converges weakly as $\varepsilon \to 0$ to the process $\hat{x}(t)$, which satisfies (2.185) with function $\hat{g}(x,v)$ in (2.182) and the merged Markov process $\hat{y}(t)$ in merged phase space $(V,\mathcal{V})$ with the generator $\hat{Q}$ given in (2.172).

2.6. Stability of Difference Equations in Averaging and Diffusion Approximation Schemes

2.6.1. STOCHASTIC STABILITY OF DIFFERENCE EQUATIONS IN AVERAGING SCHEME

In Section 2.2 we show that the solution of the following difference equation

$$X^{\epsilon}_{\nu(t/\epsilon)+1} = X^{\epsilon}_{\nu(t/\epsilon)} + \varepsilon g(X^{\epsilon}_{\nu(t/\epsilon)}, y_{\nu(t/\epsilon)+1}), X^{\epsilon}_0 = X_0 = x, t \in R^+, \tag{2.186}$$

has the limit as $\epsilon \to 0$ that satisfies the following equation

$$\frac{d\tilde{x}_t}{dt} = \tilde{g}(\tilde{x}_t), \tilde{x}_0 = x, \tag{2.187}$$

where $(y_n)_{n\in Z^+}$ is a Markov chain, $\nu(t)$ is the associated counting process, $g(x,y)$ is a bounded and continuous function on $R^d \times Y$, Y is a phase space for $(y_n)_{n\in Z^+}$,

$$\tilde{g}(x) := \int_Y p(dy)g(x,y)/m,$$

$$m := \int_Y p(dy)m(y),$$

and $\{p(A); A \in \mathcal{Y}\}$ is the stationary distribution of $(y_n)_{n\in Z^+}$, $\mathcal{Y}$ is a Borel σ-algebra on Y.

Our first purpose in this section is to state the stability property for the difference equation (2.186) under the stability property of its averaged equation (2.187). Namely, we state that if there exists a Lyapurnov function $V(x)$ and constant $\beta > 0$ such that $\tilde{g}(x)V^1_x(x) \le -\beta V(x)$, for all $x \in R$, then there exists $\epsilon_0 > 0$ so that the proces $X^{\epsilon}_{\nu(t/\epsilon)}$ in (2.186) is an asymptotically stochastically stable process provided $0 \le \epsilon \le \epsilon_0$, for some small ϵ_0.

We also consider the following difference equation in semi-Markov random media

$$X^{\epsilon}_{\nu(t/\epsilon)+1} - X^{\epsilon}_{\nu(t/\epsilon)} = \epsilon g(X^{\epsilon}_{\nu(t/\epsilon)}, y_{\nu(t/\epsilon)+1}), \quad X^{\epsilon}_0 = X_0 = x \in R, \tag{2.188}$$

here $(y_n;\theta_n)_{n\in Z^+}$ is a Markov renewal process with stochastic kernel

$$Q(y,dz,dt) := \mathcal{P}\{y_{n+1} \in dz, \theta_{n+1} \le dt | y_0 = y\} = P(y,dz)G_y(dt), \tag{2.189}$$

$\nu(t) := \sum_{k=1}^{n} \theta_k$, $\theta_0 = 0$, $y_{\nu(t)}$ is a regular semi-Markov process $g(x,y)$ is a given function on $R \times Y$, $(Y,\mathcal{Y})$ is a phase space of Markov chain $(y_n)_{n\in Z^+}$ with σ-algebra $\mathcal{Y}$ of Borel sets from Y.

We suppose that:

(i)$(y_n)_{n\in Z^+}$ is a stationary ergodic process with ergodic distribution $p(dy)$;

(ii)

$$\int_Y |g(x,y)|^2 p(dy) < +\infty, \forall x \in R; \tag{2.190}$$

(iii) $g_x(x,y)$ is bounded and continuous function of x and y;

(iv) the second moment $m_2(y) := \int_0^\infty t^2 G_y(dt)$ is uniformly integrable, where $G_y(dt)$ is defined in (2.189).

Under conditions (i) -(iv) in (2.190), the process $X^\epsilon_{\nu(t/\epsilon)}$ converges weakly as $\varepsilon \to 0$ to the averaged process $\tilde{x}_t$, where

$$\frac{d\tilde{x}_t}{dt} = \tilde{g}(\tilde{x}_t), \tilde{x}_0 = x_0 = x, \tag{2.191}$$

$$\tilde{g}(x) := \int_Y p(dy) g(x_1 y)/m, m := \int_Y p(dy) m(y), m(y) := \int_0^\infty tG_y(dt).$$

We now study the stability of zero state of equation (2.188) using the stability of zero state of equation (2.191).

Definition 1: The process $X^\epsilon_{\nu(t/\epsilon)}$ is *stochastically exponentially stable*, if there exist $\epsilon_0 > 0$, $\delta > 0$ and $\gamma > 0$ such that for each $\epsilon \in [0, \epsilon_0]$ the following inequality

$$\mathcal{P}_{x,y}\{|X^\epsilon_{\nu(t/\epsilon)}| \le \Delta_2 e^{-\gamma t}; t \ge 0\} \ge 1 - \Delta_1, \tag{2.192}$$

is satisfied for some $\Delta_1 > 0$ and $\Delta_2 > 0$, provided that $|X_0^\epsilon| = |x| < \delta$.

Definition 2: The process $X^\epsilon_{\nu(t/\epsilon)}$ is *asymptotically stochastically stable* if

$$\mathcal{P}_{x,y}\{\lim_{t\to+\infty} |X^\epsilon_{\nu(t/\epsilon)}| = 0\} = 1, \forall X \in R, y \in Y. \tag{2.193}$$

Stochastic stability was introduced by Arnold [4], *stable in probability* was used by Khasminskii [40], and *stable with probability 1* was introduced by Kushner [46].

Theorem 7: Let the conditions (i)-(iv) in (2.190) be satisfied and the following conditions hold:

(v) there exists a smooth function $V(x)$ on R such that $V(x) \to +\infty$ as $x \to +\infty$,

$$b_1|x|^{n_1} \le V(x) \le b_2|x|^{n_2},$$

for small x and some positive constants b_1 and b_2 and some $n_i \in Z^+, i = 1,2$, $V(x)$ is positively defined and $V(x) = 0 \Rightarrow x = 0$;

(vi) $g(0,y) = 0, \forall\, y \in Y$;

(vii) there exists $\beta > 0$ such that

$$\tilde{g}(x)V_x(x) \le -\beta V(x), \tag{2.194}$$

for all $x \in R$.

Then process $X^\epsilon_{\nu(t/\epsilon)}$ in (2.188) is stochastically exponentially stable and asymptotically stochastically stable.

Proof. Recall that the process $(X^\epsilon_{\nu(t/\epsilon)}, y^\epsilon_{\nu(t/\epsilon)}, \gamma(t/\epsilon))$ on $R \times Y \times R^+$ is a Markov process with the infinitesimal operator given by (see Section 1.9.3)

$$L_\varepsilon f(t,x,y) = \frac{1}{\varepsilon}Qf(t,x,y) + \frac{1}{\varepsilon}P[f(t, x+\varepsilon g(x,y), y) - f(t,x,y)],$$

$$f \in \mathbf{C}^1(R_+) \times \mathbf{C}(R) \times \mathbf{C}(Y), \tag{2.195}$$

where $\gamma(t) := t - \tau_{\nu(t)}$, $Pg(x,y) := \int_Y P(y,dz)g(x,z)$, $P(y,A)$ is defined in (2.189), $y \in Y, A \in \mathcal{Y}$,

$$Qf(t,x,y) := \frac{d}{dt}f(t,x,y) + \frac{h_y(t)}{\bar{G}_y(t)}[Pf(0,,x,y) - f(t,x,y)], \tag{2.196}$$

$$\bar{G}_y(t) := 1 - G_y(t), h_y(t) := \frac{dG_y(t)}{dt}, \tag{2.197}$$

and $G_y(t)$ is defined in (2.189).

We introduce the family of functions

$$V_\varepsilon(t,x,y) := V(x) + \epsilon V_1(t,x,y), \tag{2.198}$$

where V_1 is given by solving the equation

$$QV_1(t,x,y) + \mathbf{P}g(x,y)V_x(x) - AV(x) = 0, \tag{2.199}$$

with

$$AV(x) := \tilde{g}(x)V_x(x). \tag{2.200}$$

From (2.195)-(2.200), we obtain

$$L_\varepsilon V_\varepsilon = AV(x) + \mathbf{P}[V_1(t, x + \epsilon g(x,y), y) - V_1(t,x,y)] + 0(\epsilon), \tag{2.201}$$

where $0(\epsilon) = \frac{1}{2}\varepsilon g^2(x,y)V_{xx}(x) + o(1)$, with $o(1) \to 0$ as $\epsilon \to 0$.

Define the process

$$\begin{aligned} m_\epsilon(t) \ &:= V_\varepsilon(\gamma(t/\epsilon), X^\epsilon_{\nu(t/\epsilon)}, y_{\nu(t/\epsilon)}) \\ &-V_\varepsilon(0,x,y) - \int_0^t L_\varepsilon V_\varepsilon(\gamma(s/\epsilon), X^\epsilon_{\nu(s/\epsilon)}, y_{\nu(s/\epsilon)})ds. \end{aligned} \tag{2.202}$$

This is a right-continuous integrable $\mathcal{F}^\epsilon_t$-martingale with zero mean, where $\mathcal{F}^\epsilon_t := \sigma\{y_{\nu(s/\epsilon)}; 0 \le \epsilon \le t\}$. (See Section 1.9).

From (2.198)-(2.202) it follows that

$$\begin{aligned} &V(X^\epsilon_{\nu(t/\epsilon)}) - V(x) - \int_0^t AV(X^\epsilon_{\nu(s/\epsilon)})ds \\ &= M_\epsilon(t) + \epsilon V_1(t,x,y) + \epsilon V_1(\gamma(t/\epsilon), X^\epsilon_{\nu(t/\epsilon)}, y_{\nu(t/\epsilon)}) \\ &+ \int_0^t P[V_1(s, x + \epsilon g(X(^\epsilon_{\nu(s/\epsilon)}), y_{\nu(s/\epsilon)}) - V_1(s,x,y)]ds + 0(\epsilon). \end{aligned}$$

Therefore, $X^\epsilon_{\nu(t/\epsilon)}$ in (2.188) approximates the averaged process $\tilde{x}_t$ in (2.191).

We note that from condition (v) and boundness of g it follows that there exists $\epsilon_0 > 0$ so that for $0 < \epsilon < \epsilon_0$, we have

$$l_1 V(x) \le V_\epsilon(t,x,y) \le l_2 V(x) \tag{2.203}$$

for some positive constants l_1 and l_2.

Let $\tilde{\beta} > 0$ be a given constant. It follows from (2.198) that

$$(L_\epsilon + \hat{\beta})V_\epsilon \leq (\hat{\beta} l_2 + A + \epsilon_0 l_3)V(x), \tag{2.204}$$

for some constant $l_3 > 0$.

Choose $\hat{\beta} > 0$ so that

$$\hat{\beta} \leq \frac{\beta - \epsilon_0 \cdot l_3}{l_2} \tag{2.205}$$

for the constant $\beta > 0$ defined in (2.194). Then it follows from (2.204)-(2.205) that

$$(L_\epsilon + \hat{\beta})V_\epsilon \leq 0. \tag{2.206}$$

Rewriting (2.202) in terms of $e^{\hat{\beta}t}V_\epsilon(t, x, y)$, we obtain

$$e^{\hat{\beta}\cdot t}V_\epsilon(\gamma(t/\epsilon), X^\epsilon_{\nu(t/\epsilon)}y_{\nu(t/\epsilon)}) = V_\epsilon(0, x, y) + \int_0^t e^{\hat{\beta}\cdot s}(L_\epsilon + \hat{\beta})V_\epsilon \cdot ds + \tilde{M}_\epsilon(t), \tag{2.207}$$

where $\tilde{M}_\epsilon(t)$ is a right-continous integrable $\mathcal{F}^\epsilon_t$ - martingale with mean zero.

Using (2.203)-(2.204) and (2.206) in (2.207), we get

$$0 \leq e^{\hat{\beta}t} l_1 V(X^\epsilon_{\nu(t/\epsilon)}) \leq e^{\hat{\beta}t} V_\epsilon(\gamma(t/\epsilon), X^\epsilon_{\nu(t/\epsilon)}, y^\epsilon_{\nu(t/\epsilon)}) \leq l_2 V(x) + \tilde{M}_\epsilon(t), \tag{2.208}$$

from which it follows that $l_2 V(x) + \tilde{M}_\epsilon(t)$ is nonnegative martingale.

Applying Kolmogorov-Doob's inequality, we get for any $\tilde{\Delta}_2 > 0$ that

$$\begin{aligned} &\mathcal{P}_{x,y}\{\sup_{0\leq t\leq T} e^{\hat{\beta}t} l_1 V((X^\epsilon_{\nu(t/\epsilon)}) \geq \tilde{\Delta}_2\} \\ &\leq \mathcal{P}_{x,y}\{\sup_{0\leq t\leq T}(l_2 V(x)) + \tilde{M}_\epsilon(t)) \geq \tilde{\Delta}_2\} \leq \tfrac{l_2 V(x)}{\tilde{\Delta}_2}. \end{aligned} \tag{2.209}$$

Taking into account the existence of constants $b_1 > 0$ and $b_2 > 0$ and positive numbers n_1 and n_2 such that (see condition (v)):

$$b_1|x|^{n_1} \leq V(x) \leq b_2|x|^{n_2}, \tag{2.210}$$

we have

$$\{b_1|X^\epsilon_{\nu(t/\epsilon)}|^{n_1} \leq e^{-\hat{\beta}t}\frac{\tilde{\Delta}_2}{l_1}; t \geq 0\} \supset \{V(X^\epsilon_{\nu(t/\epsilon)}) \leq e^{-\hat{\beta}t}\frac{\tilde{\Delta}_2}{l_1}; t \geq 0\}. \tag{2.211}$$

From (2.211) we obtain:

$$\mathcal{P}_{x,y}\{|X^\epsilon_{\nu(t/\epsilon)}| \leq e^{\hat{\beta}/n_1}(\frac{\tilde{\Delta}_2}{b_1, \ell_1})^{1/n_1}; t \geq 0\} \geq 1 - \frac{l_2 V(x)}{\tilde{\Delta}_2}. \tag{2.212}$$

Now let $\Delta_1 > 0$ and $\Delta_2 > 0$ be fixed. We choose $\tilde{\Delta}_2$ so that:

$$\mathcal{P}_{x,y}\{|X^\epsilon_{\nu(t/\epsilon)}| \leq e^{-\frac{\hat{\beta}}{n_1}}\Delta_2; t \geq 0\} \geq 1 - \frac{\ell_2 V(x)}{\tilde{\Delta}_2}. \tag{2.213}$$

Hence, $\tilde{\Delta}_2 := \Delta_2^{n_1} b_1 l_1$.

Let $x \in R$ with $|x| < \delta_1$ and $V(x) < l_2^{-1}\tilde{\Delta}_2\Delta_1$. Then

$$\delta_1 = (\frac{\tilde{\Delta}_2 \cdot \Delta_1}{b_2 l_2})^{1/n^2}$$

(see (2.210)). From (2.212) and (2.213), we obtain

$$\mathcal{P}_{x,y}\{|X^{\epsilon}_{\nu(t/\epsilon)}| \geq e^{-\frac{\hat{\beta}\cdot t}{n_1}}\Delta_2; t > 0\} \geq 1 - \Delta_1.$$

This proves (2.192) with $\gamma = \frac{\hat{\beta}}{n_1}$, and thus the first assertion of the theorem 1 is proved.

To prove (2.193) and, hence, the second assertion of the theorem 7, we note that:

$$\begin{aligned} \{\lim_{t\to+\infty} |X^{\epsilon}_{\nu(t/\epsilon)}| = 0\} &= \{\lim_{t\to+\infty} V(X^{\epsilon}_{\nu(t/\epsilon)}) = 0\} \\ &\supset \{\sup_{t\geq 0} e^{\hat{\beta}t} l_1 V(X^{\epsilon}_{\nu(t/\epsilon)}) \leq D\}, \end{aligned} \tag{2.214}$$

for some constant D. Then we have from (2.209) and (2.214)that

$$\mathcal{P}_{x,y}\{\lim_{t\to+\infty} |X^{\epsilon}_{\nu(t/\epsilon)}| = 0\} = 1,$$

as $D \to \infty$. This completes the prov Δ.

2.6.2. STOCHASTIC STABILITY OF DIFFERENCE EQUATIONS IN DIFFUSION APPROXIMATION SCHEME

We have studied in Section 2.3 that if $\tilde{g}(x) = 0, \forall x \in R$ (the balance condition is fulfilled), then the following difference equation

$$X^{\epsilon}_{\nu(t/\epsilon^2)+1} = X^{\epsilon}_{\nu(t/\epsilon^2)} + \varepsilon g(X_{\nu(t/\epsilon^2)}, y^{\epsilon}_{\nu(t/\epsilon^2)+1}) \tag{2.215}$$

has the limit, as $\epsilon \to 0$, satisfying the following diffusion model:

$$d\tilde{x}(t) = \tilde{\alpha}(\tilde{x}(t)dt + \tilde{\beta}(\tilde{x}(t))dw(t), \tag{2.216}$$

where

$$\begin{aligned} \tilde{\alpha}(x) &:= \int_Y p(dy)[gRg_x + 1/2gg_x]/m, \\ \tilde{\beta}^2(x) &:= 2\int_Y p(dy)[gR_0 g + 1/2g^2]/m, \end{aligned} \tag{2.217}$$

$\mathbf{R}_0$ is a potential of Markov chain $(y_u)_{n\in Z^+}$ (see Section 1.2, Chapter 1).

Our purpose in this section is to state the stability property for difference equation (2.215) under the stability property of diffusion model (2.216). Namely, we state that if there exists a Lyapunov function $W(x)$ such that

$$\tilde{\alpha}(x)W_x + 1/2\,\tilde{\beta}(x)W_{xx} \leq -\,\gamma W(x),$$

for all $x \in R$, for some $\gamma > 0$, then the process $X^{\epsilon}_{\nu(t/\epsilon^2)}$ is asypmtotically stochastically stable process as $\epsilon \in [0, \epsilon_0]$ is fixed, where ε_0 is a small paramater.

Let the balance condition be fulfilled:

$$\tilde{g}(x) = 0, \forall x \in R. \tag{2.218}$$

Then we have to apply a diffusion approximation scheme to the difference equation (2.215) in scale of time t/ϵ^2 :

$$\begin{cases} X^{\epsilon}_{\nu(t/\epsilon^2)+1} - X^{\epsilon}_{\nu(t/\epsilon^2)} = \epsilon g(X^{\epsilon}_{\nu(t/\epsilon^2)}, y_{\nu(t/\epsilon)+1}) \\ X^{\epsilon}_0 = X_0 = x, \end{cases} \tag{2.219}$$

where all the items in (2.219) are defined in Section 2.6.1.

We suppose that the following conditions are satisfied:

(i) Condition (i) from Section 2.6.1;

$$\begin{aligned} (ii)' \quad & \textstyle\int_Y p(dy)|g(x,y)|^3 < +\infty; \\ & \textstyle\int_Y p(dy)|g_x(x,y)|^2 < +\infty, \forall x \in R; \end{aligned} \tag{2.220}$$

$(iii)' g_x(x,y)$and $g_{xx}(x,y)$ are bounded and continuous;

$(iv)'$ the third moment $m_3(y) := \int_0^\infty t^3 G_y(dt)$ is uniformly integrable.

Under conditions (2.218) and (i)', $(ii)' - (iv)'$) in (2.220), the process $X^{\epsilon}_{\nu(t/\epsilon^2)}$ in (2.219) converges weakly as $\epsilon \to 0$ to the diffusion process $\tilde{x}(t)$ such that

$$d\tilde{x}(t) = \tilde{\alpha}(\tilde{x}(t)dt + \tilde{\beta}(\tilde{x}(t))dw(t), \tag{2.221}$$

where the drift and diffusion coefficients $\alpha(x)$ and $\tilde{\beta}(x)$ are defined in (2.217), $m := \int_Y p(dy)m(y)$, $m(y) := \int_0^\infty tG_y(dt)$, and $w(t)$ is a standard Wiener process.

In the next theorem, we study the stability property of difference equation (2.219) under the stability conditions of the equation (2.221).

Theorem 8: Let conditions (2.218) and $(i)', (ii)' - (iv)'$ in (2.220) and the following conditions be satisfied:

(a) there exists a smooth function $W(x)$ on R such that $W(x) \to +\infty$ as $x \to +\infty$,

$$K_1|x|^{p_1} \le W(x) \le K_2|x|^{p_2}$$

for some positive constants K_1 and K_2 and $p_i \in Z^+, i = 1, 2$, and for small x; $W(x)$ is positively defined and $W(x) = 0 \Rightarrow x = 0$;

(b)

$$g(0, y) = 0, \forall y \in Y;$$

(c) there exists $\gamma > 0$ such that

$$\tilde{\alpha}(x)W'_x(x) + 1/2\tilde{\beta}^2(x)W''_{xx}(x) \le -\gamma W(x), \gamma > 0. \tag{2.222}$$

Then process $X^{\epsilon}_{\nu(t/\epsilon^2)}$ in (2.219) is stochastically exponentially stable. Moreover, it is asymptotically stochastically stable.

Proof. Let us consider the process $(X^{\epsilon}_{\nu(t/\epsilon^2)}, y^{\epsilon}_{\nu(t/\epsilon^2)}, \gamma_{(t/\epsilon^2)})$ on $R \times Y \times R^+$. It is a Markov process with infinitesimal operator

$$L^{\epsilon} f(t,x,y) = \frac{1}{\epsilon^2} Qf(t,x,y) + \frac{1}{\epsilon^2} P[f(t,x+\epsilon g(x,y),y) - f(t,x,y)], \qquad (2.223)$$

where Q and P are defined in (2.195)-(2.196), $f \in \mathbf{C}^1(R_+) \times \mathbf{C}(R) \times \mathbf{C}(Y)$.

We now introduce the following family of functions:

$$W^{\epsilon}(t,x,y) = W(x) + \epsilon W_1(t,x,y) + \epsilon^2 W_2(t,x,y), \qquad (2.224)$$

where V_1 is defined by the solution of the equation

$$QW_1 + Pg(x,y)W'_x(x) = 0, \qquad (2.225)$$

and V_2 is defined by the solution of the equation:

$$QW_2 + Pg(x,y)dW_1(x)/dx + 1/2Pg^2(x,y)W(x) - LW(x) = 0, \qquad (2.226)$$

where

$$\bar{L} := \tilde{\alpha}(x)\frac{d}{dx} + 1/2\tilde{\beta}^2(x)\frac{d^2}{dx^2}, \qquad (2.227)$$

$W(x)$ is defined in (a). It follows from the equations (2.225)-(2.226) that

$$L^{\epsilon} W^{\epsilon} = LW(x) + O(\epsilon)W(x), \qquad (2.228)$$

where $||O(\epsilon)W(x)|| \to 0$ as $\epsilon \to 0$, and L is defined in (2.227).

Let us define the following process:

$$m^{\epsilon}(t) := W^{\epsilon}(\gamma(t/\epsilon^2)), X^{\epsilon}_{\nu(t/\epsilon^2)}, y_{\nu(t/\epsilon^2)}) - W^{\epsilon}(0,x,y)$$
$$- \int_0^t L^{\epsilon} W^{\epsilon}(\gamma(s/\epsilon^2), X^{\epsilon}_{\nu(s/\epsilon^2)} y_{\nu(s/\epsilon^2)}) ds. \qquad (2.229)$$

The process $m^{\epsilon}(t)$ is a right-continuous integrable $\mathcal{F}^{\epsilon}_t$-martingale with

$$\mathcal{F}^{\epsilon}_t := \sigma\{\gamma_{(s/\epsilon^2)}, y_{\nu(s/\epsilon^2)}; 0 \le s \le t\}.$$

From the representations (2.223)-(2.229) it follows that the expression (229) may be rewritten as

$$W(X^{\epsilon}_{\nu(t/\epsilon^2)}) - W(x) - \textstyle\int_0^t LW(X^{\epsilon}_{\nu(s/\epsilon^2)} ds$$
$$= m^{\epsilon}(t) + \epsilon \cdot W_1(0,x,y) + \epsilon^2 W_2((0,x,y)$$
$$-\epsilon W_1(\gamma(s/\epsilon^2), X^{\epsilon}_{\nu(t/\epsilon^2)}, y_{\nu(t/\epsilon^2)})$$
$$-\epsilon^2 \cdot W_2(\gamma_{(t/\epsilon^2)}, X^{\epsilon}_{\nu(t/\epsilon^2)}, y_{\nu(t/\epsilon^2)}) + 0(\epsilon). \qquad (2.230)$$

It is also easy to see from (2.230) that the process $X^{\epsilon}_{\nu(t/\epsilon^2)}$ approximates the diffusion process $\tilde{x}(t)$ in (2.221) with infinitesimal operator $\tilde{L}$.

Using conditions of our theorem, we conclude that there exists a small number ϵ_0 such that if $0 \leq \epsilon < \epsilon_0$ then

$$C_1 W(x) \leq W^{\epsilon}(t, x, y) \leq C_2 W(x), \tag{2.231}$$

for some positive constants C_1 and C_2 and for all $x \in R$.

Let $\hat{\gamma} > 0$. It follows from (2.228) that

$$(L^{\epsilon} + \hat{\gamma}) W^{\epsilon} = \hat{\gamma} W^{\epsilon} + LW + o(\epsilon). \tag{2.232}$$

Under conditions of our theorem and using the inequality (231) we obtain from (2.232) that

$$(L^{\epsilon} + \hat{\gamma}) W^{\epsilon} \leq (\hat{\gamma} C_2 + L + \epsilon_0 C_3) W(x),$$

where C_3 is some positive constant, $x \in R$. If we choose $\hat{\gamma}$ in such a way that $C_2 \hat{\gamma} + \epsilon_0 \cdot C_3 \leq \gamma$, where γ is defined in (2.222), then we have:

$$(L^{\epsilon} + \hat{\gamma}) W^{\epsilon} \leq 0. \tag{2.233}$$

Now let us rewrite (2.229) with $e^{\hat{\gamma} \cdot t} W^{\epsilon}$:

$$\begin{aligned} & e^{\hat{\gamma} t} W^{\epsilon}(\gamma(t/\epsilon^2), X^{\epsilon}_{\nu(t/\epsilon^2)}, y_{\nu(t/\epsilon^2)}) \\ & = W^{\epsilon}(0, x, y) + \textstyle\int_0^t e^{\hat{\gamma} \cdot s} (L^{\epsilon} + \hat{\gamma}) W^{\epsilon} ds + \tilde{m}^{\epsilon}(t), \end{aligned} \tag{2.234}$$

where $\tilde{m}^{\epsilon}(t)$ is a right-continous integrable $\mathcal{F}^{\epsilon}_t$ - martingale with zero mean.

Taking into account (2.231) and (2.233), we obtain from (2.224) and (2.234) that

$$O \leq C_1 e^{\hat{\gamma} t} W(x^{\epsilon}_{\nu(t/\epsilon^2)}) \leq e^{\hat{\gamma} t} W^{\epsilon} \leq C_2 W(x) + \tilde{m}^{\epsilon}(t). \tag{2.235}$$

The inequality (2.235) means that $C_2 W(x) + \tilde{m}^{\epsilon}(t)$ is a non-negative $\Im^{\epsilon}_t$ - martingale. From Kolmogorov-Doob inequality we obtain for every $\tilde{N}_2 > 0$ that

$$\mathcal{P}_{x,y}\{\sup_{0 \leq t \leq T} C_1 e^{\hat{\gamma} t} W(X^{\epsilon}_{\nu(t/\epsilon^2)}) > \tilde{N}_2\} \leq \frac{C_2 W(x)}{\tilde{N}_2}, \tag{2.236}$$

and we have as $T \to +\infty$ the following

$$\mathcal{P}_{x,y}\{\sup_{0 \leq t \leq +\infty} C_1 e^{\hat{\gamma} t} W(X^{\epsilon}_{\nu(t/\epsilon^2)}) > \tilde{N}_2\} \leq \frac{C_2 W(x)}{\tilde{N}_2}.$$

Therefore,

$$\{K_1 |X^{\epsilon}_{\nu(t/\epsilon^2)}|^{p_1} \leq e^{\hat{\gamma} t} \frac{\tilde{N}_2}{C_1}; t \geq 0\} \supset \{W(X^{\epsilon}_{\nu(t/\epsilon^2)}) \leq e^{\hat{\gamma} t} \frac{\tilde{N}_2}{C_1}; t \geq 0\} \tag{2.237}$$

and

$$\mathcal{P}_{x,y}\{|X^{\epsilon}_{\nu(t/\epsilon^2)}| \le e^{\bar{\gamma}t}(\frac{\tilde{N}_2}{C_1 \cdot K_1})^{1/p_1}; t \ge 0\} \ge 1 - \frac{C_2 W(x)}{\tilde{N}_2}, \tag{2.238}$$

where $\bar{\gamma} := \hat{\gamma}/p_1$.

Let $N_1 > 0$ and $N_2 > 0$ be given, such that $\tilde{N}_2$ is so small that (2.238) yields the inequality:

$$\mathcal{P}_{x,y}\{|X^{\epsilon}_{\nu(t/\epsilon^2)}| \le e^{\bar{\gamma}t} N_2; t \ge 0\} \ge 1 - \frac{C_2 W(x)}{\tilde{N}_2}, \tag{2.239}$$

namely, $\tilde{N}_2 = N_2^{p_1} C_1 K_1$, Then we can take $\delta > 0$ in such a way that for $|x| < \delta$ we have:

$$W(x) < C_2^{-1} \tilde{N}_2 N_1, \tag{2.240}$$

namely, $\delta := (\frac{\tilde{N}_2 N_1}{C_2 K_2})^{1/p_2}$.

Finally, from (2.239)-(2.240) we obtain the inequality:

$$\mathcal{P}_{x,y}\{|X^{\epsilon}_{\nu(t/\epsilon^2)}| \le e^{\bar{\gamma}t} N_2; t \ge 0\} \ge 1 - N_1,$$

and thus the stochastic exponential stability is proved.

To prove the asymptotic stochastic stability, we note that

$$\begin{aligned} \{\lim_{t\to+\infty} |X^{\epsilon}_{\nu(t/\epsilon^2)}| = 0\} \quad &\supset \{\lim_{t\to+\infty} W(X^{\epsilon}_{\nu(t/\epsilon^2)}) = 0\} \\ &\supset \{\sup_{t\ge 0} C_1 \cdot e^{\bar{\gamma}\cdot t} W(X^{\epsilon}_{\nu(t/\epsilon^2)}) \le C\}, \end{aligned} \tag{2.241}$$

where C is a some positive constant.

From (2.236) and (2.241) we obtain

$$\begin{aligned} \mathcal{P}_{x,y}\{\lim_{t\to+\infty} |X^{\epsilon}_{\nu(t/\epsilon^2)}| = 0\} &\ge \mathcal{P}_{x,y}\{\sup_{t\ge 0} C_1 \cdot e^{\bar{\gamma}\cdot t} W(X^{\epsilon}_{\nu(t/\epsilon^2)}) \le C\} \\ &\ge 1 - \frac{C_2 W(x)}{C}. \end{aligned}$$

Therefore, as $C \to +\infty$, we get

$$\mathcal{P}_{x,y}\{\lim_{t\to+\infty} |X^{\epsilon}_{\nu(t/\epsilon^2)}| = 0\} = 1,$$

and hence the asymptotic stochastical stability is proved.$\triangle$

Remarks: Stochastic stability of stochastic differential equations has been studied in [70], including asymptotic and global stability. Asymptotic stability of linear stochastic systems was studied in [45]. Asymptotic stability of SDE with jumps has been studied in [31, p.325]. Asymptotic stochastic stability of stochastic systems with wide-band noise disturbances using martingale approach was studied in [12].

2.7. Limit Theorems for Vector Difference Equations in Random Media.

2.7.1. AVERAGING OF VECTOR DIFFERENCE EQUATIONS

We have considered two types of difference equations in one dimensional space: a) difference equations with Markov random pertubations as a random media and with a discrete parameter to explain how the method of random evolutions works; b) difference equations with semi-Markov random pertubations as a random media and with a continuous parameter. Here we consider the vector difference equations.

2.7.1.1. Averaging in Markov random media. We consider a system in a linear phase space X with discrete time $n \in Z^+ = \{0, 1, 2, ...\}$ which is perturbed by a Markov chain $(y_n)_{n\in Z^+}$ defined on a measurable space $(Y, \mathcal{Y})$. The system depends on a small parameter $\epsilon > 0$. Let $\mathbf{X}_n^\epsilon \in X$ denote the state of the system at time n. Throughout the remaining part of this paper we use the following phase space $X = R^d, d \geq 1$. Suppose that $\mathbf{X}_n^\epsilon$ is determined by the following recurrence relations:

$$\begin{aligned} \mathbf{X}_{n+1}^\epsilon - \mathbf{X}_n^\epsilon &= \varepsilon \mathbf{g}(\mathbf{X}_n^\epsilon, y_{n+1}), \\ \mathbf{X}^\epsilon &= \mathbf{X}_0, \end{aligned} \tag{2.242}$$

where $\mathbf{X}_0 = \mathbf{x} \in X$ is given, $\mathbf{g} : X \times Y \to X$ is a given vector-function, $\mathbf{g} := (g_1, g_1, ..., g_n)$, $\epsilon > 0$ is a small parameter. We also assume that $\mathbf{g}(\mathbf{x}, y)$ is measurable in $y \in Y$ and continuous in $\mathbf{x} \in X$. Moreover, we assume $\mathbf{g_x}(\mathbf{x}, y) := (\partial g_i(x, y)/\partial x_j)_{i,j=1}^n$ is bounded and continuous as a function from $X \times Y \to X$, and that

$$\int_Y \mathbf{g}^T(\mathbf{x}, y)\mathbf{g}(\mathbf{x}, y)p(dy) < +\infty \text{ for all fixed } \mathbf{x} \in \text{X}. \tag{2.243}$$

We also suppose here that the noise process $(y_n)_{n\in Z_+}$ is a stationary ergodic process with ergodic distribution $p(dy)$. Namely, for any function $f(y) : Y \to R$ with $\int_Y |f(y)|p(dy) < +\infty$, we have

$$P\{\lim_{n\to+\infty} \frac{1}{n} \sum_{k=1}^n f(y_k) = \int_Y f(y)p(dy)\} = 1. \tag{2.244}$$

Our goal is to investigate the asymptotic behaviours of the system as $\varepsilon \to 0$ and $n \to \infty$.

Theorem 9: Under conditions (2.243) and (2.244) the process $\mathbf{X}_{[t/\epsilon]}^\epsilon$ in (2.242) converges weakly to the process $\hat{\mathbf{x}}_t$ as $\epsilon \to 0$, and $\hat{\mathbf{x}}_t$ satisfies the equation:

$$\left\{ \begin{aligned} \frac{d\hat{\mathbf{x}}_t}{dt} &= \hat{\mathbf{g}}(\hat{\mathbf{x}}_t) \\ \hat{\mathbf{x}}_0 &= \mathbf{x}, \end{aligned} \right. \tag{2.245}$$

where

$$\hat{\mathbf{g}}(x) := \int_Y p(dy)\mathbf{g}(\mathbf{x}, y).$$

Proof. We define the family of operators on the space $\mathbf{C}^1(R^d)$:

$$\mathcal{D}^\epsilon(\mathbf{x}, y)f(\mathbf{x}) := f(\mathbf{x} + \epsilon\mathbf{g}(\mathbf{x}, y)),$$

where $\mathbf{x} \in R^d$. Then the proof follows directly from the averaging theorem for random evolutions (see [81], Chapter 4 and [82], section 1) with the operator $\mathcal{D}^\epsilon(\mathbf{x}, y)$.

Remark 6: For the process

$$\mathbf{X}_t^\epsilon = \mathbf{X}_{[t/\epsilon]}^\epsilon + (t/\epsilon - [t/\epsilon])(\mathbf{X}_{[t/\epsilon]+1}^\epsilon - \mathbf{X}_{[t/\epsilon]}^\epsilon) = \mathbf{X}_{[t/\epsilon]}^\epsilon + \epsilon(t/\epsilon - [t/\epsilon])\mathbf{g}(\mathbf{X}_{t/\epsilon]}^\epsilon, Y_{[t/\epsilon]+1}),$$

we can also obtain from Theorem 9 that $\mathbf{X}_t^\epsilon$ converges weakly to the process $\hat{\mathbf{x}}_t$ as $\epsilon \to 0$.

Remark 7: From Theorem 9 it follows that $\sup_{\mathrm{n}\epsilon\leq t}||\mathbf{X}_\mathrm{n}^\epsilon - \hat{\mathbf{x}}(\epsilon \mathrm{n})||\epsilon \to 0$ for any $t > 0$, where $\mathbf{X}_n^\epsilon$ and $\hat{\mathbf{x}}(t)$ are defined in (2.242) and (2.244), respectively.

Remark 8: Our Theorem 9 includes the general result in [76] (Theorem 1, pp. 466) as a special case.

2.7.1.2. Averaging in semi-Markov random media. We now consider Markov renewal process [79] $(y_n; \theta_n)_{n\in Z^+}$ with stochastic kernel

$$Q(y, dz, dt) := \mathcal{P}\{y_{n+1} \in dz, \theta_{n+1} \leq dt | y_0 = y\} = P(y, dz)G_y(dt). \tag{2.246}$$

Let

$$\nu(t) = \max\{\mathrm{n} : \tau_\mathrm{n} \leq \mathrm{t}\} \tag{2.247}$$

be a counting process, where

$$\tau_n = \sum_{k=1}^{n} \theta_k, \quad \theta_0 = 0.$$

We consider the following difference equation in semi-Markov random media:

$$\begin{cases} \mathbf{X}_{\nu(t/\epsilon)+1}^\epsilon - \mathbf{X}_{\nu(t/\epsilon)}^\epsilon &= \epsilon\, \mathbf{g}(\mathbf{X}_{\nu(t/\epsilon)}^\epsilon, y_{\nu(t/\epsilon)+1}), \\ \mathbf{X}_0^\epsilon &= \mathbf{X}_0 = \mathbf{x} \in X, \end{cases} \tag{2.248}$$

where $y_{\nu(t)}$ is a semi-Markov process, $\nu(t)$ is defined in (2.247).

We note, that if $t \in [\varepsilon\tau_n, \varepsilon\tau_{n+1})$ with τ_n being defined in (16), then $\mathbf{X}_{\nu(t/\epsilon)}$ satisfies the equation:

$$\mathbf{X}_{n+1}^\varepsilon - \mathbf{X}_n^\varepsilon = \varepsilon \mathbf{g}(\mathbf{X}_n^\epsilon, y_{n+1}), \quad t \in [\varepsilon\tau_n, \varepsilon\tau_{n+1}).$$

In what follows, we consider a regular semi-Markov process, namely, we assume

$$\mathcal{P}\{\nu(t) < +\infty\} = 1, \forall t \in R^+. \tag{2.249}$$

We assume that

$$m_2(y) := \int_0^\infty t^2 G_y(dt) \tag{2.250}$$

is uniformly integrable, where $G_y(dt) := \mathcal{P}\{\theta_{n+1} \leq t | y_0 = y\}$.

Let $\tilde{\mathbf{x}}_t$ satisfy the following equation

$$\frac{d\tilde{\mathbf{x}}_t}{dt} = \tilde{\mathbf{g}}(\tilde{\mathbf{x}}_t), \quad \tilde{\mathbf{x}}_0 = \mathbf{X}_0 = \mathbf{x}, \tag{2.251}$$

where

$$\tilde{\mathbf{g}}(x) = \int_Y p(dy)\mathbf{g}(\mathbf{x}, y)/m,$$

$$m := \int_Y m(y)p(dy), \quad m(y) = \int_0^\infty tG_y(dt). \tag{2.252}$$

We obtain the following

Theorem 10: Under conditions (2.243), (2.244), (2.249) and (2.250), the process $\mathbf{X}^\epsilon_{\nu(t/\epsilon)}$ in (2.252) converges weakly to the process $\tilde{\mathbf{x}}_t$ in (2.251) as $\epsilon \to 0$.

Proof. Theorem 10 follows directly from Theorem 9 and renewal theorem for counting process $\nu(t)$ in (2.247) [81,82].

Remark 9: Consider the process

$$\begin{aligned} \mathbf{X}^\epsilon(t) &:= \mathbf{X}^\epsilon_{\nu(t/\epsilon)} + (t/\epsilon - \tau_{\nu(t/\epsilon)})(\mathbf{X}^\epsilon_{\nu(t/\epsilon)+1} - \mathbf{X}^\epsilon_{\nu(t/\epsilon)}) \\ &= \mathbf{X}^\epsilon_{\nu(t/\epsilon)} + \varepsilon(t/\epsilon - \tau_{\nu(t/\epsilon)})\mathbf{g}(\mathbf{X}^\epsilon_{\nu(t/\epsilon)}, y_{\nu(t/\epsilon)+1}). \end{aligned} \tag{2.253}$$

This process also converges weakly to the process $\tilde{\mathbf{x}}_t$ in (2.249) as $\epsilon \to 0$. This conclusion follows from Theorem 10, representation (2.253) and the following result [79, p. 163]:

For every bounded and continuous function $f(y)$ of y and for any $t \in [0, T]$

$$\lim_{\epsilon\to 0} E_y[(t/\epsilon - \tau_{\nu(t/\epsilon)})f(y_{\nu(t/\epsilon)})] = t\int_Y p(dy)m_2(y)f(y)/2m,$$

where m and $m_2(y)$ are defined in (2.241) and (2.239), respectively.

Remark 10: If $\mathbf{g} \equiv \mathbf{g}^\epsilon(\mathbf{x}, y) := \mathbf{g}(\mathbf{x}, y) + \epsilon\mathbf{g}_1(\mathbf{x}, y)$ then the conditions of Theorems 9 and 10 remain true, since in equations (2.242) and (2.248) the term $\epsilon^2\mathbf{g}_1(x, y)$ vanishes under averaging in scale of time t/ϵ [79].

2.7.2. DIFFUSION APPROXIMATION OF VECTOR DIFFERENCE EQUATIONS

In this section we consider the same types of equations as in Section 2.3.1. Under *balance condition* and using the method of random evolutions in diffusion approximation scheme, we obtain diffusion approximation of difference equations in Markov and semi-Markov random media.

2.7.2.1. Diffusion approximation in Markov media. We suppose in this section that the *balance condition* is fulfilled:

$$\hat{\mathbf{g}}(\mathbf{x}) := \int_Y p(dy)\mathbf{g}(\mathbf{x}, y) = 0, \forall \mathbf{x} \in X. \tag{2.254}$$

Under condition (2.254) $\hat{\mathbf{x}}_t = \tilde{\mathbf{x}}_t = \mathbf{x}_0$, where $\hat{\mathbf{x}}_t$ and $\tilde{\mathbf{x}}_t$ are defined in (2.245) and (2.251), respectively.

In this case we can study the behavior of solutions of the equations (2.242) and (2.248), by rescaling the time via $n^1 = \epsilon^2 n$ and $t^1 = \epsilon^2 t$, respectively.

We consider the following difference equation:

$$\mathbf{X}^\epsilon_{n+1} = \mathbf{X}^\epsilon_n + \epsilon\mathbf{g}(\mathbf{X}^\epsilon_n, y_{n+1}), \quad \mathbf{X}^\epsilon_0 = \mathbf{X}_0 = \mathbf{x} \in X, \tag{2.255}$$

where $(y_n)_{n\in Z_+}$ and $\mathbf{g}(\mathbf{x}, y)$ are defined in Section 2.7.1.1.

Let the following conditions be satisfied: the third moment of $G_y(dt)$

$$m_3(y) := \int_0^{\infty} t^3 G_y(dt) \tag{2.256}$$

is uniformly integrable and, in addition,

$$\begin{cases} \int_Y p(dy)\mathbf{g}^T(\mathbf{x}, y)\mathbf{g}(\mathbf{x}, y) < +\infty, \\ \int_Y p(dy)|\partial g_i(\mathbf{x}, y)/\partial x_j|^2 < +\infty, \quad \forall x \in X, \quad i, j = 1, 2, ..., n. \end{cases} \tag{2.257}$$

Let

$$\begin{aligned} \alpha(\mathbf{x}) &:= (\int_Y p(dy)\mathbf{g}^T R_0 \mathbf{g}_{\mathbf{x}})^T, \\ \beta^2(\mathbf{x}) &:= 2\int_Y p(dy)[\mathbf{g}R_0\mathbf{g}^T + 1/2\mathbf{g}\mathbf{g}^T]. \end{aligned} \tag{2.258}$$

where R_0 is a potential of Markov chain $(y_n)_{n\in Z^+}$ [49, 81, 82].

Let the process $\bar{\mathbf{X}}_t$ satisfy the following stochastic differential equation

$$d\bar{\mathbf{X}}_t = \alpha(\bar{\mathbf{X}}_t)dt + \beta(\bar{\mathbf{X}}_t)d\mathbf{w}_t, \tag{2.259}$$

where w_t is a standard n-dimensional Wiener process, and drift vector-coefficient $\alpha(\mathbf{x})$ and diffusion matrix $\beta(\mathbf{x})$ are defined in (2.258).

The following result is true.

Theorem 11: Under conditions (2.254), (2.256), (2.257), the process $X^{\epsilon}_{[t/\epsilon^2]}$ converges weakly to the process $\bar{\mathbf{X}}_t$ given in (2.259) as $\epsilon \to 0$ with drift vector-coefficient $\alpha(\mathbf{x})$ and diffusion matrix $\beta(\mathbf{x})$ given in (2.258).

Proof. We define the family of operators on the space $\mathbf{C}^2(R^d)$:

$$\mathcal{D}^{\epsilon}(\mathbf{x}, y)f(\mathbf{x}) := f(\mathbf{x} + \epsilon\mathbf{g}(\mathbf{x}, y)),$$

where $\mathbf{x} \in X$. Then the proof follows directly from the diffusion approximation theorem for random evolutions (see [49], Chapter 4 and [81], section 1) with the operators $\mathcal{D}^{\epsilon}(\mathbf{x}, y)$.

Corrolary 1: Consider the vector-function $\mathbf{g}$, depending on a parameter ϵ, as

$$\mathbf{g} \equiv \mathbf{g}^{\epsilon}(\mathbf{x}, y) := \mathbf{g}(\mathbf{x}, y) + \epsilon\mathbf{g}_1(\mathbf{x}, y).$$

Then from Theorem 3 it follows that the process $\mathbf{X}^{\epsilon}_{[t/\epsilon^2]}$ (see (2.253)) converges weakly to the process $\bar{\mathbf{X}}_t$ given in (2.257) as $\epsilon \to 0$, when the drift vector-coefficient is

$$\alpha(\mathbf{x}) := (\int_Y p(dy)\mathbf{g}^T R_0 \mathbf{g}_{\mathbf{x}} + \mathbf{g}_1^T)^T$$

and the same diffusion matrix $\beta(\mathbf{x})$ given in (2.258).

2.7.2.2. Diffusion approximation in semi-Markov random media. We now consider the Markov renewal process $(y_n, \theta_n)_{n \in Z_+}$ with stochastic kernel $Q(y, dz, dt)$ given in (2.246), counting process given in (2.247) with regular condition (2.249). We suppose also that the *balance condition* (2.254) is satisfied.

In this subsection, we study the following difference equation:

$$\begin{cases} \mathbf{X}^{\epsilon}_{\nu(t/\epsilon^2)+1} - \mathbf{X}^{\epsilon}_{\nu(t/\epsilon^2)} = \epsilon \mathbf{g}(\mathbf{X}^{\epsilon}_{\nu(t/\epsilon^2)}, y_{\nu(t/\epsilon^2)+1}), \\ \mathbf{X}^{\epsilon}_0 = \mathbf{X}_0 = \mathbf{x} \in X. \end{cases} \tag{2.260}$$

Note that if $t \in [\epsilon^2 \tau_n, \epsilon^2 \tau_{n+1})$, then $\mathbf{X}^{\epsilon}_{\nu(t/\epsilon^2)+1}$ satisfies the equation (2.255).

To study the asymptotic behavior of the system (2.260) as $\epsilon \to 0$, we rewrite equation (2.260) as

$$\mathbf{X}^{\epsilon}_{\nu(t/\epsilon^2)+1} = \mathbf{X}_0 + \epsilon \sum_{k=1}^{\nu(t/\epsilon^2)} \mathbf{g}(\mathbf{X}^{\epsilon}_k, y_{k+1}). \tag{2.261}$$

Let

$$\tilde{\alpha}(\mathbf{x}) := \frac{\alpha(\mathbf{x})}{m}, \qquad \tilde{\beta}^2(\mathbf{x}) := \frac{\beta^2(\mathbf{x})}{m}, \tag{2.262}$$

where $\alpha(\mathbf{x}), \beta^2(\mathbf{x})$ are defined in (2.258).

Let the process $\tilde{\mathbf{X}}(t)$ satisfy the following stochastic differential equation:

$$d\tilde{\mathbf{X}}(t) = \tilde{\alpha}(\tilde{\mathbf{X}}(t))dt + \tilde{\beta}(\tilde{\mathbf{X}}(t))d\mathbf{w}(t), \tag{2.263}$$

where $\mathbf{w}(t)$ is an n-dimensional standard Wiener process, and $\tilde{\alpha}(\mathbf{x}), \tilde{\beta}(\mathbf{x})$ are defined in (2.262).

The following result is true.

Theorem 12: Under conditions of Theorem 11 and condition (2.249) the process $\mathbf{X}^{\epsilon}_{\nu(t/\epsilon^2)}$ in (2.260) converges weakly as $\epsilon \to 0$ to the diffusion process $\tilde{\mathbf{X}}(t)$ in (2.263) with the drift coefficient drift $\tilde{\alpha}(\mathbf{x})$ and diffusion matrix $\tilde{\beta}(\mathbf{x})$ given in (2.262).

Proof. Theorem 12 follows directly from Theorem 11 and renewal theorem (see [49,79]) for counting process $\nu(t)$ in (2.249).

Corollary 2: Consider the vector-function $\mathbf{g}$, depending on a parameter ϵ, as

$$\mathbf{g} \equiv \mathbf{g}^{\epsilon}(\mathbf{x}, y) := \mathbf{g}(\mathbf{x}, y) + \epsilon \mathbf{g}_1(\mathbf{x}, y).$$

Then from Theorem 12 it follows that the process $\mathbf{X}^{\epsilon}_{[t/\epsilon^2]}$ (see (2.260)) converges weakly to the process $\bar{\mathbf{X}}_t$ given in (2.263) as $\epsilon \to 0$, when the drift vector-coefficient is

$$\tilde{\alpha}(\mathbf{x}) := (\int_Y p(dy)\mathbf{g}^T R_0 \mathbf{g}_{\mathbf{x}} + \mathbf{g}_1^T)^T / m$$

and and the diffusion matrix $\tilde{\beta}(\mathbf{x})$ is the same and given in (2.262).

2.7.3. NORMAL DEVIATIONS OF VECTOR DIFFERENCE EQUATIONS

We consider a system in a linear vector space X with discrete parameter of time $n \in Z_+$ (the set of nonnegative integers) which is perturbed by a Markov chain

$(y_n)_{n\in Z_+}$ defined on a measurable space $(Y,\mathcal{Y})$. The system depends on a small parameter $\varepsilon>0$, and its state at time n is determined by X_n^ε.

We suppose that $\mathbf{X}_n^\varepsilon$ is determined by the recurrence relations:

$$\mathbf{X}_{n+1}^\varepsilon-\mathbf{X}_n^\varepsilon=\epsilon\mathbf{g}(\mathbf{X}_n^\varepsilon,y_{n+1}),\quad \mathbf{X}_0^\varepsilon=\mathbf{x}_0=\mathbf{x}\in X, \tag{2.264}$$

where $\mathbf{g}:X\times Y\to X$ is given function and $X=R^d, d\geq 1$. We also assume that $\mathbf{g}(\mathbf{x},y)$ is assumed to be measurable in y and continuous in $\mathbf{x}$. We suppose that the process $(y_n)_{n\in Z_+}$ is a stationary ergodic Markov chain in $(Y,\mathcal{Y})$ with ergodic distribution $p(A), A\in\mathcal{Y}$, function $\mathbf{g}(\mathbf{x},y)$ has 2nd order continuous derivatives with respect to $\mathbf{x}\in X$, and $\int_Y ||\mathbf{g}(\mathbf{x},y)||^2p(dy)<+\infty$, for all $\mathbf{x}\in X$.

In Section 2.7.1, we proved that under above conditions the process $\mathbf{X}_{[t/\varepsilon]}^\varepsilon$ converges as $\varepsilon>0$ to the process $\tilde{\mathbf{x}}_t$ such that

$$\frac{d\tilde{\mathbf{x}}_t}{dt}=\tilde{\mathbf{g}}(\tilde{\mathbf{x}}_t),\tilde{\mathbf{x}}_0=\mathbf{x}, \tag{2.265}$$

where

$$\tilde{\mathbf{g}}(\mathbf{x}):=\int_Y p(dy)\mathbf{g}(\mathbf{x},y). \tag{2.266}$$

Furthermore, we proved that if $\mathcal{P}\{\nu(t)<+\infty)\}=1$, for all $t\in R^+:=[0,+\infty)$, where $\nu(t)$ is a counting process, $\nu(t):=\max\{n:\tau_n\leq t\},\tau_n:=\sum_{k=1}^n\theta_k$, and $(\theta_n)_{n\in Z_+}$ is a sojourn time with distribution function $G_y(dt)$ such that $\int_0^\infty t^2G_y(dt):=m_2(y)$ is uniformly integrable, then the process $\mathbf{X}_{\nu(t/\varepsilon)+1}^\varepsilon$ converges weakly as $\varepsilon>0$ to the process $\hat{\mathbf{x}}(t)$ such that:

$$\frac{d\hat{\mathbf{x}}(t)}{dt}=\hat{\mathbf{g}}(\hat{\mathbf{x}}(t)),\hat{\mathbf{x}}(0)=\mathbf{x}_0=\mathbf{x}, \tag{2.267}$$

where

$$\hat{\mathbf{g}}(\mathbf{x}):=\tilde{\mathbf{g}}(\mathbf{x})/m,\quad m:=\int_Y p(dy)m(y),$$

$$m(y):=\int_0^\infty tG_y(dt), \tag{2.268}$$

and $\hat{\mathbf{g}}(\mathbf{x})$ is defined in (2.266).

Define the following non-random sequence $\tilde{\mathbf{X}}_n^\varepsilon$:

$$\hat{\mathbf{X}}_{n+1}^\varepsilon-\hat{\mathbf{X}}_n^\varepsilon=\epsilon\hat{\mathbf{g}}(\hat{\mathbf{X}}_n^\varepsilon), \tag{2.269}$$

where $\hat{\mathbf{g}}(\mathbf{x})$ is defined in (2.267). We assume that $\hat{\mathbf{g}}(\mathbf{x})\neq\mathbf{0}$, for $\mathbf{x}\in X$.

In this section, we study normal deviations of the solution of the perturbed system (2.264) from the solution of the averaged system (2.269).

Let

$$\mathbf{Z}_n^\varepsilon:=[\mathbf{X}_n^\varepsilon-\hat{\mathbf{X}}_{n]}^\varepsilon]/\sqrt{\varepsilon} \tag{2.270}$$

where $\mathbf{X}_n^\varepsilon$ is defined in (2.264) and $\hat{\mathbf{X}}_n^\varepsilon$ is defined in (2.269).

We want to show that under some conditions, $\mathbf{Z}_n^\varepsilon$ converges weakly to a diffusion process.

Let the following conditions be satisfied:

(C_1) there exists a measurable function $h : Y \to R_+$ such that

$$\int_Y h(y)p(dy) := \hat{h} < +\infty \text{ and for } y \in \mathcal{Y}, x, x' \in R^d : \tag{2.271}$$

$$||\mathbf{g}(\mathbf{x}, y) - \mathbf{g}(\mathbf{x}', y)|| \le h(y)||\mathbf{x} - \mathbf{x}'||; \tag{2.272}$$

(C_2) *strong mixing condition:*

$$\sum_{k=1}^{+\infty} \sup_{y \in Y, A \in \mathcal{Y}} |P_k(y, A) - p(A)| < +\infty,$$

where $P_k(y, A)$ are transitions probabilities of the k-step of the Markov chain $(y_n)_{n\in Z^+}$.

We note that $\mathbf{Z}_n^\varepsilon$ in (2.270) satisfies

$$\begin{aligned} \mathbf{Z}_n^\varepsilon &= \sqrt{\varepsilon}\textstyle\sum_{k=0}^n[\mathbf{g}(\mathbf{X}_k^\varepsilon, y_{k+1}) - \hat{\mathbf{g}}(\hat{\mathbf{X}}_k^\varepsilon)] \\ &= \sqrt{\varepsilon}\textstyle\sum_{k=0}^n[\mathbf{g}(\mathbf{X}_k^\varepsilon, y_{k+1}) - \mathbf{g}(\hat{\mathbf{X}}_n^\varepsilon, y_{k+1})] + \sqrt{\varepsilon}\textstyle\sum_{k=0}^n[\mathbf{g}(\hat{\mathbf{X}}_k^\varepsilon, y_{k+1}) - \hat{\mathbf{g}}(\hat{\mathbf{X}}_k^\varepsilon)]. \end{aligned} \tag{2.273}$$

From condition (C_1) it follows that

$$||\mathbf{Z}_n^\varepsilon|| \le \epsilon \sum_{k=0}^n h(y_{k+1})||\mathbf{Z}_k^\varepsilon|| + \sqrt{\varepsilon}\sum_{k=0}^n[\mathbf{g}(\hat{\mathbf{X}}_k^\varepsilon, y_{k+1}) - \hat{\mathbf{g}}(\hat{\mathbf{X}}_k^\varepsilon)],$$

and

$$\sup_{k\le n} ||\mathbf{Z}_k^\varepsilon|| \le \sup_{k\le n} \sqrt{\varepsilon}||S_k^\varepsilon||\exp\{\varepsilon \cdot \sum_{k=0}^n h(y_{k+1})\}, \tag{2.274}$$

where

$$S_n^\varepsilon := \sum_{k=0}^n[\mathbf{g}(\hat{\mathbf{X}}_k^\varepsilon, y_{k+1}) - \hat{\mathbf{g}}(\hat{\mathbf{X}}_k^\varepsilon)]. \tag{2.275}$$

We note that

$$\lim_{\varepsilon\to 0} \varepsilon \sum_{\varepsilon n \le t_0} h(y_{n+1}) = t_0 \int_Y h(y)p(y), \tag{2.276}$$

due to ergodicity of $(y_n)_{n\in Z_+}$, and that

$$\lim_{\varepsilon\to 0} \varepsilon \sum_{n\le\nu(t/\varepsilon)} h(y_{u+1}) = t_0 \int_Y h(y)p(dy)/m, \tag{2.277}$$

due to (2.276) and renewal theorem [27, Ch. XI] where m is defined in (2.68).

From (2.274)-(2.277) it follows that $\sup_{n\le\nu_{(t/\varepsilon)}} ||\mathbf{Z}_n^\varepsilon||$ is bounded in probability as $\varepsilon \to 0$ if $\sup_{n\le\nu_{(t/\varepsilon)}} \sqrt{\varepsilon}||S_n^\varepsilon||$ converges.

Let us consider firstly the expression $\sqrt{\varepsilon}S_n^\varepsilon$ which is the second term on the righthand side of (2.273).

We note that the function $\mathbf{G}(\mathbf{x}, y) := (\mathbf{g}(\mathbf{x}, y) - \hat{\mathbf{g}}(\mathbf{x}))$ satisfies balance condition

$$\int_Y p(dy)\mathbf{G}(\mathbf{x}, y) = \int_Y p(dy)(\mathbf{g}(\mathbf{x}, y) - \tilde{\mathbf{g}}(\mathbf{x})) = 0. \tag{2.278}$$

From the theory of random evolutions (see section 2.1., and also [49,79,81]) it follows that under condition (2.278) the process $\sqrt{\varepsilon}\sum_{k=0}^{\nu(t/\varepsilon)}[g(\hat{\mathbf{X}}_k^\varepsilon, y_{k+1}) - \hat{g}(\hat{\mathbf{X}}_k^\varepsilon)]$ converges weakly as $\varepsilon \to 0$ to the stochastic Ito integral

$$\int_0^t \beta(\hat{\mathbf{x}}_s)d\mathbf{w}_s, \tag{2.279}$$

with diffusion matrix $\beta(\mathbf{x})$ given by

$$\begin{aligned} \beta^2(\mathbf{x}) &:= \int_y p(dy)[(\mathbf{g}(\mathbf{x},y) - \hat{\mathbf{g}}(\mathbf{x}))R_0(\mathbf{g}(\mathbf{x},y) - \hat{\mathbf{g}}(\mathbf{x}))^T \\ &+ (\mathbf{g}(\mathbf{x},y) - \hat{\mathbf{g}}(\mathbf{x}))(\mathbf{g}(\mathbf{x},y) - \hat{\mathbf{g}}(\mathbf{x}))^T/2]/m, \end{aligned} \tag{2.280}$$

where $\mathbf{w}_s$ is a standard n-dimensional Wiener process and R_0 is a potential of Markov chain $(y_n)_{n\in Z_+}$.

We now consider the following family of operators $D^\varepsilon(y)$ on $B := C^1(R^d)$:

$$D^\varepsilon(y)f(\mathbf{x}) := f(\mathbf{x} + \varepsilon\mathbf{g}(\mathbf{x},y)).$$

We note that operators $D^\varepsilon(y)$ are linear contractive uniformly by y and admit the representation:

$$D^\varepsilon(y)f(\mathbf{x}) = f(\mathbf{x}) + \varepsilon\mathbf{g}^T(\mathbf{x},y)f_{\mathbf{x}}(\mathbf{x}) + o(\varepsilon)f(\mathbf{x}) \quad \text{as } \varepsilon \to 0,$$

where $||o(\varepsilon)f||/\epsilon \to_{\varepsilon\to 0} 0, ||\cdot||$ is a norm in $C^1(R^d)$.

Therefore, the second term in the righthand side of (2.273) converges weakly as $\varepsilon \to 0$ to the integral (2.279).

Now, consider the first term on the righthand side of (2.273). By Taylor formula we obtain

$$\begin{aligned} \mathbf{g}(\mathbf{X}_k^\varepsilon, y_{k+1}) - \mathbf{g}(\hat{\mathbf{X}}_k^\varepsilon, y_{k+1}) &= \sqrt{\varepsilon}\mathbf{g}_{\mathbf{x}}(\hat{\mathbf{X}}_k^\varepsilon, y_{k+1})\mathbf{Z}_n^\varepsilon + 1/2\varepsilon\mathbf{g}_{\mathbf{xx}}(\hat{\mathbf{X}}_k^\varepsilon \\ &+ \sqrt{\varepsilon}\theta\mathbf{Z}_n^\varepsilon, y_{k+1})(\mathbf{Z}_n^\varepsilon)(\mathbf{Z}_n^\varepsilon), \end{aligned} \tag{2.281}$$

where $0 < \theta < 1$, $\mathbf{g}_{\mathbf{x}}$ is the Jacobian matrix and

$$\begin{aligned} \mathbf{g}_{\mathbf{xx}}(\mathbf{Z}_n^\varepsilon)(\mathbf{Z}_n^\varepsilon) &:= \sum_{i=1}^d\sum_{j=1}^d D_iD_j\mathbf{g}(\mathbf{x})(Z_{ni}^\varepsilon)(Z_{nj}^\varepsilon) \\ &= [\sum_{i=1}^d\sum_{j=1}^d D_iD_jg_1(\mathbf{x})(Z_{ni}^\varepsilon)(Z_{nj}^\varepsilon), \\ &\dots \ , \sum_{i=1}^d\sum_{j=1}^d D_iD_jg_d(\mathbf{x})(Z_{ni}^\varepsilon)(Z_{nj}^\varepsilon)], \end{aligned}$$

where $D_i := \partial/\partial x_i$.

So, the first term on the righthand side of (2.273) is

$$\varepsilon\sum_{k=0}^n \mathbf{g}_{\mathbf{x}}(\hat{\mathbf{X}}_k^\varepsilon, y_{k+1})\mathbf{Z}_k^\varepsilon + 1/2(\varepsilon)^{3/2}\sum_{k=0}^n \mathbf{g}_{\mathbf{xx}}(\hat{\mathbf{X}}_k^\varepsilon + \sqrt{\varepsilon}\theta\mathbf{Z}_k^\varepsilon, y_{k+1})(\mathbf{Z}_n^\varepsilon)(\mathbf{Z}_n^\varepsilon). \tag{2.282}$$

The second term in (2.282) converges weakly to zero as $\varepsilon \to 0$ due to the ergodicity of $(y_n)_{n\in Z_+}$ and continuosity of $\mathbf{g_{xx}}$ by $\mathbf{x}$.

We now consider the following process:

$$\mathbf{Z}_t^\varepsilon := \sum_{k=1}^{\infty} \mathbf{Z}_k^\varepsilon \mathbf{1}\{\tau_k \le t/\varepsilon < \tau_{k+1}\}. \tag{2.280}$$

This process is tight in $D[0,T]$ (see [5, 6]), since

$$E||\mathbf{Z}_{t_1}^\varepsilon - \mathbf{Z}_{t_2}^\varepsilon||^4 \le K \cdot |t_1 - t_2|^4,$$

that follows from the strong mixing condition (C_2), where the constant K does not depend on n. Hence, the sequence $\mathbf{Z}_t^{\varepsilon_n}$ converges weakly in $D[0,T]$ to some process $\tilde{\mathbf{Z}}_t$ in $C[0,T]$.

For the first term in (2.282) we obtain

$$\varepsilon \sum_{k=0}^{\nu(t/\varepsilon)} \mathbf{g_x}(\hat{\mathbf{X}}_k^\varepsilon, y_{k+1})\mathbf{Z}_k^\varepsilon \to_{\varepsilon\to 0} \int_0^t \hat{\mathbf{g}}_\mathbf{x}(\hat{\mathbf{x}}_s)\tilde{\mathbf{Z}}_s ds, \tag{2.284}$$

where

$$\hat{\mathbf{g}}_\mathbf{x} := \int_Y p(dy)\mathbf{g_x}(\mathbf{x}, y)/m., \tag{2.285}$$

From (2.282) and (2.284) we obtain that the first term on the righthand side of (2.273) converges weakly as $\varepsilon \to 0$ to the limit in (2.285).

From (2.278)-(2.280) and (2.281)-(2.284), we finally obtain the following result:

Theorem 13: Under conditions $(C_1)-(C_2)$, the process $\mathbf{Z}_t^\varepsilon$ in (2.283) converges weakly as $\varepsilon \to 0$ to the process $\tilde{\mathbf{Z}}_t$ which satisfies the following stochastic differential equation:

$$\tilde{\mathbf{Z}}_t = \int_0^t \hat{\mathbf{g}}_\mathbf{x}(\hat{\mathbf{x}}_s)\tilde{\mathbf{Z}}_s ds + \int_0^t \beta(\hat{\mathbf{x}}_s)d\mathbf{w}_s. \tag{2.286}$$

Remark 11: The result of Theorem 1 remains true for the function $\mathbf{g} \equiv \mathbf{g}^\epsilon(\mathbf{x}, y) := \mathbf{g}(\mathbf{x}, y) + \epsilon \mathbf{g}_1(\mathbf{x}, y)$, since in equations (2.281) and (2.282), the term $\epsilon^{3/2}\mathbf{g}_1(x, y)$ vanishes.

2.7.4. MERGING OF VECTOR DIFFERENCE EQUATIONS

We consider a Markov renewal process $(y_n^\varepsilon, \theta_n)_{n\in Z_+}$ in phase space $(Y, \mathcal{Y})$ with the semi-markov kernel

$$Q_\varepsilon(y, dz, t) := P_\varepsilon(y, dz)G_y(t), \tag{2.287}$$

where the stochastic kernel $P_\varepsilon(y, dz)$, which defines the transition probabilities of perturbed Markov chain $(y_n^\varepsilon)_{n\in Z_+}$, given by

$$P_\varepsilon(y, B) := P(y, B) - \varepsilon P_1(y, B), y \in Y, B \subset Y, \tag{2.280}$$

here $P(y, B)$ is the transition probability of the basic non-perturbed Markov chain $(y_n)_{n\in Z_+}$, and $P_1(y, B)$ is a certain probability measure.

Our basic assumptions is that the stochastic kernel $P(y, B)$ is co-ordinated with the given decomposition of phase space $(Y, \mathcal{Y})$:

$$Y = \bigcup_{v \epsilon V} Y_v, Y_v \bigcap Y'_v = \phi, v \neq v', \tag{2.289}$$

by the following way:

$$P(y, Y_v) = \mathbf{1}_v(y) := \begin{cases} 1, & ify \in Y_v, \\ 0, & ify \notin Y_v. \end{cases} \tag{2.290}$$

In each class $Y_v, v\epsilon V$, the basic non-perturbed Markov chain is uniformly ergodic with respect to $v\epsilon V$ with stationary distribution $p_v(A), v \in V, A \in \mathcal{Y} : p_v(B) = \int_{Y_v} p_v(dy)P(y, B), B \subset Y_v, p_v(Y_v) = \mathbf{1}$. The decomposition (2.289) defines a merging function below

$$v(y) = v, \text{if } y \in Y_v, v \in V.$$

Here, $(V, \mathcal{V})$ is a measurable merged phase space.

We introduce the following notations:

$$m_v := \int_{Y_v} p_v(dy)m(y), \hat{P}_1(v, H) := \int_{Y_v} p_v(dy)P_1(y, Y_H), v\epsilon V, H\epsilon V,$$

$$Y_H := \bigcup_{v \epsilon H} Y_v \epsilon Y, H \subset V. \tag{2.291}$$

It is known from [5], that the kernel

$$\hat{Q}(v, H) := \hat{P}_1(v, H)/m_v, v \notin H, \tag{2.292}$$

with the function $q(v)$ satisfying

$$0 < q(v) := -\hat{P}_1(v)/m_v = \int_{Y_v} p_v(dy)P_1(y, Y_v)/m_v, v \notin H, \tag{2.293}$$

defines a jump Markov process $\hat{y}(t)$ in the phase space $(V, \mathcal{V})$ with the stochastic kernal

$$\hat{Q}(v, H, t) := \hat{P}(v, H) \cdot (1 - e^{-q(v)t}), \hat{P}(v, H) := \hat{Q}(v, H)/g(v), \tag{2.294}$$

where $\hat{Q}(v, H)$ and $q(v)$ are defined in (2.292) and (2.293), respectively.

Namely, the semi-Markov process $y^{\varepsilon}_{\nu(t/\varepsilon)}$ converges weakly under conditions (2.287)-(2.294) as $\varepsilon \to 0$ to the jump Markov process $\hat{y}(t)$. This Markov process $\hat{y}(t)$ is called a *merged Markov process* and the phase space $(V, \mathcal{V})$ is called a *merged phase space*. The infinitessimal operator $\hat{Q}$ of the merged Markov process $\hat{y}(t)$ is defined by

$$\hat{Q}\hat{b}(v) := q(v) \int_V \hat{P}(v, dv')[\hat{f}(v') - \hat{f}(v)],$$

$$\hat{f}(v) := \int_{Y_v} p_v(dy)f(y). \tag{2.295}$$

We now consider the behaviour of the solution of the following difference equation, as $\varepsilon \to 0$:

$$\mathbf{X}^{\varepsilon}_{\nu(t/\varepsilon)+1)} - \mathbf{X}^{\varepsilon}_{\nu(t/\varepsilon)} = \epsilon \mathbf{g}(\mathbf{X}^{\varepsilon}_{\nu(t/\varepsilon)}, y^{\varepsilon}_{\nu(t/\varepsilon)+1)}), \tag{2.296}$$

where $(y^{\varepsilon}_n)_{n\in Z_+}$ is a perturbed Markov chain in the phase space $(Y, \mathcal{Y})$ with the transition probability $P_{\varepsilon}(y, B)$ in (2.291).

We consider the following family of operators $D^{\varepsilon}(y)$ on $B := C^1(R^d)$:

$$D^{\varepsilon}(y)f(\mathbf{x}) := f(\mathbf{x} + \varepsilon \mathbf{g}(\mathbf{x}, y)). \tag{2.297}$$

We note that these operators $D^{\varepsilon}(y)$ are linear contractive uniformly with respect to y and admit the representation:

$$D^{\varepsilon}(y)f(\mathbf{x}) = f(\mathbf{x}) + \varepsilon \mathbf{g}^T(\mathbf{x}, y)\frac{d}{d\mathbf{x}}f(\mathbf{x}) + o(\varepsilon)f(\mathbf{x}) \quad \text{as } \varepsilon \to 0, \tag{2.298}$$

where $||o(\varepsilon)f|| \to_{\varepsilon\to 0} 0, ||\cdot||$ is a norm in $C^1(R^d)$.

Put

$$D_1(y)f(x) := \mathbf{g}^T(\mathbf{x}, y)f_{\mathbf{x}}(\mathbf{x}), \quad \forall f \in C^1(R^d). \tag{2.299}$$

Also, consider the operator

$$\hat{D}(v) := \int_Y p_v(dy)D_1(y)/m_v, \tag{2.300}$$

where $D_1(y)$ and m_v are defined in (2.299) and (2.300), respectively. Define the opeator $\hat{V}(t)$ as a solution of the following equation:

$$\hat{V}(t)f - f - \int_0^t \hat{D}(\hat{y}(s))\hat{V}(s)f ds = 0, \forall f \epsilon B, \tag{2.301}$$

where operator $\hat{D}(y)$ is defined in (2.300).

From (2.296) we get that

$$\mathbf{X}^{\varepsilon}_{\nu(t/\varepsilon)+1)} = \mathbf{X}_0 + \varepsilon \sum_{k=0}^{\nu(t/\varepsilon)} \mathbf{g}(\mathbf{X}^{\varepsilon}_k, y^{\varepsilon}_{k+1}). \tag{2.302}$$

Note that for $f \in C^1(R^d)$, we have from (2.296) and (2.297) that

$$\begin{aligned} f(\mathbf{X}^{\varepsilon}_{\nu(t/\varepsilon)}) &= f(\mathbf{x}_0 + \varepsilon \cdot \textstyle\sum_{k=0}^{\nu(t/\varepsilon)} \mathbf{g}(\mathbf{X}^{\varepsilon}_k, y^{\varepsilon}_{k+1})) \\ &= \textstyle\prod_{k=0}^{\nu(t/\varepsilon)} D^{\varepsilon}(y^{\varepsilon}_{k+1})f(\mathbf{x}) := V^{\varepsilon}_{\nu(t/\varepsilon)}f(\mathbf{x}). \end{aligned} \tag{2.303}$$

The operator process $V^{\varepsilon}_{\nu(t/\varepsilon)}$ is a semi-Markov random evolution. Therefore, from the theory of random evolutions (see [49,79,81,82]) it follows that under conditions of Theorem 13 and conditions (2.287)-(2.294), the family of random evolutions $V^{\epsilon}_{\nu(t/\varepsilon)}$ converges weakly as $\varepsilon \to 0$ to the *merged Markov random evolution* $\hat{V}(t)$ which is defined by the solution of the equation (43):

$$V^{\varepsilon}_{\nu(t/\varepsilon)} \Longrightarrow_{\varepsilon\to 0} \hat{V}(t). \tag{2.304}$$

We now calculate the operator $\hat{D}(v)$ in (2.300), using (2.298)-(2.299), as follows

$$\begin{aligned}\hat{D}(v)f(\mathbf{x}) &= \int_{Y_v} p_v(dy)D(y)/m_v \\ &= \int_{Y_v} p_v(dy)\mathbf{g}^T(\mathbf{x},y)f_{\mathbf{x}}(\mathbf{x})/m_v =: \hat{\mathbf{g}}^T(\mathbf{x},v)f_{\mathbf{x}}(\mathbf{x}).\end{aligned}$$

Namely,

$$\hat{D}(v)f(\mathbf{x}) = \hat{\mathbf{g}}(\mathbf{x},v)f_{\mathbf{x}}(\mathbf{x}) = \int_{Y_v} p_v(dy)\mathbf{g}(\mathbf{x},y)f_{\mathbf{x}}(\mathbf{x})/m_v. \tag{2.305}$$

From (2.303) and (2.304), we obtain

$$f(\mathbf{X}^{\varepsilon}_{\nu(t/\varepsilon)}) = V^{\varepsilon}_{\nu(t/\varepsilon)}f(\mathbf{x}) \Longrightarrow_{\varepsilon\to 0} \hat{V}(t)f(\mathbf{x}) := f(\hat{\mathbf{x}}(t)), \tag{2.306}$$

where $\hat{\mathbf{x}}(t)$ is a limit for the $\mathbf{X}^{\varepsilon}_{\nu(t/\varepsilon)}$ process such that

$$f(\hat{\mathbf{x}}(t)) - f(\mathbf{x}) - \int_0^t \hat{\mathbf{g}}^T(\hat{\mathbf{x}}(s),\hat{y}(s))f_{\mathbf{x}}(\hat{\mathbf{x}}(s))ds = 0, \tag{2.307}$$

which follows directly from (2.301), (2.305) and (2.306).

Therefore, $\hat{x}(t)$ satisfies the initial alue problem:

$$\begin{cases} \frac{d\hat{\mathbf{x}}(t)}{dt} &= \hat{\mathbf{g}}(\hat{\mathbf{x}}(t),\hat{y}(t)), \\ \hat{\mathbf{x}}(0) &= \mathbf{x}_0 = \mathbf{x}. \end{cases} \tag{2.308}$$

In summary, we obtain the following

Theorem 14: Under conditions (2.287)-(2.299), the random process $\mathbf{X}^{\varepsilon}_{\nu(t/\varepsilon)}$ converges weakly as $\varepsilon \to 0$ to the process $\hat{\mathbf{x}}(t)$ which satisfies equation (2.308) with function $\hat{\mathbf{g}}(\mathbf{x},v)$ given in (2.305) and the merged Markov process $\hat{y}(t)$ in the merged phase space $(V,\mathcal{V})$ with generator $\hat{Q}$ given in (2.295).

Remark 12: The same conclusion remains valid if $\mathbf{g} \equiv \mathbf{g}^{\epsilon}(\mathbf{x},y) := \mathbf{g}(\mathbf{x},y) + \epsilon\mathbf{g}_1(\mathbf{x},y)$, since in equation (2.299) the term $\epsilon^{3/2}\mathbf{g}_1(x,y)$ vanishes.

CHAPTER 3: EPIDEMIC MODELS

In this Chapter, we study *epidemic models*. We start with deterministic models and then concentrated on their stochastic analogue (in random media), and consider limit theorems and stochastic stability in averaging and diffusion approximation schemes.

3.1. Deterministic Epidemic Models

The study of epidemics has a long history with a vast variety of models and explanations for the spread and cause of epidemic outbreaks [64]. Even simple models can pose important questions with regard to the underlying process and possible means of control of the disease or epidemic. One such case study is the model proposed by Capasso and Paveri-Fontana [13] for the 1973 cholera epidemic in Bari, southern Italy.

An interesting early mathematical model, involving a non-linear ordinary differential equation, proposed by Bernoulli [11], considered the effect of cow-pox innoculation on the spread of smallpox. The article has some interesting data on child mortality at the time, and it is probably the first time that a mathematical model was used to assess the practical advantages of a vaccination control programme. Further development has shown that mathematical models can be extremely useful in estimating the level of vaccination for the control of directly transmitted infectious diseases. See, for example, [2,3]. The series of papers on epidemic models by Kermack and McKendrick [44] had a major influence in the development of mathematical models, and the Reed-Frost model (see Bailey [6]) is the basic one in the theory of epidemics. Therefore, we will start from these models.

A good introduction and survey of various problems and models for the spread and control of infectious diseases is given by the books Bailey [7] and Hoppensteadt [37,39,40]. The survey by Wickwire [93] and the collection of articles on the population dynamics of infectious diseases edited by Anderson [1] are also good references.

3.1.1. REED-FROST MODEL: $S \to I \to R$

A model of susceptible-infective interaction was introduced by L.J. Reed and W.H. Frost in 1928. To introduce the model, we first select an appropriate time unit. This is usually related to the mean length of the infectious period of the disease. Then for the nth period, we can define S_n as the number of susceptibles (who can catch the disease) and I_n as the number of infectives (who have and can transmit the disease). Let R_n be the number of those removed in the nth period. That is, R_n denotes the number of individuals who have passed through the infectious stage by the nth period (who have either had the disease, or are recovered, immune or isolated until recovered). The population is assumed to be constant, say N, and randomly mixing. That is, the probability of effective contact between any susceptible and any infective is the same for the entire population.

Let p be a probability of an effective contact between susceptible and an infective in the nth time period. At the end of each time period, the population will be examined and the new numbers of susceptibles, infectives, and those removed will be determined. The probability that a given susceptible will avoid contact with all infectives is $(1-p)^{I_n}$ and equals to probability of a given susceptible not contacting any infective in the nth time period.

We therefore have the following *Reed-Frost model*

$$\mathcal{P}\{S_{n+1}=k, I_{n+1}=m|S_n, I_n\} = \binom{S_n}{k}(1-p)^{I_n \cdot k}[1-(1-p)^{I_n}]^m. \tag{3.1}$$

That is, the probability that $S_{n+1}=k$ and $I_{n+1}=m$, given S_n and I_n, is the number of ways k individuals can be selected from among S_n times the probability that k miss contact and m make contact. If $S_{n+1}=k$ and $I_{n+1}=m$, then $R_{n+1}=N-k-m$, because the total population remains constant.

3.1.2. KERMACK-MCKENDRICK MODEL

Since $E(S_{n+1}|S_n, I_n) = S_n(1-p)^{I_n}$ and $E(I_{n+1}|S_n, I_n) = S_n(1-(1-p)^{I_n})$, we can consider the system of equations $S_{n+1} = S_n \cdot (1-p)^{I_n}, I_{n+1} = S_n \cdot (1-(1-p)^{I_n})$ as a deterministic analog of the *Reed-Frost model.* It is convenient to rewrite this deterministic system as

$$\begin{cases} S_{n+1} &= \exp(-\alpha I_n)S_n, \\ I_{n+1} &= S_n(1-\exp(-\alpha I_n)), \end{cases} \tag{3.2}$$

where $\alpha = -\ln(1-p)$. This is called *the Kermack-McKendrick model,* see [39].

Model (3.2) can be extended to

$$\begin{cases} S_{n+1} &= \exp(-\alpha I_n)S_n, \\ I_{n+1} &= \beta I_n + (1-\exp(-\alpha I_n))S_n, \\ R_{n+1} &= (1-\beta)I_n + R_n, \end{cases} \tag{3.3}$$

here β is the probability that an infective survives one time period as an infective. Note that if the time periods are chosen equal to the length of the infective period, then $\beta = 0$ as in (3.2). Note also the third equation is decoupled from the first two equations.

The *threshold phenomenon of epidemics* can now be illustrated by the Kermack-McKendrick model (3.3) as follows (see [39]):

Threshold Theorem. Suppose that $R_0 = 0$. Then we have:

(i) the susceptible population approaches a limiting value: $S_n \to S_\infty$ as $n \to +\infty$;
(ii) $R_n \to R_\infty = N - S_\infty$ as $n \to +\infty$;
(iii) if $F = S_\infty / S_0$, then

$$F = \exp\Big(\Big(-\frac{\alpha S_0}{1-\beta} \cdot \Big(1 + \frac{I_0}{S_0} - F\Big)\Big). \tag{3.4}$$

Proof: As $S_{n+1} = \exp(-\alpha I_n) S_n \leq S_n$ and $S_n \geq 0$ for all $n \in Z^+$ $\lim_{n\to\infty} S_n := S_{+\infty}$ exists. This proves (i). As $R_{n+1} = (1-\beta) \cdot I_n + R_n \geq R_n$ and $R_n \leq N$, $\lim_{n\to+\infty} R_n$ exists. Moreover, since $I_n := R_{n+1} - R_n$, we have that $\lim_{n\to+\infty} I_n = 0$ and so, $R_\infty = N - S_\infty$. As $S_{n+1} = \exp(-\alpha I_n) S_n$, we have that

$$\begin{aligned} S_n &= S_0 \exp\Big(-\alpha \cdot \textstyle\sum_{k=0}^{n-1} I_k\Big) \\ &= S_0 \exp\Big(-\big(\tfrac{\alpha}{1-\beta}\big) \textstyle\sum_{k=0}^{n-1} (R_{k+1} - R_k)\Big) \\ &= S_0 \exp\big(-\tfrac{\alpha}{1-\beta} \cdot R_{n-1}\big). \end{aligned}$$

Passing to the limit as $n \to +\infty$, we obtain

$$S_\infty = S_0 \exp\big(-\frac{\alpha}{1-\beta}(N - S_\infty)\big).$$

Setting $F = S_\infty / S_0$ gives

$$F = \exp\big(-\big(\frac{\alpha S_0}{1-\beta}\big)\big(1 + \frac{I_0}{S_0} - F\big)\big).$$

This completes the proof of the theorem.

Remark. From the above Theorem, we conclude that if I_0/S_0 is small, then $T := \frac{1-\beta}{\alpha}$ is the threshold of susceptible population size. In other words, if $S_0 < T$, then $F \sim 1$ and few susceptibles are infected, if $S_0 > T$ then an epidemic occurs.

F is a measure of the epidemic's final size and of its severity. If $T \gg 1$, then $F \sim 0$ and a large epidemic occurs. But if $T \sim 1$, then $F \sim 1$, and only a few susceptibles are exposed. As T passes through the critical value $T = 1$, there is a dramatic change (referred to as a bifurcation) in the epidemic's final size. This is a basic result in epidemiology.

3.1.3. CONTINUOUS EPIDEMIC MODELS

Let $S(t), I(t)$ and $R(t)$ be the number of individuals in each class of susceptibles, infectives and those removed, respectively. Assume that: (i) the gain in the infective class is at a rate proportional to the number of infectives and susceptibles (that is $\lambda S \cdot I$), the susceptibles are lost at the same rate; (ii) the rate of removal of infectives to the removal class is proportional to the number of infectives (that is λI, where $\lambda > 0$ is a constant); (iii) the incubation period is short enough to be negligible (that is, a susceptible who contracts the diseases is infective right away). We further assume, as in the previous sections, that various classes are uniformly mixed (that is, every pair of individuals has equal probability of coming into contact with each other). Then we obtain the following

$$\begin{cases} \frac{dS}{dt} &= -rSI, \quad S(0) > 0, \\ \frac{dI}{dt} &= rSI - \lambda I, \quad I(0) > 0, \\ \frac{dR}{dt} &= \beta I, \quad R(0) = 0, \end{cases} \tag{3.5}$$

where $r > 0$ is the infection rate and $\lambda > 0$ the removal rate of infectives. This is the *continuous-time Kermack-McKendrick model* (see, [34,59]).

We now show that model (3.5) can be obtained from discrete-time model (3.3). Suppose that $\alpha_n = \varepsilon \cdot r, \beta_n = 1 - \varepsilon\lambda$ and let $S_n = S(n\varepsilon), I_n = I(n\varepsilon)$ and $R_n = R(n\varepsilon)$, where S, I and R are smooth functions. Since $S(n\varepsilon + \varepsilon) - S(n\varepsilon) = -rS(n\varepsilon) \cdot I(n\varepsilon)\varepsilon$, by setting $t = \varepsilon n$, we obtain as $\varepsilon \to 0$ the first equation in (3.5). More explicitely, $S(n\varepsilon+\varepsilon) - S(n\varepsilon) = S_{n+1} - S_n = -\alpha_n I_n S_n + 0(\varepsilon^2) = -\varepsilon r S(n\varepsilon) \cdot I(n\varepsilon) + 0(\varepsilon^2)$, where we used the first equation in (3.3) and the Taylor expansion for $\exp(-\varepsilon\alpha_n I_n)$. The second equation in (3.5) can be obtain from the second equation of (3.3): $I(n\varepsilon+\varepsilon) - I(n\varepsilon) = I_{n+1} - I_n = -\varepsilon\lambda I_n + \alpha_n I_n \cdot S_n + 0(\varepsilon^2) = -\varepsilon\lambda I(n\varepsilon) + \varepsilon r I(\varepsilon n) \cdot S(n\varepsilon) + 0(\varepsilon^2)$. Therefore, if setting $t = \varepsilon n$, as $\varepsilon \to 0$ we obtain the second equation in (3.5):

$$\lim_{\varepsilon \to 0} \frac{I(n\varepsilon + \varepsilon) - I(n\varepsilon)}{\varepsilon} = \frac{dI(t)}{dt} = rI(t) \cdot S(t) - \lambda I(t).$$

The equation for $R(t)$ follows in the same way.

We can solve system (3.5) for I as function of S. From the first and the second equation in (3.5) we obtain

$$\frac{dI(t)}{dS(t)} = -1 + \frac{\lambda}{r} S(t), \tag{3.6}$$

and hence,

$$I(t) = I(0) + \frac{\lambda}{r} \ln(S(t)/S(0)) - (S(t) - S(0)). \tag{3.7}$$

Since $S(t) \geq 0$ and $dS(t)/dt \leq 0, S(t)$ approaches a limit, say $S(\infty)$, as $t \to +\infty$. Also, since $R(t) \leq S(0) + I(0)$ and $dR(t)/dt \geq 0, R(t)$ approaches a limit denoted by $R(+\infty)$. Thus, $dR(t)/dt \to 0$ as $t \to +\infty$, and from the third equation in (3.5) we obtain that $I(t) = \frac{1}{\lambda} dR(t)/dt \to 0$ as $t \to +\infty$. Using this and the formula (3.7), we have

$$S(+\infty) = S(0) + I(0) + \frac{\lambda}{r} \ln(S(+\infty)/S(0)), \tag{3.8}$$

or equivalently,

$$\frac{S(+\infty)}{S(0)} = \exp\left\{\frac{r}{\lambda S(0)}[(\frac{S(+\infty)}{S(0)}) - 1 - \frac{I(0)}{S(0)}]\right\}. \tag{3.9}$$

In most epidemics, it is difficult to determine how many new infectives each day since only those that are removed, via medical aid or whatever, can be counted. So, to apply the model to actual epidemic situations in general, we need to know the number removed per unit time, namely, $dR(t)/dt$ as a function of time. This can be done as follows. First of all, we note, that from (3.5) it follows that

$$\frac{dS(t)}{dt} + \frac{dI(t)}{dt} + \frac{dR(t)}{dt} = 0, \tag{3.10}$$

that is,

$$S(t) + I(t) + R(t) = N = \text{const}. \tag{3.11}$$

From the first and the third equations in (3.5), we obtain

$$\frac{dS(t)}{dR(t)} = -\frac{S(t)}{p}, \tag{3.12}$$

where $p := \frac{\lambda}{r}$ and

$$S(t) = S(0)\ \exp\,[-\frac{R(t)}{p}] \geq S(0)\ \exp\,[-N/p] > 0.$$

Hence, $0 < S(+\infty) \leq N$.

From the third equation in (3.5) and (3.10)-(3.12), we obtain an equation for $R(t)$

$$\begin{aligned} \tfrac{dR(t)}{dt} &= \lambda I = \lambda(N - R(t) - S(t)) \\ &= \lambda\big[N - R(t) - S_0\ \exp\big(-\tfrac{R(t)}{p}\big)\big], R(0) = 0, \end{aligned} \tag{3.13}$$

for which an explicit formula for the solution can be obtained but is complicated, though knowing all the parameters $\lambda, r, S(0)$ and N, computing the solution numerically is staightforward. However, in many cases, we don not know all the parameters, and so we have to carry out a best fit procedure assuming, of course, the epidemic is reasonably described by such a model. In practice, it is often the case that if the epidemic is not large, R/p is small; certainly $R/p < 1$.

Following Kermack and McKendrick [39], we can then approximate (3.13) by the equation

$$\frac{dR(t)}{dt} = \lambda[N - S(0) + (\frac{S(0)}{p} - 1)R(t) - \frac{S(0)R^2(t)}{2p^2}]. \tag{3.14}$$

Factoring the right hand side quadratic in $R(t)$, we can interprete this equation to get (see [59, p. 615]):

$$\frac{dR(t)}{dt} = \frac{\lambda d^2 p^2}{2S(0)} \cdot \ \text{sech}^2(\frac{\mu\lambda t}{2} - \phi),$$

where

$$\mu := \big[(\frac{S(0)}{p} - 1)^2 + \frac{2S(0)/(N - S(0))}{p^2}\big]^{1/2},$$

and

$$\phi := \frac{\tanh^{-1}(\frac{S(0)}{p} - 1)}{\mu}.$$

We now describe the *threshold phenomenon* for the continuous time epidemic model (3.5). From the second equation in (3.5), we obtain

$$\frac{dI(t)}{dt}|_{t=0} = I(0)(rS(0) - \lambda) \begin{cases} > 0, & \text{if} \quad S(0) > \lambda/r =: p, \\ \leq 0, & \text{if} \quad S(0) < p. \end{cases} \tag{3.15}$$

Since, from the first equation in (3.5), $dS(t)/dt < 0, S(t) \leq S(0)$, we have, if $S(0) < \lambda/p$,

$$\frac{dI(t)}{dt} = I(t) \cdot (r \cdot S(t) - a) \leq 0, \forall\, t \geq 0, \tag{3.16}$$

in which case $I(0) > I(t) \to 0$ as $t \to +\infty$ and so *the infection dies out:* (that is, *no epidemic can occur*). On the other hand, if $S(0) > \lambda/r$ then $I(t)$ initially increases and *we have an epidemic.* The term "epidemic" here means that $I(t) > I(0)$ for some $t > 0$. We thus have *a threshold phenomenon.* The critical parameter $p = \lambda/r$ is sometimes called *the relative removal rate* and its reciprocal $\sigma := r/\lambda$ is called *the infection's contact rate.* We write

$$R_0 := \frac{rS(0)}{\lambda},$$

and regard R_0 as the basic reproduction rate of the infection, that is, the number of secondary infections produced by one primary infection in a wholly susceptible population. Here $1/\lambda$ is the average infectious period. If more than one secondary infection is produced from one primary infection, that is $R_0 > 1$, clearly an epidemic results. A mathematical introduction to the subject is given by Waltman [88].

Let us now derive some other useful analytical results from the continuous time epidemic model. Integrating equation (3.6) gives the (I, S) phase trajectories as

$$I(t) + S(t) - p \cdot \ln S(t) = \text{ const } = I(0) + S(0) - p \ \ln S(0), \tag{3.17}$$

where the initial conditions in (3.5) are used. If an epidemic exists, it is natural to know how severe it will be. From (3.6) the maximim $I(t), I(\max)$, occurs at $S = p$, where $dI(t)/dt = 0$. From (3.17), with $S = p$, we obtain:

$$\begin{aligned} I_{\max} &= p \ln p - p + I(0) + S(0) - p \ln S(0) \\ &= I(0) + (S(0) - p) + p \ln(\tfrac{p}{S(0)}) = N - p + p \ln(\tfrac{p}{S(0)}). \end{aligned} \tag{3.18}$$

We note that with initial conditions in (3.5), all initial values $S(0)$ and $I(0)$ satisfy $I(0) + S(0) = N$ since $R(0) = 0$ and so for $t > 0, 0 \leq S(t) + I(t) < N$. For any initial value $I(0)$ and $S(0) > p$, the phase trajectory starts with $S > p$ and $I(t)$ increases from $I(0)$, and hence an epidemic results. It is not necessarily a severe epidemic, though if $I(0)$ is close to $I_{\max}$. If $S(0) < p$, then $I(t)$ decreases from $I(0)$, and no epidemic occurs.

We have seen from (3.12) that $0 < S(+\infty) \leq N$. In fact, $0 < S(+\infty) < p$. Since $I(+\infty) = 0$, (3.10)-(3.11) implies that $R(+\infty) = N - S(+\infty)$. Thus, from (3.12) we have

$$S(+\infty) = S_0 \exp\left[\frac{R(+\infty)}{p}\right] = S_0 \exp\left[-\frac{N - S(+\infty)}{p}\right], \tag{3.19}$$

and so $S(+\infty)$ is the positive root $0 < z < p$ of the transcendental equation

$$S_0 \exp\left[-\frac{N - z}{p}\right] = z. \tag{3.20}$$

We then get the total number of susceptibles who catch the disease in the course of the epidemic as

$$I_{\text{total}} = I(0) + S(0) - S(+\infty), \tag{3.21}$$

where $S(+\infty)$ is the positive solution z of (3.20).

An important implication of this analysis, namely that $I(t) \to 0$ and $S(t) \to S(+\infty) > 0$, is that *the disease dies out from a lack of infectives* and *not from a lack of susceptibles.*

3.2. Stochastic Epidemic Model (EM in Random Media)

Here we consider an epidemic model, introduced in subsection 3.1, in random media.

We consider an discrete-time epidemic model in Markov random environment, and a continuous-time epidemic model in semi-Markov random environment.

3.2.1. DISCRETE EPIDEMIC MODEL IN MARKOV RANDOM MEDIA.

Let $(y_n)_{n\in Z^+}$ be a homogeneous Markov chain in a measurable space $(X, \mathcal{X})$ with transition probabilities $P(y, A), y \in X, A \in \mathcal{X}$, and ergodic distribution $p(A), A \in \mathcal{X}$ (see Chapter 1).

Let $\beta(y)$ and $d(y)$ be non-negative, bounded measurable functions defined on Y. *A discrete epidemic model in Markov random environment* is defined

$$\begin{cases} S_{n+1} &= S_n \exp\left(-\alpha(y_{n+1}) I_n\right), \\ I_{n+1} &= \left(1 - \exp\left(-\alpha(y_{n+1}) I_n\right)\right) S_n + \beta(y_{n+1}) I_n, \\ R_{n+1} &= \left(1 - \beta(y_{n+1})\right) I_n + R_n. \end{cases} \tag{3.22}$$

As above S_n is a number of susceptible in the nth period (who can catch the disease); I_n is a number of infective in the nth period (who have the disease and can transmit it); R_n is a number of those removed in the nth period. We note that all the functions S_n, I_n and R_n are random processes.

Function $\beta(y)$ describes the removal rate of infectives for each fixed $y \epsilon Y$, and function $\alpha(y)$ describes the infection rate for each fixed $y \epsilon Y$.

In such a way, we have an epidemic model in Markov random environment Y. The evolution of this epidemic model is the following: the number S_n of susceptibles in the nth period defined as the state y_n of random evironment through the infection rate $\alpha(y_n)$; the number I_n of infectives in the nth period also depends on the state y_n of random enviroment through the infection rate $\alpha(y_n)$ and also by the removal rate $\beta(y_n)$; the number of those removed in the nth period depends on the state y_n of random environment via the removal rate $\beta(y_n)$.

The difference between classical Kermack-McKendrick epidemic model (3.3) and discrete epidemic model in Markov random environment (3.22) is the following: in the first model (3.3) the infection rate d and the removal rate β are constant for all periods $n \geq 1$; in the second model (3.22) the infection rates $\alpha(y_n)$ and the removal rates $\beta(y_n)$ are changed from one period to another with respect to the states of Markov chain $(y_n)_{n \in Z^+}$.

Let us state some analog of threshold theorem for discrete epidemic model.

Analogue of Threshold Theorem. Suppose that $R_0 = 0$. Then:

a) the susceptible population approaches a limiting value: $S_n \to \bar{S}_\infty$ as $n \to +\infty$ ($\mathcal{P}$-almost sure (a.s));

b) $\bar{R}_\infty = N - \bar{S}_\infty$ ($\mathcal{P}$-a.s).

Proof: a) as $S_{n+1} = \exp\,(-\alpha(y_{n+1})I_n)S_n \leq S_n$ $\mathcal{P}$-a.s. and $S_n \geq 0$ $\mathcal{P}$-a.s. $\forall n \in Z^+$, we see that $\lim_{n\to\infty} S_n$ exists $\mathcal{P}$-a.s. Denote this by $\bar{S}_\infty := \lim_{n\to\infty} S_n$ exists $\mathcal{P}$-a.s.

b) As $R_{n+1} = (1 - \beta(y_{n+1}))I_n + R_n \geq R_n$ ($\mathcal{P}$-a.s.) and $R_n \leq N$ ($\mathcal{P}$-a.s.), where N is the total size of the population, we see that $\lim_{n\to\infty} R_n$ exists $\mathcal{P}$-a.s. Let $\bar{R}_\infty = \lim_{n\to+\infty} R_n$. As $I_n = R_{n+1} - R_n$, we have that $\lim_{n\to\infty} I_n = 0$ $\mathcal{P}$-a.s. and so $\bar{R}_\infty = N - \bar{S}_\infty$.

As we see, the result so far is the same as a) and b) in the Threshold Theorem. The third part of the Threshold Theorem will be different from its stochastic analogue, and this will be studied later.

3.2.2. CONTINUOUS EPIDEMIC MODEL IN MARKOV RENEWAL RANDOM MEDIA

Let $(y_n;\theta_n)_{n\in Z^+}$ be a Markov renewal process in the phase space $(Y \times R_+, \mathcal{Y} \times \mathcal{R}_+)$ with stochastic kernel $Q(y, dz, t), y \in Y, dz \in \mathcal{Y}, t \in R_+$ (see Section 1.3, Chapter 1).

Let $\nu(t) := \max\,\{n : \tau_n \leq t\}$ be a counting process, $\tau_n := \sum_{k=0}^{n} \theta_k, \theta_0 = 0$.

Let also $\alpha(y)$ and $\beta(y)$ be non-negative, bounded measurable functions defined on Y.

A continuous epidemic model in Markov renewal random environment is defined as

$$\begin{cases} S_{\nu(t)+1} &= S_{\nu t}\exp\,(-\alpha(y_{\nu(t)+1})I_{\nu(t)}), \\ I_{\nu(t)+1} &= (1-\exp\,(-\alpha(y_{\nu(t)+1})I_{\nu(t)}))S_{\nu(t)} + \beta(y_{\nu(t)+1})\cdot I_{\nu(t)}, \\ R_{\nu(t)+1} &= (1-\beta(y_{\nu(t)+1}))I_{\nu(t)} + R_{\nu(t)}. \end{cases} \tag{3.23}$$

It is important to note that S, I and R in (3.23) are random processes. Here, $S_{\nu(t)}$ is a number of susceptible in the $\nu(t)$th period. It means that if $t\epsilon[\tau_n, \tau_{n+1}]$, where τ_n are renewal times, then $S_{\nu(t)} \equiv S_n$, as $\nu(t) = n$, for $t\epsilon[\tau_n, \tau_{n+1}]$; $I_{\nu(t)}$ and $R_{\nu(t)}$ are the numbers of infective, and of those removed in the $\nu(t)$th period, respectively.

In such a way, if $t\epsilon[\tau_n, \tau_{n+1}]$, then our system (3.23) is the same as (3.22). However, the length of the infectious period of the disease is random and is given by $\theta_n := \tau_n - \tau_{n-1}, n \geq 1$.

The number of periods on the interval $[0, t]$ is defined by the counting process $\nu(t) := \max\,\{n; \tau_n \leq t\}$. Since our MRP is a regular one; i.e., $P(\nu(t) < +\infty, \forall\ t\epsilon R_+) = 1$, we have a finite number of periods on the finite interval $[0, t]$.

We now show that an analog of Threshold Theorem is also true for the system (3.23).

Analog of Threshold Theorem. Suppose that $R_0 = 0$. Then:

a) the susceptible population approaches a limiting value: $S_{\nu(t)} \to \hat{S}_\infty$ as $t \to +\infty$ ($\mathcal{P}$-a.s.);

b) $\hat{R}_\infty = N - \hat{S}_\infty$ ($\mathcal{P}$-a.s.).

Proof: a) As $S_{\nu(t)+1} = \exp\,(-\alpha(y_{\nu(t)+1})I_{\nu(t)}S_{\nu(t)} \leq S_{\nu(t)}$ $\mathcal{P}$-a.s. and $S_{\nu(t)} \geq 0$ $\mathcal{P}$-a.s. $\forall\ t\epsilon R_+$, we see that $\lim_{t\to+\infty} S_{\nu(t)}$ exists $\mathcal{P}$-a.s., as $\nu(t) \to +\infty$. Denote this by $\hat{S}_\infty := \lim_{t\to+\infty} S_{\nu(t)}$ ($\mathcal{P}$-a.s.).

b) We note that $R_{\nu(t)} \leq N$ ($\mathcal{P}$-a.s.). We see that $\lim_{t\to+\infty} R_{\nu(t)}$ exists $\mathcal{P}$-a.s. Let $\hat{R}_\infty := \lim R_{\nu(t)}$. As $I_{\nu(t)} = R_{\nu(t)+1} - R_{\nu(t)}$, we have that $\lim_{t\to+\infty} I_{\nu(t)} = 0$. $\mathcal{P}$-a.s. and so $\hat{R}_\infty = N - \hat{S}_\infty$.

3.2.3. CONTINUOUS EPIDEMIC MODEL IN SEMI-MARKOV RANDOM MEDIA

Let $y(t) := y_{\nu(t)}$ be a semi-Markov process, which is defined by a Markov renewal process $(y_n; \theta_n)_{n\in Z^+}$, and functions $\beta(y)$ and $\alpha(y)$ be the same as in sections 3.1.1-3.1.2.

A continuous epidemic model in semi-Markov random environment is defined as

$$\begin{cases} \frac{dS(t)}{dt} &= -\alpha(y(t))S(t))I(t), \quad S(0) > 0, \\ \frac{dI(t)}{dt} &= \alpha(y(t))S(t)I(t) - \beta(y(t))I(t), \quad I(0) > 0, \\ \frac{dR(t)}{dt} &= \beta(y(t))I(t), \quad R(0) = 0, \end{cases} \tag{3.24}$$

The model (3.24) differs from the classical continuous-time Kermack-McKendrick model (3.5) where coefficients r and λ are constants. We note that we cannot obtain the model (3.24) from any discrete-time epidemic model, neither from (3.22) nor from (3.23), as it was in the classical case of Kermack-McKendrick model (see Section 3.1.2).

If $y(t)$ in (3.24) is a Markov process, we can obtain this model from the *merging theorem for epidemic model* (to be described later). We note that $S(t), I(t)$ and $R(t)$ in (3.24) are random processes.

We now state the result for the measurability of epidemic models in RM from the general results in Section 1.10.2.

The discrete-time epidemic model (3.22) in Markov random environment is $\mathcal{F}_{n+1}/\mathcal{R}$ measurable. It means that all the processes S_{n+1}, I_{n+1} and R_{n+1} are $\mathcal{F}^Y_{n+1}/\mathcal{R}$-measurable, since, for example, in the expression

$$S_{n+1} = S_n \exp\left(-\alpha(y_{n+1})I_n\right),$$

S_n is $\mathcal{F}^Y_n$-measurable and I_n is $\mathcal{F}^Y_n$-measurable, hence, S_n and I_n are also $\mathcal{F}^Y_{n+1}$-measurable, $(\mathcal{F}^Y_n \subset \mathcal{F}^Y_{n+1})$, $\alpha(y_{n+1})$ is $\mathcal{F}^Y_{n+1}$-measurable, and $\mathcal{F}^Y_{n+1} \subset \mathcal{F}_{n+1}$.

The same is true for I_{n+1} and R_{n+1} in (3.22).

The continuous-time epidemic model in (3.23) in Markov renewal random environment is $\mathcal{G}_t/\mathcal{R}$-measurable, since, for example, $S_{\nu(t)}$ and $I_{\nu(t)}$ are $\mathcal{H}_t/\mathcal{R}$-measurable, $G_t \subset G_t$ and $d(y_{\nu(t)+1})$ is $\mathcal{G}_t/\mathcal{R}$-measurable.

The continuous-time epidemic model in (3.24) in semi-Markov random environment is $\mathcal{H}_t/\mathcal{R}$-measurable, as $\alpha(y(t))$ is $\mathcal{H}_t/\mathcal{R}$-measurable and $\beta(y(t))$ is $\mathcal{H}_t/\mathcal{R}$-measurable.

3.3. Averaging of the Epidemic Model in Random Media

In this section, we apply the results from section 2.7, namely Theorems 9 and 10 to the Epidemic Model in Random Media. In this case, $X = R^3$, $\mathbf{x} = [S, I, R]^T$ and the vector-function $\mathbf{g}(\mathbf{x}, y) = [-\alpha(y)SI, \alpha(y)SI - \beta(y)I, \beta(y)I]^T$.

Let us consider a disease in random environment which, after recovery, confers immunity. The population can be then divided into three distinct classes: the susceptibles, S^ε, who can catch the disease; the infectives, I^ε, who have the disease and can transmit it, and the removed class, R^ε, namely, those who have either had the disease, or are recovered, immune or isolated until recovered, where ε is a small positive parameter.

We suppose that these classes $S^\varepsilon, I^\varepsilon, R^\varepsilon$ satisfy the *Kermack-McKendrick model* in randon environment in series scheme:

$$\begin{cases} S^\varepsilon_{\nu(t/\varepsilon)+1} &= S^\varepsilon_{\nu(t/\varepsilon)} e^{-\varepsilon\alpha(y_{\nu(t/\varepsilon)+1}) I^\varepsilon_{\nu(t/\varepsilon)}}, \\ I^\varepsilon_{\nu(t/\varepsilon)+1} &= (1 - e^{-\varepsilon\alpha(y_{\nu(t/\varepsilon)+1}) I^\varepsilon_{\nu(t/\varepsilon)}}) S^\varepsilon_{\nu(t/\varepsilon)} \\ &+ (1 - \varepsilon\beta(y_{\nu(t/\varepsilon)+1})) I^\varepsilon_{\nu(t/\varepsilon)}, \\ R^\varepsilon_{\nu(t/\varepsilon)+1} &= R^\varepsilon_{\nu(t/\varepsilon)} + \varepsilon \cdot \beta(y_{\nu(t/\varepsilon)+1}) I^\varepsilon_{\nu(t/\varepsilon)}, \\ S^\varepsilon_0 &= S_0, \quad I^\varepsilon_0 = I_0, \quad R^\varepsilon_0 = 0. \end{cases} \tag{3.25}$$

The above is called a *continuous epidemic model* in *Markov renewal random environment*. We can rewrite system (3.25) in the following form:

$$\begin{cases} S^\varepsilon_{\nu(t/\varepsilon)+1} - S_{\nu(t/\varepsilon)} &= -S^\varepsilon_{\nu(t/\varepsilon)} \cdot (1 - e^{-\varepsilon\alpha(y_{\nu(t/\varepsilon)+1}) \cdot I^\varepsilon_{\nu(t/\varepsilon)}}), \\ I^\varepsilon_{\nu(t/\varepsilon)+1} - I_{\nu(t/\varepsilon)} &= S^\varepsilon_{\nu(t/\varepsilon)} \cdot (1 - e^{-\varepsilon\alpha(y_{\nu(t/\varepsilon)+1}) \cdot I^\varepsilon_{\nu(t/\varepsilon)}}) \\ &- \varepsilon\beta \cdot (y_{\nu(t/\varepsilon)+1}) \cdot I^\varepsilon_{\nu(t/\varepsilon)}, \\ R^\varepsilon_{\nu(t/\varepsilon)+1} - R^\varepsilon_{\nu(t/\varepsilon)} &= \varepsilon\beta(y_{\nu(t/\varepsilon)+1}) \cdot I^\varepsilon_{\nu(t/\varepsilon)}, \\ S^\varepsilon_0 &= S_0, \quad I^\varepsilon_0 = I_0, \quad R^\varepsilon_0 = 0. \end{cases} \tag{3.26}$$

Using the Taylor expansion for exponential functions, we obtain the following representations:

$$\begin{cases} S^\varepsilon_{\nu(t/\varepsilon)+1} - S^\varepsilon_{\nu(t/\varepsilon)} &= -\varepsilon \cdot \alpha(y_{\nu(t/\varepsilon)+1}) S^\varepsilon_{\nu(t/\varepsilon)} \cdot I^\varepsilon_{\nu(t/\varepsilon)} + o(\varepsilon) \\ I^\varepsilon_{\nu(t/\varepsilon)+1} - I^\varepsilon_{\nu(t/\varepsilon)} &= \varepsilon(\alpha(y_{\nu(t/\varepsilon)+1} S^\varepsilon_{\nu(t/\varepsilon)}) \cdot I^\varepsilon_{\nu(t/\varepsilon)}) \\ &- \beta(y_{\nu(t/\varepsilon)+1}) \cdot I^\varepsilon_{\nu(t/\varepsilon)} + o(\varepsilon), \\ R^\varepsilon_{\nu(t/\varepsilon)+1} - R^\varepsilon_{\nu(t/\varepsilon)} &= \varepsilon\beta(y_{\nu(t/\varepsilon)+1}) \cdot I^\varepsilon_{\nu(t/\varepsilon)}. \\ S^\varepsilon_0 &= S_0, \quad I^\varepsilon_0 = I_0, \quad R^\varepsilon_0 = 0. \end{cases} \tag{3.27}$$

We note that sequences $S^\varepsilon_{\nu(t/\varepsilon)}$, $I^\varepsilon_{\nu(t/\varepsilon)}$ and $R^\varepsilon_{\nu(t/\varepsilon)}$ are bounded by N- total number (size) of population. It means that all the sequences have the limiting points as $\varepsilon \to 0$. Let us denote these limit points by $\hat{S}_t, \hat{I}_t$ and $\hat{R}_t$.

Applying Theorem 10 (Subsection 2.7.1, Chapter 2) with functions $\mathbf{g}(\mathbf{x}, y)$ given by $\mathbf{g}(\mathbf{x}, y) = [-\alpha(y)SI, \alpha(y)SI - \beta(y)I, \beta(y)I]^T$ we obtain that

$$\lim_{\varepsilon \to 0} \mathcal{P}\{(|S^\varepsilon_{\nu(t/\varepsilon)} - \hat{S}_t| + |I^\varepsilon_{\nu(t/\varepsilon)} - \hat{I}_t| + |R^\varepsilon_{\nu(t/\varepsilon)} - \hat{R}_t|) > \delta\} = 0 \tag{3.28}$$

for any $\delta > 0$ and $t \epsilon R^+$, where $\hat{S}_t$, $\hat{I}_t$ and $\hat{R}_t$ satisfy the following system of equations:

$$\left\{\begin{array}{rcl} \frac{d\hat{S}_t}{dt} & = & -\hat{\alpha}\hat{S}_t \cdot \hat{I}_t \\ \frac{d\hat{I}_t}{dt} & = & \hat{\alpha} \cdot \hat{S}_t \hat{I}_t - \hat{\beta} \cdot \hat{I}_t \\ \frac{d\hat{R}_t}{dt} & = & \hat{\beta} \cdot \hat{I}_t \\ \hat{S}_0 & = & S_0, \quad \hat{I}_0 = I_0, \quad \hat{R}_0 = 0, \end{array}\right. \tag{3.29}$$

where

$$\hat{\alpha} := \int_Y p(dy)\alpha(y)/m,$$
$$\hat{\beta} := \int_Y p(dy)\beta(y)/m. \tag{3.30}$$

Put

$$\hat{r} := \hat{\beta}/\hat{\alpha} \text{ and } \hat{\sigma} := \hat{\alpha}/\hat{\beta} = \frac{r}{\hat{r}}. \tag{3.31}$$

$\hat{r}$ is called *average relative removal rate* and $\hat{\sigma}$ is called *the infection's average contact rate.* $\hat{\sigma} \cdot S_0 := \hat{\gamma}$ is the *basic average reproduction rate* of the infection, and $1/\hat{\beta}$ is the *average infection period.*

Integrating the system (3.29) gives

$$\hat{I}_t = S_0 + I_0 - \hat{S}_t + \hat{r} \cdot \ln \frac{\hat{S}_t}{S_0},$$

here we have used that

$$\frac{d\hat{I}_t}{d\hat{S}_t} = -1 + \frac{\hat{r}}{\hat{S}_t},$$

which follows from (3.29), and $\hat{R}_0 = 0$.

If $\hat{r} > S_0$, then $\hat{I}_t \downarrow 0$ and $\hat{S}_t \downarrow \hat{S}_{\min}$ as $t \to +\infty$, where $\hat{S}_{\min}$ satisfies the equation

$$\hat{S}_{\min} = S_0 + I_0 + \hat{r} \cdot \ln \frac{\hat{S}_{\min}}{S_0}. \tag{3.32}$$

On the other hand, if $\hat{r} < S_0$ then $\hat{I}_t$ increases for $\hat{r} < \hat{S}_t < S_0$, and $\hat{I}_t$ decreases for $\hat{S}_{\min} < \hat{S}_t < \hat{r}$. The maximum of $\hat{I}_t = \hat{I}_{\max}$ is determined from the equation

$$\hat{I}_{\max} = S_0 + I_0 - \hat{r} + \hat{r} \cdot \ln \frac{\hat{r}}{S_0}. \tag{3.33}$$

Therefore, the dimensionless parameter $\hat{r}/S_0$ indicates the vulnerability of the susceptible population to supporting an epidemic. This is referred to as the analogue of Kermack-McKendrick *average threshold theorem.*

We note that from (3.29) and $\frac{d\hat{S}}{d\hat{R}} = -\frac{\hat{S}}{\hat{r}}$, and from $\hat{S}_t + \hat{I}_t + \hat{R}_t = N$, we have the equation for $\hat{R}_t$:

$$d\hat{R}_t/dt = \hat{\beta}\cdot(N - \hat{R}_t - S_0\cdot\exp(-\frac{\hat{R}_t}{\hat{r}})). \tag{3.34}$$

From (3.28), (3.32) and (3.33) we obtain that

$$\lim_{\varepsilon\to 0}\mathcal{P}\{|S^\varepsilon_{\min} - \hat{S}_{\min}| + |I^\varepsilon_{\max} - \hat{I}_{\max}|) > \delta\} = 0,$$

where

$$S^\varepsilon_{\min} := \lim_{t\to+\infty} S^\varepsilon_{\nu(t/\varepsilon)} \text{ and } I^\varepsilon_{\max} := \lim_{t\to+\infty} I^\varepsilon_{\nu(t/\varepsilon)}.$$

Now we prove the result (3.29) on averaging of the epidemic model by another method, based on the renewal theorem and an averaging result for integral functionals of semi-Markov process.

We note that

$$\triangle(S^\varepsilon_{\nu(t/\varepsilon)} + I^\varepsilon_{\nu(t/\varepsilon} + R^\varepsilon_{\nu(t/\varepsilon)}) = 0, \forall t\epsilon R^+,$$

and hence,

$$S^\varepsilon_{\nu(t/\varepsilon)} + I^\varepsilon_{\nu(t/\varepsilon)} + R^\varepsilon_{\nu(t/\varepsilon)} = N = \text{const}, \tag{3.35}$$

where N is the total size of the population, and $\Delta S^\varepsilon_{\nu(t/\varepsilon)} := S^\varepsilon_{\nu(t/\varepsilon)+1} - S_{\nu(t/\varepsilon)}$. Hence, each term $S^\varepsilon_{\nu(t/\varepsilon)}, I^\varepsilon_{\nu(t/\varepsilon)}$ and $R^\varepsilon_{\nu(t/\varepsilon)}$ in (3.25) is bounded and we can choose a subsequence for which each $S^\varepsilon, I^\varepsilon$ and R^ε is convergent.

Let us prove that $S^\varepsilon \to \hat{S}_t, I^\varepsilon \to \hat{I}_t, R^\varepsilon \to \hat{R}_t$, as $\epsilon \to 0$, and $\hat{S}_t, \hat{I}_t, \hat{R}_r$ are defined in (3.29).

We note that

$$\begin{aligned} R^\varepsilon_{\nu(t/\varepsilon)} - \hat{R}_t &= \varepsilon\textstyle\sum_{k=1}^{\nu(t/\varepsilon)}\beta(y_{k+1})I^\varepsilon_k - \hat{\beta}\cdot\int_0^t \hat{I}_s ds \\ &= \varepsilon\textstyle\sum_{k=1}^{\nu(t/\varepsilon)}\beta(y_{k+1})I^\varepsilon_k - \varepsilon\hat{\beta}\cdot\int_0^{t/\varepsilon}\hat{I}_{\varepsilon s}ds \\ &= \varepsilon\cdot(\textstyle\sum_{k=1}^{\nu(t/\varepsilon)}\beta(y_{k+1})I^\varepsilon_k - \hat{\beta}\cdot\sum_{k=1}^{\nu(t/\varepsilon)}\hat{I}_{\varepsilon\alpha_k} \\ &-\hat{\beta}\textstyle\int_{\nu(t/\varepsilon}^{t/\varepsilon}\hat{I}_{\varepsilon s}ds), \end{aligned} \tag{3.36}$$

where we used the mean theorem and $k \le \varepsilon\alpha_k < k+1$.

We note that

$$\int_{\nu(t/\varepsilon}^{t/\varepsilon}\hat{I}_{\varepsilon s}ds \to 0,$$

as $\epsilon \to 0$, $\forall t\epsilon R^+$, since $t/\epsilon - \nu(t/\epsilon) \to 0$ as $\epsilon \to 0$. Further,

$$\begin{aligned} &\varepsilon\textstyle\sum_{k=1}^{\nu(t/\varepsilon)}\beta(y_{k+1})I^\varepsilon_k - \hat{\beta}\sum_{k=1}^{\nu(t/\varepsilon)}\hat{I}_{\varepsilon\alpha_k} \\ &= \varepsilon\cdot\textstyle\sum_{k=1}^{\nu(t/\varepsilon)}(\beta(y_k/m) - \hat{\beta})I^\varepsilon_k + \varepsilon\cdot\hat{\beta}\cdot\sum_{k=1}^{\nu(t/\varepsilon)}(I^\varepsilon_k - \hat{I}_{\varepsilon\alpha_k}), \end{aligned} \tag{3.37}$$

and $|I^\varepsilon_k| \le N, \forall_k \ge 1, \forall\varepsilon > 0$.

Using the renewal theorem (see Chapter 1), we get

$$\varepsilon \cdot \nu(t/\varepsilon) \to t/m,$$

as $\epsilon \to 0$, $\forall t \geq 0$, and using the ergodicity of Markov chain $(y_n)_{n\in Z^+}$ (see Chapter 1), we get

$$\varepsilon \sum_{k=1}^{[t/\varepsilon]} \beta(y_n) \to t \cdot \hat{\beta}.$$

Therefore,

$$|\varepsilon \sum_{k=1}^{\nu(t/\varepsilon)} (\beta(y_k) - \hat{\beta}) \cdot I_k^\varepsilon| \leq N \cdot |\varepsilon \sum_{k=1}^{\nu(t/\varepsilon)} (\beta(y_n) - \hat{\beta})| \to 0, \quad \forall t \epsilon R^+, \tag{3.38}$$

as $\epsilon \to 0$. Also, since $\hat{I}_t$ is continuous with respect to t and $I^\varepsilon_{\nu(t/\varepsilon)}$ converges to $\hat{I}_t$ as $\varepsilon \to 0$, we obtain that

$$\begin{aligned} |\varepsilon\hat{\beta} \textstyle\sum_{k=1}^{\nu(t/\varepsilon)} (I_k^\varepsilon - \hat{I}_{\varepsilon\alpha_k})| &\leq \varepsilon \cdot \delta\hat{\beta} \cdot \nu(t/\varepsilon) \\ &\leq \hat{\beta} \cdot \delta \cdot \tfrac{ct}{m} \to_{\delta\to 0} 0, \end{aligned} \tag{3.39}$$

for some constant $c > 0$ and $\forall t \epsilon R^+$.

It means, taking into account (3.35)-(3.39), that

$$R^\varepsilon_{\nu(t/\varepsilon)} \to \hat{R}_t, \forall t \epsilon R^+ \tag{3.40}$$

as $\epsilon \to 0$. We now consider the convergence of the sequence $S^\varepsilon_{\nu(t/\varepsilon)}$. From (3.25) it follows that

$$S^\varepsilon_{n+1} = S^\varepsilon_0 \cdot \prod_{k=1}^{n} e^{-\varepsilon\alpha(y_{k+1})\cdot I_k^\varepsilon},$$

or

$$\ln \frac{S^\varepsilon_{n+1}}{S^\varepsilon_0} = -\varepsilon \cdot \sum_{k=1}^{n} \alpha(y_{k+1}) \cdot I_k^\varepsilon. \tag{3.41}$$

From (3.35) and (3.38)-(3.39) we know that

$$\varepsilon \cdot \sum_{k=1}^{\nu(t/\varepsilon)} \alpha(y_{k+1}) \cdot I_k^\varepsilon \to \hat{\alpha} \int_0^t \hat{I}_s ds \tag{3.42}$$

as $\epsilon \to 0$. Hence,

$$\ln \frac{S^\varepsilon_{\nu(t/\varepsilon)+1}}{S^\varepsilon_0} \to \ln \frac{\hat{S}_t}{S_0} = -\hat{\alpha} \cdot \int_0^t \hat{I}_s ds \tag{3.43}$$

as $\epsilon \to 0$. Namely, from (3.41)-(3.43), we finally get

$$\hat{S}_t = S_0 e^{-\hat{\alpha}\cdot \int_0^t \hat{I}_s ds},$$

or

$$\begin{cases} \frac{d\hat{S}_t}{dt} & = & -\hat{\alpha} \cdot \hat{I}_t \cdot \hat{S}_t, \\ \hat{S}_0 & = & S_0. \end{cases} \tag{3.44}$$

It remains to prove the convergence for $I^\varepsilon_{\nu(t/\varepsilon)}$ in (3.25).

From (3.25) we obtain

$$I^\varepsilon_{\nu(t/\varepsilon)+1} = I^\varepsilon_0 + \sum_{k=1}^{\nu(t/\varepsilon)} (1 - e^{-\varepsilon\alpha(y_{k+1})} I^\varepsilon_k) S^\varepsilon_k - \varepsilon \cdot \sum_{k=1}^{\nu(t/\varepsilon)} \beta(y_{k+1}) I^\varepsilon_k. \tag{3.45}$$

From previous arguments we already know that

$$\varepsilon \sum_{k=1}^{\nu(t/\varepsilon)} \beta(y_{k+1}) I^\varepsilon_k \to \hat{\beta} \cdot \int_0^t \hat{I}_s ds \tag{3.46}$$

as $\epsilon \to 0$.

We now claim that the second term in (3.45) converges to $\hat{\alpha} \cdot \hat{I}_t \cdot \hat{S}_t$ as $\varepsilon \to 0$. In fact, we have

$$|\sum_{k=1}^{\nu(t/\varepsilon)} (1 - e^{-\varepsilon\alpha(y_{k+1})} \cdot I^\varepsilon_k) S^\varepsilon_k - \hat{\alpha} \int_0^t \hat{I}_s \cdot \hat{S}_s ds|$$

$$\leq |\varepsilon \sum_{k=1}^{\nu(t/\varepsilon)} \alpha(y_{k+1}) \cdot I^\varepsilon_k S^\varepsilon_k - \hat{\alpha} \cdot \int_0^t \hat{I}_s \cdot \hat{S}_s ds| + \varepsilon^2 | \sum_{k=1}^{\nu(t/\varepsilon)} \frac{\alpha^2(y_{k+1})}{2} (I^\varepsilon_k)^2 S^\varepsilon_k|. \tag{3.47}$$

The second term in (3.47) tends to zero as $\varepsilon \to 0$, since

$$\varepsilon \cdot \sum_{k=1}^{\nu(t/\varepsilon)} \frac{\alpha^2(y_{k+1})}{2} \to \frac{\hat{\alpha}^2}{2},$$

that follows from renewal theorem and ergodicity of Markov chain $(y_n)_{n \in Z^+}$, and

$$\varepsilon^2 | \sum_{k=1}^{\nu(t/\varepsilon)} \frac{\alpha^2(y_{k+1})}{2} (I^\varepsilon_k)^2 \cdot S^\varepsilon_k| \leq \varepsilon \cdot \frac{\hat{\alpha}^2}{2} \cdot N^3 \to 0 \tag{3.48}$$

as $\epsilon \to 0$. For the first term in (3.47), we have

$$\begin{aligned} & |\varepsilon^2 \textstyle\sum_{k=1}^{\nu(t/\varepsilon)} \alpha(y_{k+1}) I^\varepsilon_k S^\varepsilon_k - \hat{\alpha} \cdot \hat{I}_t \cdot \hat{S}_t| \\ & \leq |\varepsilon \textstyle\sum_{k=1}^{\nu(t/\varepsilon)} (\alpha(y_{k+1}) - \hat{\alpha}) I^\varepsilon_k \cdot S^\varepsilon_k| + |\varepsilon\hat{\alpha} \cdot \textstyle\sum_{k=1}^{\nu(t/\varepsilon)} (I^\varepsilon_k - \hat{I}_{\varepsilon\alpha_k}) \cdot S^\varepsilon_k| \\ & + |\varepsilon\hat{\alpha} \cdot \textstyle\sum_{k=1}^{\nu(t/\varepsilon)} \hat{I}_{\varepsilon\alpha_k} \cdot (S^\varepsilon_k - \hat{S}_{\varepsilon\alpha_k})| + \varepsilon\hat{\alpha} \cdot | \int_{\nu(t/\varepsilon)}^{t/\varepsilon} \hat{I}_s \hat{S}_s ds| \to 0, \end{aligned} \tag{3.49}$$

as $\epsilon \to 0$, for all $t \in R_+$, using the same arguments as in (3.36)-(3.39).

From (3.45)-(3.49) we finally obtain that

$$I^{\varepsilon}_{\nu(t/\varepsilon)} \to \hat{I}_t, \quad \forall t \epsilon R_+$$

as $\epsilon \to 0$. In this way, we conclude that all the processes $S^{\varepsilon}_{\nu(t/\varepsilon)}, I^{\varepsilon}_{\nu(t/\varepsilon)}$ and $R^{\varepsilon}_{\nu(t/\varepsilon)}$ in (3.25) have the limiting points $\hat{S}_t, \hat{I}_t$ and $\hat{R}_t$, respectively.

Let us now consider the *discrete-time epidemic model in Markov random environment* in series scheme:

$$\begin{cases} S^{\varepsilon}_{[t/\varepsilon]+1} &= S^{\varepsilon}_{[t/\varepsilon]}\exp(-d(y_{[t/\varepsilon]+1}) \cdot I_{[t/\varepsilon]}, \\ I^{\varepsilon}_{[t/\varepsilon]+1} &= (1-\exp(-dy_{[t/\varepsilon]+1}) \cdot I^{\varepsilon}_{[t/\varepsilon]})S^{\varepsilon}_{[t/\varepsilon]+1} + \beta(y_{[t/\varepsilon]} \cdot I_{[t/\varepsilon]}, \\ R^{\varepsilon}_{[t/\varepsilon]+1} &= (1-\beta(y_{[t/\varepsilon]+1}))I^{\varepsilon}_{[t/\varepsilon]} + R^{\varepsilon}_{[t/\varepsilon]}, \\ S^{\varepsilon}_0 &= S_0, \quad I^{\varepsilon}_0 = I_0, \quad R^{\varepsilon}_0 = 0. \end{cases}$$

From Theorem 9 (Subsection 2.7.1, Chapter 2) it follows that

$$(S_{[t/\varepsilon]}, I^{\varepsilon}_{[t/\varepsilon]}, R^{\varepsilon}_{[t/\varepsilon]}) \to (\tilde{S}_t, \tilde{I}_t, \tilde{R}_t)$$

as $\epsilon \to 0$, where

$$\begin{cases} \frac{d\tilde{S}_t}{dt} &= -\tilde{\alpha} \cdot \tilde{S}_t \cdot \tilde{I}_t, \\ \frac{d\tilde{I}_t}{dt} &= \tilde{\alpha} \cdot \tilde{S}_t \cdot \tilde{I}_t - \tilde{\beta} \cdot \tilde{I}_t, \\ \frac{d\tilde{R}_t}{dt} &= \tilde{\beta} \cdot \tilde{I}_t, \end{cases}$$

and

$$\tilde{\alpha} := \int_Y p(dy)d(y),$$

$$\tilde{\beta} := \int_Y p(dy)\beta(y).$$

The averaged system (3.29) for epidemic model differs from the above averaged system by a factor m^{-1} in the expression for $\hat{\alpha}$ and $\hat{\beta}$.

The presence of scaling factor m in the averaging formulas (3.29) and (3.30) can be explained as follows: the discrete influence of a random factor on the original evolution of epidemics model in (3.25) transforms into the continuous component of the limiting averaged evolution of epidemic model in (3.29). In this case, m^{-1} is the mean stationary intensity of the jumps of random environment.

We consider finally the continuous-time epidemic model in semi-Markov random environment in series scheme:

$$\begin{cases} \frac{dS^\varepsilon(t)}{dt} &= -\alpha(y(t/\varepsilon))S^\varepsilon(t)\cdot I^\varepsilon(t), \\ \frac{dI^\varepsilon(t)}{dt} &= \alpha(y(t/\varepsilon))S^\varepsilon(t)\cdot I^\varepsilon(t) - \beta(y(t/\varepsilon))I^\varepsilon(t), \\ \frac{dR^\varepsilon(t)}{dt} &= \beta(y(t/\varepsilon))I^\varepsilon(t), \\ S^\varepsilon(0) &= S_0, \quad I^\varepsilon(0) = I_0, \quad R^\varepsilon(0) = 0. \end{cases}$$

From the Theorem 10 (Subsection 2.7.1, Chapter 2) it follows that

$$(S^\varepsilon(t), I^\varepsilon(t), R^\varepsilon(t)) \to (\hat{S}(t), \hat{I}(t), \hat{R}(t))$$

as $\epsilon \to 0$, where

$$\begin{cases} \frac{d\hat{S}(t)}{dt} &= -\bar{\alpha}\cdot\hat{S}(t)\cdot\hat{I}(t), \\ \frac{d\hat{I}(t)}{dt} &= \bar{\alpha}\cdot\hat{S}(t)\cdot\hat{I}(t) - \bar{\beta}\hat{I}(t), \\ \frac{d\hat{R}(t)}{dt} &= \bar{\beta}\hat{I}(t), \\ \hat{S}(t) &= S_0, \quad \hat{I}(t) = I_0, \quad \hat{R}(t) = 0, \end{cases}$$

and

$$\bar{\alpha} := \int_Y p(dy)m(y)\cdot\alpha(y)/m,$$

$$\bar{\beta} := \int_Y p(dy)m(y)\beta(y)/m,$$

$$m(y) := \int_0^\infty tG_y(dt), \quad m := \int_Y p(dy)m(y).$$

Here $m(y)$ is the lifetime in the state y of the continuous-time epidemic model in semi-Markov random environment, and m is the mean stationary lifetime in the states of the epidemic model.

Remark. The same result on averaging for discrete-time epidemic model in Markov random environment was obtained in [41].

3.4. Merging of Epidemic Models in Random Media

Here we apply the results from Section 2.7.4, Chapter 2, namely, Theorem 14, to the epidemic models in random media with $X = R^3$, $\mathbf{x} = [S, I, R]^T$ and vector-function

$$\mathbf{g}(\mathbf{x}, y) = [-\alpha(y)SI, \alpha(y)SI - \beta(y)I, \beta(y)I]^T.$$

Suppose that classes $S^\varepsilon, I^\varepsilon$ and R^ε satisfy the Kermack-McKendrick model in random environment $y^\varepsilon_{\nu(t/\varepsilon)}$ in series scheme:

$$\left\{\begin{array}{rcl} S^{\varepsilon}_{\nu(t/\varepsilon)+1} & = & S^{\varepsilon}_{\nu(t/\varepsilon)}\exp\left(-\varepsilon\alpha(y^{\varepsilon}_{\nu(t/\varepsilon)})I^{\varepsilon}_{\nu(t/\varepsilon)}\right), \\ I^{\varepsilon}_{\nu(t/\varepsilon)+1} & = & (1-\exp\left(-\varepsilon\alpha(y^{\varepsilon}_{\nu(t/\varepsilon)})I^{\varepsilon}_{\nu(t/\varepsilon)}\right)S^{\varepsilon}_{\nu(t/\varepsilon)} \\ & + & (1-\varepsilon\beta(y^{\varepsilon}_{\nu(t/\varepsilon)})I^{\varepsilon}_{\nu(t/\varepsilon)}, \\ R^{\varepsilon}_{\nu(t/\varepsilon)+1} & = & R^{\varepsilon}_{\nu(t/\varepsilon)} + \varepsilon\beta(y^{\varepsilon}_{\nu(t/\varepsilon)}\cdot I^{\varepsilon}_{\nu(t/\varepsilon)}, \\ S^{\varepsilon}(0) & = & S_0, \quad I^{\varepsilon}(0) = I_0, \quad R^{\varepsilon}(0) = 0, \end{array}\right. \tag{3.50}$$

where $(y^{\varepsilon}_n)_{n\in Z^+}$ is a perturbed Markov chain in $(Y, \mathcal{Y})$ with transition probabilities $P_{\varepsilon}(y, A)$, described in Section 1.9.4 of Chapter 1. This system can be written as

$$\left\{\begin{array}{rcl} S^{\varepsilon}_{\nu(t/\varepsilon)+1} - S_{\nu(t/\varepsilon)} & = & -S^{\varepsilon}_{\nu(t/\varepsilon)}\cdot(1-e^{-\varepsilon\alpha(y_{\nu(t/\varepsilon)+1})\cdot I^{\varepsilon}_{\nu(t/\varepsilon)}}), \\ I^{\varepsilon}_{\nu(t/\varepsilon)+1} - I_{\nu(t/\varepsilon)} & = & S^{\varepsilon}_{\nu(t/\varepsilon)}\cdot(1-e^{-\varepsilon\alpha(y_{\nu(t/\varepsilon)+1})\cdot I^{\varepsilon}_{\nu(t/\varepsilon)}}) - \varepsilon\beta\cdot(y_{\nu(t/\varepsilon)+1})\cdot I^{\varepsilon}_{\nu(t/\varepsilon)}, \\ R^{\varepsilon}_{\nu(t/\varepsilon)+1} - R^{\varepsilon}_{\nu(t/\varepsilon)} & = & \varepsilon\beta(y_{\nu(t/\varepsilon)+1})\cdot I^{\varepsilon}_{\nu(t/\varepsilon)}, \\ S^{\varepsilon}_0 & = & S_0, \quad I^{\varepsilon}_0 = I_0, \quad R^{\varepsilon}_0 = 0. \end{array}\right.$$

Using the Taylor expansion for exponential functions, we obtain the following

$$\left\{\begin{array}{rcl} S^{\varepsilon}_{\nu(t/\varepsilon)+1} - S^{\varepsilon}_{\nu(t/\varepsilon)} & = & -\varepsilon\cdot\alpha(y_{\nu(t/\varepsilon)+1})S^{\varepsilon}_{\nu(t/\varepsilon)}\cdot I^{\varepsilon}_{\nu(t/\varepsilon)} + o(\varepsilon), \\ I^{\varepsilon}_{\nu(t/\varepsilon)+1} - I^{\varepsilon}_{\nu(t/\varepsilon)} & = & \varepsilon(\alpha(y_{\nu(t/\varepsilon)+1}S^{\varepsilon}_{\nu(t/\varepsilon)})\cdot I^{\varepsilon}_{\nu(t/\varepsilon)}) - \epsilon\beta(y_{\nu(t/\varepsilon)+1})\cdot I^{\varepsilon}_{\nu(t/\varepsilon)} + o(\varepsilon), \\ R^{\varepsilon}_{\nu(t/\varepsilon)+1} - R^{\varepsilon}_{\nu(t/\varepsilon)} & = & \varepsilon\beta(y_{\nu(t/\varepsilon)+1})\cdot I^{\varepsilon}_{\nu(t/\varepsilon)}. \\ S^{\varepsilon}_0 & = & S_0, \quad I^{\varepsilon}_0 = I_0, \quad R^{\varepsilon}_0 = 0. \end{array}\right.$$

We note that sequences $S^{\varepsilon}_{\nu(t/\varepsilon)}$, $I^{\varepsilon}_{\nu(t/\varepsilon)}$ and $R^{\varepsilon}_{\nu(t/\varepsilon)}$ are bounded by $N = S_0 + I_0$: the total number (size) of population. As shown in Section 1.9.4 of Chapter 1, under merging conditions the perturbed semi-Markov process $y^{\varepsilon}(\nu(t/\varepsilon))$ is merged to the jump Markov process $\hat{y}(t)$ in phase space $(V, \mathcal{V})$ with stochastic kernel

$$\hat{Q}(v, H, t) := \hat{P}(v, H)\cdot(1-e^{-q(v)\cdot t}), \tag{3.51}$$

where $\hat{P}(v, H)$ and $q(v)$ are defined in Section 1.9.4 of Chapter 1, $v\epsilon V, H\epsilon\mathcal{V}$. Here, the process $\hat{y}(t)$ is called the *merged Markov process.*

From the Theorem 14 (Chapter 2) it follows that $(S^{\varepsilon}, I^{\varepsilon}, R^{\varepsilon})$ in (3.50) converges weakly as $\varepsilon\to 0$ to the triplet $(\tilde{S}, \tilde{I}, \tilde{R})$ which satisfies the following system:

$$\begin{cases} \frac{d\tilde{S}(t)}{dt} &= -\tilde{\alpha}(\hat{y}(t))\tilde{S}(t)\tilde{I}(t), \\ \frac{d\tilde{I}(t)}{dt} &= \tilde{\alpha}(\hat{y}(t))\tilde{I}(t)\tilde{S}(t) - \tilde{\beta}(\hat{y}(t))\tilde{I}(t), \\ \frac{d\tilde{R}(t)}{dt} &= \tilde{\beta}(\hat{y}(t))\tilde{I}(t), \\ \tilde{S}(0) &= S_0, \quad \tilde{I}(0) = I_0, \quad \tilde{R}(0) = 0, \end{cases} \tag{3.52}$$

where

$$\begin{aligned} \tilde{\alpha}(v) &:= \int_{Y_v} p_v(dy)\alpha(y)/m_v, \\ \tilde{\beta}(v) &:= \int_{Y_v} p_v(dy)\beta(y)/m_v, \\ m_v &:= \int_{Y_v} p_v(dy)m(y). \end{aligned} \tag{3.53}$$

The same result can be obtained by the method developed in Section 3.3. We only need the following known result (see Chapter 1):

$$\varepsilon \sum_{k=1}^{\nu(t/\varepsilon)} \alpha(y_k^\varepsilon) \to \int_0^t \tilde{\alpha}(\hat{y}(s))ds, \tag{3.54}$$

as $\epsilon \to 0$, where $(y_n^\varepsilon)_{n\in Z^+}$ is a perturbed Markov chain, $\hat{y}(t)$ is a merged Markov process in the phase space $(V, \mathcal{V})$ with stochastic kernel $\hat{Q}$ in (3.51), $\tilde{\alpha}(v)$ is defined in (3.53).

The system (3.52) for the limiting epidemic model $(\tilde{S}, \tilde{I}, \tilde{R})$ looks like the continuous-time epidemic model in semi-Markov random environment considered in Section 3.1.3, but only in merged Markov random environment $\hat{y}(t)$.

3.5. Diffusion Approximation of Epidemic Models in Random Media

Here, we apply the results from Section 2.7.2, Chapter 2, namely, Theorems 11 and 12, to the epidemic models in random media with $X = R^3$, $\mathbf{x} = [S, I, R]^T$ and vector-function

$$\mathbf{g}(\mathbf{x}, y) = [-\alpha(y)SI, \alpha(y)SI - \beta(y)I, \beta(y)I]^T.$$

Suppose now that the averaged infection rate $\hat{\alpha}$ and averaged removal rate of infectives $\hat{\beta}$ are equal to zero. Namely,

$$\hat{\alpha} := \int_Y p(dy)\alpha(y) = 0, \tag{3.55}$$

$$\hat{\beta} := \int_Y p(dy)\beta(y) = 0,$$

where $\alpha(g)$ and $\beta(y)$ are bounded and continuous functions on Y defined in Section 3.1. Therefore, the right hand-side of the averaged system (3.29) for $(\hat{S}, \hat{I}, \hat{R})$ in

Section 3.3, is equal to zero and the solutions of the system for time of order $1/\varepsilon$ will not exit from the neighbourhood of initial conditions. In this case, we can obtain more precise information on the behaviors of the solution to original stochastic model in series scheme, considering time scale t/ε^2.

In this section, under the additional *balance condition* (3.55), we show that the initial system (3.56) in series scheme in scale of time t/ε^2 admits an approximation described by a system of stochastic differential equations, involving some diffusion processes.

Let us consider the following system of equations for $S^\varepsilon, I^\varepsilon, R^\varepsilon$:

$$\begin{cases} S^\varepsilon_{\nu(t/\varepsilon^2)+1} &= S^\varepsilon_{\nu(t/\varepsilon^2)} \cdot e^{-\varepsilon\cdot\alpha(y_{\nu(t/\varepsilon^2)+1})\cdot I^\varepsilon_{\nu(t/\varepsilon^2)}}, \\ I^\varepsilon_{\nu(t/\varepsilon^2)+1} &= (1-e^{-\varepsilon\cdot\alpha(y_{\nu(t/\varepsilon^2)+1})I^\varepsilon_{\nu(t/\varepsilon^2)}})S^\varepsilon_{\nu(t/\varepsilon^2)} \\ &+ (1-\varepsilon\beta(y_{\nu(t/\varepsilon^2)+1})\cdot I_{\nu(t/\varepsilon^2)}), \\ R^\varepsilon_{\nu(t/\varepsilon^2)+1)} &= R^\varepsilon_{\nu(t/\varepsilon^2)} + \varepsilon\beta(y^\varepsilon(\nu(t/\varepsilon^2)+1)\cdot I^\varepsilon(\nu(t/\varepsilon^2)), \\ S^\varepsilon_0 &= S_0, \quad I^\varepsilon_0 = I_0, \quad R_0 = 0. \end{cases} \tag{3.56}$$

Here, function $\mathbf{g}^\varepsilon(\mathbf{x}, y)$ is given by (see Corollary 1, Chapter 2)

$$\mathbf{g}^\varepsilon(\mathbf{x}, y) := \mathbf{g}(\mathbf{x}, y) + \epsilon\mathbf{g}_1(\mathbf{x}, y),$$

where $\mathbf{g}(\mathbf{x}, y) = [-\alpha(y)SI, \alpha(y)SI - \beta(y)I, \beta(y)I]^T$ and $\mathbf{g}_1(\mathbf{x}, y) = [\alpha^2(y)SI^2, \alpha^2(y)SI^2, 0]$

Applying Theorem 12 and Corollary 1 (see Section 2.7.2) with the above function $\mathbf{g}^\varepsilon([S, I]^T, y)$, we obtain that drift coefficient $\alpha(\mathbf{x})$ and diffusion matrix $\beta(\mathbf{x})$ are equal

$$\tilde{\alpha}(\mathbf{x}) = [\tilde{\alpha}_1(\tilde{S}, \tilde{I}, \tilde{R}), \tilde{\alpha}_2(\tilde{S}, \tilde{I}, \tilde{R}), \tilde{\alpha}_3(\tilde{S}, \tilde{I}, \tilde{R})]^T$$

and

$$\tilde{\beta}(\mathbf{x}) = \tilde{\beta}(\tilde{S}, \tilde{I}, \tilde{R}) := \begin{pmatrix} \beta_{11}(\tilde{S}, \tilde{I}, \tilde{R}) & \beta_{12}(\tilde{S}, \tilde{I}, \tilde{R}) & \beta_{13}(\tilde{S}, \tilde{I}, \tilde{R}) \\ \beta_{21}(\tilde{S}, \tilde{I}, \tilde{R}) & \beta_{22}(\tilde{S}, \tilde{I}, \tilde{R}) & \beta_{23}(\tilde{S}, \tilde{I}, \tilde{R}) \\ \beta_{31}(\tilde{S}, \tilde{I}, \tilde{R}) & \beta_{32}(\tilde{S}, \tilde{I}, \tilde{R}) & \beta_{33}(\tilde{S}, \tilde{I}, \tilde{R}) \end{pmatrix},$$

where

$$\begin{aligned}
\tilde{\alpha}_1(\tilde{S},\tilde{I},\tilde{R}) &= \textstyle\int_Y p(dy)[\alpha(y)R_0\alpha(y)\tilde{S}\tilde{I}^2 + (\alpha(y)\tilde{S}-\beta(y))R_0\alpha(y)\tilde{I}^2 \\
&+ 2^{-1}\alpha^2(y)\tilde{S}\tilde{I}^2]/m, \\
\tilde{\alpha}_2(\tilde{S},\tilde{I},\tilde{R}) &= \textstyle\int_Y p(dy)[\alpha(y)R_0\alpha(y)\tilde{S}^2\tilde{I} + (\alpha(y)\tilde{S}-\beta(y))R_0(\alpha(y)\tilde{S}-\beta(y))\tilde{I} \\
&+ \beta(y)R_0\beta(y)\tilde{I} + 2^{-1}\alpha^2(y)\tilde{S}\tilde{I}^2]/m, \\
\tilde{\alpha}_3(\tilde{S},\tilde{I},\tilde{R}) &= 0,
\end{aligned} \tag{3.57}$$

and matrix entries are equal

$$\begin{aligned}
\beta_{11} &= 2\textstyle\int_Y p(dy)[\alpha(y)R_0\alpha(y)\tilde{S}^2\tilde{I}^2 + \alpha^2(y)\tilde{S}^2\tilde{I}^2]/m, \\
\beta_{12} &= 2\textstyle\int_Y p(dy)[\alpha(y)R_0(\beta(y)-\alpha(y)\tilde{S})\tilde{S}\tilde{I}^2 + \alpha(y)(\beta(y)-\alpha(y)\tilde{S})\tilde{S}\tilde{I}^2]/m, \\
\beta_{13} &= 2\textstyle\int_Y p(dy)[\alpha(y)\beta(y)\tilde{S}\tilde{I}^2 + \alpha(y)R_0\beta(y)\tilde{S}\tilde{I}^2]/m, \\
\beta_{21} &= 2\textstyle\int_Y p(dy)[(\beta(y)-\alpha(y)\tilde{S})R_0\alpha(y)\tilde{S}\tilde{I}^2 + \alpha(y)(\beta(y)-\alpha(y)\tilde{S})\tilde{S}\tilde{I}^2]/m, \\
\beta_{22} &= 2\textstyle\int_Y p(dy)[(\alpha(y)\tilde{S}-\beta(y))R_0(\alpha(y)\tilde{S}-\beta(y))^2\tilde{I}^2 + (\alpha(y)\tilde{S}-\beta(y))^2\tilde{I}^2]/m, \\
\beta_{23} &= 2\textstyle\int_Y p(dy)[(\alpha(y)\tilde{S}-\beta(y))R_0\beta(y)\tilde{I}^2 + \beta(y)(\alpha(y)\tilde{S}-\beta(y))\tilde{I}^2]/m, \\
\beta_{31} &= 2\textstyle\int_Y p(dy)[\beta(y)R_0\alpha(y)\tilde{S}\tilde{I}^2 + \alpha(y)\beta(y)\tilde{S}\tilde{I}^2]/m, \\
\beta_{32} &= 2\textstyle\int_Y p(dy)[\beta(y))R_0(\alpha(y)\tilde{S}-\beta(y))\tilde{I}^2 + \beta(y)(\alpha(y)\tilde{S}-\beta(y))\tilde{I}^2]/m, \\
\beta_{33} &= 2\textstyle\int_Y p(dy)[\beta(y)R_0\beta(y)\tilde{I}^2 + \beta^2(y)\tilde{I}^2]/m.
\end{aligned} \tag{3.58}$$

We obtain that $(S^\varepsilon_{\nu(t/\varepsilon^2)}, I^\varepsilon_{\nu(t/\varepsilon^2)}, R^\varepsilon_{\nu(t/\varepsilon^2)})$ converges weakly as $\varepsilon \to 0$ to the process $(\tilde{S}(t),\tilde{I}(t),\tilde{R}(t))$ such that satisfy the following stochastic differential equations:

$$\begin{cases}
d\tilde{S}(t) = \tilde{\alpha}_1 dt + \beta_{11}dw_1(t) + \beta_{12}dw_2(t) + \beta_{13}dw_3(t), \\
d\tilde{I}(t) = \tilde{\alpha}_2 dt + \beta_{21}dw_1(t) + \beta_{22}dw_2(t) + \beta_{23}dw_3(t), \\
d\tilde{R}(t) = \beta_{31}dw_1(t) + \beta_{32}dw_2(t) + \beta_{33}dw_3(t),
\end{cases} \tag{3.59}$$

where $\tilde{\alpha}_1 := \tilde{\alpha}_1(\tilde{S},\tilde{I},\tilde{R})$, $\tilde{\alpha}_2 := \tilde{\alpha}_2(\tilde{S},\tilde{I},\tilde{R})$ are defined in (3.57), and β_{ij}, $i,j = 1,2,3$ are defined in (3.58). Here $w_i(t)$, $i = 1,2,3$, are independent standard Wiener processes.

3.6. Normal Deviations of Epidemic Model in Random Media

Here, we apply the results from Section 2.7.3, Chapter 2, namely, Theorem 13, to the epidemic model in random media with $X = R^3$, $\mathbf{x} = (S, I, R)^T$ and vector-function

$$\mathbf{g}(\mathbf{x}, y) = (-\alpha(y)SI, \alpha(y)SI - \beta(y)I, \beta(y)I)^T.$$

Recall, that we consider the transmission of a disease in random environment, which, after recovery, confers immunity. The population can be divided into three distinct classes: the susceptibles S^ε, who can catch the disease; the infectives I^ε, who have the disease and can transmit it, and the removed class R^ε, who have either had the disease, or are recovered, immune or isolated until recovered, ε is a small positive parameter.

We suppose that these classes $S^\varepsilon, I^\varepsilon, R^\varepsilon$ satisfy the *Kermack-McKendrick model* in randon environment in series scheme:

$$\begin{cases} S^\varepsilon_{\nu(t/\varepsilon)+1} &= S^\varepsilon_{\nu(t/\varepsilon)} e^{-\varepsilon\alpha(y_{\nu(t/\varepsilon)+1})I^\varepsilon_{\nu(t/\varepsilon)}}, \\ I^\varepsilon_{\nu(t/\varepsilon)+1} &= (1 - e^{-\varepsilon\alpha(y_{\nu(t/\varepsilon)+1})I^\varepsilon_{\nu(t/\varepsilon)}})S^\varepsilon_{\nu(t/\varepsilon)} + (1 - \varepsilon\beta(y_{\nu(t/\varepsilon)+1}))I^\varepsilon_{\nu(t/\varepsilon)}, \\ R^\varepsilon_{\nu(t/\varepsilon)+1} &= R^\varepsilon_{\nu(t/\varepsilon)} + \varepsilon\beta(y_{\nu(t/\varepsilon)+1})I^\varepsilon_{\nu(t/\varepsilon)}, \\ S^\varepsilon_0 &= S_0, \quad I^\varepsilon_0 = I_0, \quad R^\varepsilon_0 = 0, \end{cases}$$

and we call this a *continuous epidemic model* in *Markov renewal random environment.*

This system can be written as

$$\begin{cases} S^\varepsilon_{\nu(t/\varepsilon)+1} - S_{\nu(t/\varepsilon)} &= -S^\varepsilon_{\nu(t/\varepsilon)}(1 - e^{-\varepsilon\alpha(y_{\nu(t/\varepsilon)+1})\cdot I^\varepsilon_{\nu(t/\varepsilon)}}), \\ I^\varepsilon_{\nu(t/\varepsilon)+1} - I_{\nu(t/\varepsilon)} &= S^\varepsilon_{\nu(t/\varepsilon)}(1 - e^{-\varepsilon\alpha(y_{\nu(t/\varepsilon)+1})\cdot I^\varepsilon_{\nu(t/\varepsilon)}}) - \varepsilon\beta\cdot(y_{\nu(t/\varepsilon)+1})\cdot I^\varepsilon_{\nu(t/\varepsilon)}, \\ R^\varepsilon_{\nu(t/\varepsilon)+1} - R^\varepsilon_{\nu(t/\varepsilon)} &= \varepsilon\beta(y_{\nu(t/\varepsilon)+1})I^\varepsilon_{\nu(t/\varepsilon)}, \\ S^\varepsilon_0 &= S_0, \quad I^\varepsilon_0 = I_0, \quad R^\varepsilon_0 = 0. \end{cases}$$

Using the Taylor expansion for exponential functions, we obtain the following

$$\begin{cases} S^\varepsilon_{\nu(t/\varepsilon)+1} - S^\varepsilon_{\nu(t/\varepsilon)} &= -\varepsilon\alpha(y_{\nu(t/\varepsilon)+1})S^\varepsilon_{\nu(t/\varepsilon)}\cdot I^\varepsilon_{\nu(t/\varepsilon)} + o(\varepsilon), \\ I^\varepsilon_{\nu(t/\varepsilon)+1} - I^\varepsilon_{\nu(t/\varepsilon)} &= \varepsilon(\alpha(y_{\nu(t/\varepsilon)+1}S^\varepsilon_{\nu(t/\varepsilon)})\cdot I^\varepsilon_{\nu(t/\varepsilon)}) - \epsilon\beta(y_{\nu(t/\varepsilon)+1})I^\varepsilon_{\nu(t/\varepsilon)} + o(\varepsilon), \\ R^\varepsilon_{\nu(t/\varepsilon)+1} - R^\varepsilon_{\nu(t/\varepsilon)} &= \varepsilon\beta(y_{\nu(t/\varepsilon)+1})I^\varepsilon_{\nu(t/\varepsilon)}. \\ S^\varepsilon_0 &= S_0, \quad I^\varepsilon_0 = I_0, \quad R^\varepsilon_0 = 0. \end{cases}$$

It is easy to note that sequences $S^\varepsilon_{\nu(t/\varepsilon)}$, $I^\varepsilon_{\nu(t/\varepsilon)}$ and $R^\varepsilon_{\nu(t/\varepsilon)}$ are bounded by $N = S_0 + I_0$: the total number (size) of the population.

We now use Theorem 13 with a vector-function $\mathbf{g}((S,I,R)^T, y)$ given by

$$\mathbf{g}((S,I,R)^T, y) := ((-\alpha(y)SI), (\alpha(y)SI - \beta(y)I), (\beta(y)I))^T,$$

(here $\mathbf{x} = ((x_1, x_2, x_3))^T = (S,I,R)^T$, and T means transpose) to obtain

$$\lim_{\varepsilon\to 0} \mathcal{P}\{(|S^\varepsilon_{\nu(t/\varepsilon)} - \hat{S}_t| + |I^\varepsilon_{\nu(t/\varepsilon)} - \hat{I}_t| + |R^\varepsilon_{\nu(t/\varepsilon)} - \hat{R}_t|) > \delta\} = 0$$

for any $\delta > 0$ and $t\epsilon R_+$, where $\hat{S}_t$, $\hat{I}_t$ and $\hat{R}_t$ satisfy the following system of equations:

$$\begin{cases} \frac{d\hat{S}_t}{dt} &= -\hat{\alpha}\hat{S}_t\hat{I}_t \\ \frac{d\hat{I}_t}{dt} &= \hat{\alpha}\hat{S}_t\hat{I}_t - \hat{\beta}\hat{I}_t \\ \frac{d\hat{R}_t}{dt} &= \hat{\beta}\hat{I}_t \\ \hat{S}_0 &= S_0, \quad \hat{I}_0 = I_0, \quad \hat{R}_0 = 0, \end{cases}$$

with

$$\hat{\alpha} := \int_y p(dy)\alpha(y)/m,$$
$$\hat{\beta} := \int_y p(dy)\beta(y)/m.$$

We note, that function $g^\epsilon(\mathbf{x}, y)$ is given by (see Remark 11, Chapter 2)

$$\mathbf{g}^\epsilon(\mathbf{x}, y) := \mathbf{g}(\mathbf{x}, y) + \epsilon \mathbf{g}_1(\mathbf{x}, y),$$

where $\mathbf{g}(\mathbf{x}, y) = (-\alpha(y)SI, \alpha(y)SI - \beta(y)I, \beta(y)I)^T$ and $\mathbf{g}_1(\mathbf{x}, y) = (\alpha^2(y)SI^2, \alpha^2(y)SI^2, 0)^T$.

Applying Theorem 13 and Remark 11 (see Section 2.7.3, Chapter 2) with the above function $\mathbf{g}^\epsilon([S,I,R]^T, y)$, we obtain that the drift coefficient $\tilde{\alpha}(\mathbf{x}) := \tilde{\mathbf{g}}_{\mathbf{x}}(\hat{\mathbf{x}})\tilde{\mathbf{Z}}$ and the diffusion matrix $\beta(\hat{\mathbf{x}})$ are given by ($\hat{\mathbf{x}} := (\hat{S}, \hat{I}, \hat{R})^T$ and $\tilde{Z} := (\tilde{S}, \tilde{I}, \tilde{R})^T$):

$$\tilde{\alpha}(\mathbf{x}) = (\alpha_1(\hat{S}, \hat{I}, \tilde{S}, \tilde{I}), \alpha_2(\hat{S}, \hat{I}, \tilde{S}, \tilde{I}), \alpha_3(\hat{S}, \hat{I}, \tilde{S}, \tilde{I}))^T$$

and

$$\tilde{\beta}(\hat{\mathbf{x}}) = \begin{pmatrix} \beta_{11}(\hat{S}, \hat{I}) & \beta_{12}(\hat{S}, \hat{I}) & \beta_{13}(\hat{S}, \hat{I}) \\ \beta_{21}(\hat{S}, \hat{I}) & \beta_{22}(\hat{S}, \hat{I}) & \beta_{23}(\hat{S}, \hat{I}) \\ \beta_{31}(\hat{S}, \hat{I}) & \beta_{32}(\hat{S}, \hat{I}) & \beta_{33}(\hat{S}, \hat{I}) \end{pmatrix},$$

where

$$\begin{aligned} \alpha_1(\hat{S}, \hat{I}, \tilde{S}, \tilde{I}) &= -\hat{\alpha}(\tilde{S}\hat{I} + \hat{S}\tilde{I}), \\ \alpha_2(\hat{S}, \hat{I}, \tilde{S}, \tilde{I}) &= \hat{\alpha}(\tilde{S}\hat{I} + \hat{S}\tilde{I}) - \hat{\beta}\tilde{I}, \\ \alpha_3(\hat{S}, \hat{I}, \tilde{S}, \tilde{I}) &= \hat{\beta}\tilde{I}, \end{aligned} \tag{3.60}$$

and matrix entries are given below

$$
\begin{aligned}
\beta_{11}(\hat{S},\hat{I}) = {} & 2 \quad \textstyle\int_Y p(dy)[(\alpha(y)-\hat{\alpha})R_0(\alpha(y)-\hat{\alpha})\hat{S}^2\hat{I}^2 \\
& + \quad (\alpha(y)-\hat{\alpha})(\alpha(y)-\hat{\alpha})\hat{S}^2\hat{I}^2]/m, \\
\beta_{12}(\hat{S},\hat{I}) = {} & 2 \quad \textstyle\int_Y p(dy)[(\alpha(y)-\hat{\alpha})R_0(\alpha(y)-\hat{\alpha})\hat{S}^2\hat{I}^2 + (\hat{\alpha}-\alpha(y))R_0(\hat{\beta}-\beta(y))\hat{S}\hat{I}^2 \\
& + \quad (\alpha(y)-\hat{\alpha})(\alpha(y)-\hat{\alpha})\hat{S}^2\hat{I}^2 + (\hat{\alpha}-\alpha(y))(\hat{\beta}-\beta(y))\hat{S}\hat{]}/m, \\
\beta_{13}(\hat{S},\hat{I}) = {} & 2 \quad \textstyle\int_Y p(dy)[(\hat{\alpha}-\alpha(y))R_0(\beta(y)-\hat{\beta})\hat{S}\hat{I}^2 \\
& + \quad (\hat{\alpha}-\alpha(y))(\beta(y)-\hat{\beta})\hat{S}\hat{I}^2]/m, \\
\beta_{21}(\hat{S},\hat{I}) = {} & 2 \quad \textstyle\int_Y p(dy)[(\alpha(y)-\hat{\alpha})R_0(\alpha(y)-\hat{\alpha})\hat{S}^2\hat{I}^2 + (\hat{\beta}-\beta(y))R_0(\hat{\alpha}-\alpha(y))\hat{S}\hat{I}^2 \\
& + \quad (\alpha(y)-\hat{\alpha})(\alpha(y)-\hat{\alpha})\hat{S}^2\hat{I}^2 + (\hat{\beta}-\beta(y))(\hat{\alpha}-\alpha(y))\hat{S}\hat{I}^2]/m, \\
\beta_{22}(\hat{S},\hat{I}) = {} & 2 \quad \textstyle\int_Y p(dy)[(\alpha(y)-\hat{\alpha})R_0(\alpha(y)-\hat{\alpha})\hat{S}^2\hat{I}^2 \\
& + \quad (\hat{\alpha}-\alpha(y))R_0(\hat{\beta}-\beta(y))\hat{S}\hat{I}^2 + (\hat{\beta}-\beta(y))R_0(\hat{\alpha}-\alpha(y))\hat{S}\hat{I}^2 \\
& + \quad (\hat{\beta}-\beta(y))R_0(\hat{\beta}-\beta(y))\hat{I}^2 + (\alpha(y)-\hat{\alpha})(\alpha(y)-\hat{\alpha})\hat{S}^2\hat{I}^2 \\
& + \quad (\hat{\alpha}-\alpha(y))(\hat{\beta}-\beta(y))\hat{S}\hat{I}^2 + (\hat{\beta}-\beta(y))(\hat{\alpha}-\alpha(y))\hat{S}\hat{I}^2 \\
& + \quad (\hat{\beta}-\beta(y))(\hat{\beta}-\beta(y))\hat{I}^2]/m, \\
\beta_{23}(\hat{S},\hat{I}) = {} & 2 \quad \textstyle\int_Y p(dy)[(\hat{\alpha}-\alpha(y))R_0(\hat{\beta}-\beta(y))\hat{S}\hat{I}^2 + (\hat{\beta}-\beta(y))R_0(\hat{\beta}-\beta(y))\hat{I}^2 \\
& + \quad (\hat{\alpha}-\alpha(y))(\hat{\beta}-\beta(y))\hat{S}\hat{I}^2 + (\hat{\beta}-\beta(y))^2\hat{I}^2]/m, \\
\beta_{31}(\hat{S},\hat{I}) = {} & 2 \quad \textstyle\int_Y p(dy)[(\hat{\beta}-\beta(y))R_0(\hat{\alpha}-\alpha(y))\hat{I}^2 \\
& + \quad (\hat{\beta}-\beta(y))(\hat{\alpha}-\alpha(y))\hat{I}^2]/m, \\
\beta_{32}(\hat{S},\hat{I}) = {} & 2 \quad \textstyle\int_Y p(dy)[(\hat{\beta}-\beta(y))R_0(\hat{\alpha}-\alpha(y))\hat{S}\hat{I}^2 + (\hat{\beta}-\beta(y))R_0(\hat{\beta}-\beta(y))\hat{I}^2 \\
& + \quad (\hat{\beta}-\beta(y))(\hat{\alpha}-\alpha(y))\hat{S}\hat{I}^2 + (\hat{\beta}-\beta(y))^2\hat{I}^2]/m, \\
\beta_{33}(\hat{S},\hat{I}) = {} & 2 \quad \textstyle\int_Y p(dy)[(\hat{\beta}-\beta(y))R_0(\hat{\beta}-\beta(y))\hat{I}^2 + (\hat{\beta}-\beta(y))^2\hat{I}^2]/m.
\end{aligned}
\tag{3.61}
$$

We conclude that $\mathbf{Z}_t^\epsilon$, where $\mathbf{Z}_n^\epsilon := [\mathbf{X}_n^\epsilon - \hat{\mathbf{X}}_n^\epsilon]/\sqrt{\epsilon} = (S^\varepsilon_{\nu(t/\varepsilon)} - \hat{S}^\varepsilon_{\nu(t/\varepsilon)}, I^\varepsilon_{\nu(t/\varepsilon)} - \hat{I}^\varepsilon_{\nu(t/\varepsilon)}, R^\varepsilon_{\nu(t/\varepsilon)} - R^\varepsilon_{\nu(t/\varepsilon)})/\sqrt{\epsilon}$ converges weakly as $\varepsilon \to 0$ to the process $\tilde{\mathbf{Z}}_t := (\tilde{S}(t), \tilde{I}(t), \tilde{R}(t))$ such that satisfies the following stochastic differential equations:

$$\begin{cases} d\tilde{S}(t) &= \alpha_1(\hat{S},\hat{I},\tilde{S},\tilde{I})dt + \beta_{11}(\hat{S},\hat{I})dw_1(t) + \beta_{12}(\hat{S},\hat{I})dw_2(t) + \beta_{13}(\hat{S},\hat{I})dw_3(t), \\ d\tilde{I}(t) &= \alpha_2(\hat{S},\hat{I},\tilde{S},\tilde{I})dt + \beta_{21}(\hat{S},\hat{I})dw_1(t) + \beta_{22}(\hat{S},\hat{I})dw_2(t) + \beta_{23}(\hat{S},\hat{I})dw_3(t), \\ d\tilde{R}(t) &= \alpha_3(\hat{S},\hat{I},\tilde{S},\tilde{I})dt + \beta_{31}(\hat{S},\hat{I})dw_1(t) + \beta_{32}(\hat{S},\hat{I})dw_2(t) + \beta_{33}(\hat{S},\hat{I})dw_3(t), \end{cases} \tag{3.62}$$

where $\alpha_i := \alpha_i(\hat{S},\hat{I},\tilde{S},\tilde{I})$, $i = 1,2,3$, are defined in (3.60), and $\beta_{ij}(\hat{S},\hat{I})$, $i,j = 1,2,3$, are defined in (3.61). Here $w_i(t)$, $i = 1,2,3$, are independent standard Wiener processes.

3.7. Stochastic Stability of Epidemic Model

In this section, we study the stochastic stability of stochastic epidemic models in series scheme in averaging and diffusion approximation schemes

3.7.1. STOCHASTIC STABILITY OF EPIDEMIC MODEL IN AVERAGING SCHEME

Let us consider the epidemic model in series scheme:

$$\begin{cases} S^\varepsilon_{\nu(t/\varepsilon)+1} &= S^\varepsilon_{\nu(t/\varepsilon)} \cdot e^{-\varepsilon\alpha(y_{\nu(t/\varepsilon)+1})I^\varepsilon_{\nu(t/\varepsilon)}}, \\ I^\varepsilon_{\nu(t/\varepsilon)+1} &= (1 - e^{-\varepsilon\alpha(y_{\nu(t/\varepsilon)+1})} \cdot I^\varepsilon_{\nu(t/\varepsilon)}) \cdot S^\varepsilon_{\nu(t/\varepsilon)} \\ &+ (1 - \varepsilon\beta(y_{\nu(t/\varepsilon)+1}))I^\varepsilon_{\nu(t/\varepsilon)}, \\ R^\varepsilon_{\nu(t/\varepsilon)+1} &= R^\varepsilon_{\nu(t/\varepsilon)} + \varepsilon \cdot \beta(y_{\nu(t/\varepsilon)+1}) \cdot I^\varepsilon_{\nu(t/\varepsilon)}, \\ S^\varepsilon_0 &= S_0 > 0, \quad I^\varepsilon_0 = I_0 > 0, \quad R^\varepsilon_0 = 0, \end{cases} \tag{3.63}$$

and the averaged epidemic model (considered in Section 3.3):

$$\begin{cases} d\hat{S}_t &= -\hat{\alpha} \cdot \hat{S}_t \cdot \hat{I}_t, \\ \frac{d\hat{I}_t}{dt} &= \hat{\alpha} \cdot \hat{S}_t \cdot \hat{I}_t - \hat{\beta} \cdot \hat{I}_t, \\ \frac{d\hat{R}_t}{dt} &= \hat{\beta} \cdot \hat{I}_t, \\ \hat{S}_0 &= S_0, \quad \hat{I}_0 = I_0, \quad \hat{R}_0 = 0. \end{cases} \tag{3.64}$$

We note that for the second equation in (3.64) function $g(x,y)$ (in stability Theorem 7 for averaging scheme) satisfies the condition:

$$g(0,y)=0$$

since $g(x,y)=(\alpha(y)S-\beta(y))x$ for the second equation, $S:=S_0$.

Using the same argument in the proof of Theorem 7 for the stability in averaging scheme (see subsection 2.7), we can show that there exist a smooth Lyapunov function $W(x)$ on R such that:

$$(\hat{\alpha}\hat{S}_0-\hat{\beta})xW_x\le -\gamma W(x), \tag{3.65}$$

for some $\gamma>0$, where $\hat{\beta}:=\int_Y p(dy)\beta(y)/m$. Then the process $I^\varepsilon_{\nu(t/\varepsilon)}$ in (3.63) is stochastically exponentially stable. Moreover, it is asymptotically stochastically stable.

Indeed, take $W(x)=x^2$, then

$$(\hat{\alpha}\hat{S}_0-\hat{\beta})x2x=2(\hat{\alpha}\cdot\hat{S}_0-\hat{\beta})x^2,$$

and the inequality (3.65) holds for $0<\gamma<2(\hat{\beta}-\hat{\alpha}\cdot\hat{S}_0)$. In this case $S_0\le\hat{\beta}/\hat{\alpha}$.

Let us study the behavior of $R^\varepsilon_{\nu(t/\varepsilon)}$:

$$R^\varepsilon_{\nu(t/\varepsilon)+1}-R^\varepsilon_{\nu(t/\varepsilon)}=\varepsilon\beta(y_{\nu(t/\varepsilon)+1})I^\varepsilon_{\nu(t/\varepsilon)}\ge R^\varepsilon_{\nu(t/\varepsilon)} \tag{3.66}$$

and $R^\varepsilon_{\nu(t/\varepsilon)}\le N$, we see that $\lim_{t\to\infty}R^\varepsilon_{\nu(t/\varepsilon)}$ exists.

As

$$I^\varepsilon_{\nu(t/\varepsilon)}\ge\varepsilon\beta(y_{\nu(t/\varepsilon)+1})I^\varepsilon_{\nu(t/\varepsilon)}\ge R^\varepsilon_{\nu(t/\varepsilon)+1}-R^\varepsilon_{\nu(t/\varepsilon)}$$

we have that

$$\lim_{t\to+\infty}R^\varepsilon_{\nu(t/\varepsilon)}=N\lim_{t\to+\infty}I^\varepsilon_{\nu(t/\varepsilon)}-\lim_{t\to+\infty}S^\varepsilon_{\nu(t/\varepsilon)}=N-\lim_{t\to+\infty}S^\varepsilon_{\nu(t/\varepsilon)},$$

for fixed $\varepsilon:0\le\varepsilon\le\varepsilon_0$, since $\lim_{t\to+\infty}I^\varepsilon_{\nu(t/\varepsilon)}=0$.

From equation (3.64) it follows that:

$$\hat{S}_\infty=S_0\exp\,[-\frac{\hat{\alpha}\hat{R}_\infty}{\hat{\beta}}]=S_0\exp\,[-\frac{(N-\hat{S}_\infty)\hat{\alpha}}{\hat{\beta}}],$$

and so $\hat{S}_\infty$ is the positive root $0<z<\frac{\hat{\beta}}{\hat{\alpha}}$ of the transcendental equation:

$$S_0\exp\,[-\frac{(N-z)\hat{\alpha}}{\hat{\beta}}]=z.$$

As

$$|R^\varepsilon_\infty-\hat{R}_\infty|\le\varepsilon\cdot C_T$$

and

$$|S^\varepsilon_\infty-\hat{S}_\infty|\le\varepsilon\cdot C^1_T,$$

Then from here we obtain:

$$\hat{R}^{\varepsilon}_{\infty} \simeq N - (\hat{S}_{\infty} + \varepsilon C^1_T) = \hat{R}_{\infty} + \varepsilon C_T; \tag{3.67}$$

It means that:

$$\hat{R}_{\infty} \simeq N - \hat{S}_{\infty} - \varepsilon(C^1_T + C_T),$$

that approximates $\hat{R}_{\infty}$ by $\hat{S}_{\infty}$. Formula (3.67) gives approximation R^{ε}_{∞} by $\hat{S}_{\infty}$.

3.7.2. STABILITY OF EPIDEMIC MODEL IN DIFFUSION APPROXIMATION SCHEME.

Let us consider the epidemic model in series scheme in scale of time t/ε^2 (see section 3.5):

$$\begin{cases} S^{\varepsilon}_{\nu(t/\varepsilon^2)+1} &= S^{\varepsilon}_{\nu(t/\varepsilon^2)} e^{-\varepsilon\alpha(y_{\nu(t/\varepsilon^2)+1}) I^{\varepsilon}_{\nu(t/\varepsilon^2)}} \\ I^{\varepsilon}_{\nu(t/\varepsilon^2)+1} &= (1 - e^{-\varepsilon\alpha(y_{\nu(t/\varepsilon^2)+1})} I^{\varepsilon}_{\nu(t/\varepsilon^2)}) S^{\varepsilon}_{\nu(t/\varepsilon^2)} \\ R^{\varepsilon}_{\nu(t/\varepsilon^2)+1} &= R^{\varepsilon}_{\nu(t/\varepsilon^2)} + \varepsilon \cdot \beta(y_{\nu(t/\varepsilon^2)+1}) \cdot I^{\varepsilon}_{\nu(t/\varepsilon^2)}, \\ S^{\varepsilon}_0 &= S_0, \quad I^{\varepsilon}_0 = I_0, \quad R^{\varepsilon}_0 = 0. \end{cases} \tag{3.68}$$

We have stated in subsection 3.5, that under balance condition

$$\begin{aligned} \hat{\alpha} &:= \textstyle\int_Y p(dy)\alpha(y) = 0 \\ \hat{\beta} &:= \textstyle\int_Y p(dy)\beta(y) = 0, \end{aligned} \tag{3.69}$$

$(S^{\varepsilon}_{\nu(t/\varepsilon^2)}, I^{\varepsilon}_{\nu(t/\varepsilon^2)}, R^{\varepsilon}_{\nu(t/\varepsilon^2)})$ in (3.67) converges weakly as $\varepsilon \to 0$ to the triplet-process $(\tilde{S}(t), \tilde{I}(t), \tilde{R}(t))$ such that satisfies the following stochastic differential equations:

$$\begin{cases} d\tilde{S}(t) &= \tilde{\alpha}_1 dt + \beta_{11} dw_1(t) + \beta_{12} dw_2(t) + \beta_{13} dw_3(t), \\ d\tilde{I}(t) &= \tilde{\alpha}_2 dt + \beta_{21} dw_1(t) + \beta_{22} dw_2(t) + \beta_{23} dw_3(t), \\ d\tilde{R}(t) &= \beta_{31} dw_1(t) + \beta_{32} dw_2(t) + \beta_{33} dw_3(t), \end{cases} \tag{3.70}$$

where $\tilde{\alpha}_1 := \tilde{\alpha}_1(\tilde{S}, \tilde{I}, \tilde{R})$, $\tilde{\alpha}_2 := \tilde{\alpha}_2(\tilde{S}, \tilde{I}, \tilde{R})$ are defined in (3.57), and β_{ij}, $i, j = 1, 2, 3$ are defined in (3.58). Here $w_i(t)$, $i = 1, 2, 3$, are independent standard Wiener processes.

Let us consider first the stability of mean value of vector $(\tilde{S}(t), \tilde{I}(t), \tilde{R}(t))$. For this purpose we make a linearization of the coefficients $\tilde{\alpha}_i$, $i = 1, 2$, near the point $(0, 0, 0)$, which is the solution of the system (3.70). We can expand the functions $\tilde{\alpha}_1$ and $\tilde{\alpha}_2$ in Taylor series near $(0, 0, 0)$. We obtain that

$$d\tilde{\alpha}_i/dx|_{(0,0,0,)} = 0, \tag{3.71}$$

when $i \neq 2$ and $x \neq \tilde{I}$. For $i = 2$ and $x = \tilde{I}$ we have

$$d\tilde{\alpha}_2/d\tilde{I}|_{(0,0,0)} = 2\int_Y p(dy)\beta(y)R_0\beta(y)/m. \tag{3.72}$$

Finally, from (3.70), (3.71) and (3.72) we obtain that $E\tilde{I}(t)$ satisfies the equation

$$dE\tilde{I}(t)/dt = 2\int_Y p(dy)\beta(y)R_0\beta(y)/mE\tilde{I}(t)$$

and has the following look:

$$E\tilde{I}(t) = I_0 e^{2t\int_Y p(dy)\beta(y)R_0\beta(y)/m}. \tag{3.73}$$

We note that error of approximation is $O(r^2)$, where $r := \sqrt{S^2 + I^2 + R^2}$.

From (3.73) we have that mean value of infective goes to infinity when $t \to +\infty$, since $\int_Y p(dy)\beta(y)R_0\beta(y)/m > 0$. We note that $\tilde{S}(0) = S_0$ and $\tilde{R}(0) = 0$. It means that $E\tilde{S}(t) = S_0$ and $E\tilde{R}(t) = 0$.

We can conclude that under balance condition (3.69) number of infectives $\tilde{I}(t)$ in mean, namely, $E\tilde{I}(t)$ goes to infinity in diffusion approximation scheme, while number of susceptible in mean $E\tilde{S}(t)$ is constant S_0 and number of removed is zero.

We note that since vector $\mathbf{X}^\epsilon(t) := (S^\varepsilon_{\nu(t/\varepsilon^2)}, I^\varepsilon_{\nu(t/\varepsilon^2)}, R^\varepsilon_{\nu(t/\varepsilon^2)})^T$ in (3.68) converges weakly to the vector $(\tilde{S}(t), \tilde{I}(t), \tilde{R}(t))^T$, then we conclude that $E(S^\varepsilon_{\nu(t/\varepsilon^2)}, I^\varepsilon_{\nu(t/\varepsilon^2)}, R^\varepsilon_{\nu(t/\varepsilon^2)})^T$ converges to the $E\tilde{\mathbf{X}}(t) := E(\tilde{S}(t), \tilde{I}(t), \tilde{R}(t))^T$.

And stability of the epidemic model in (3.68) is completely defined by the stability of mean value $E(\tilde{S}(t), \tilde{I}(t), \tilde{R}(t))$.

3.7.3. STABILITY OF EPIDEMIC MODEL IN NORMAL DEVIATIONS SCHEME.

In this section we study the stability of mean value of the limiting vector process $(\tilde{S}(t), \tilde{I}(t), \tilde{R}(t))$ (see (3.62)) in normal deviations scheme (see section 3.6).

We use the linearization of the coefficients α_i, $i = 1, 2, 3$ near the point $(0, 0, 0)$. We remind that these coefficients are equal to

$$\begin{aligned}
\alpha_1(\hat{S}, \hat{I}, \tilde{S}, \tilde{I}) &= -\hat{\alpha}(\tilde{S}\hat{I} + \hat{S}\tilde{I}), \\
\alpha_2(\hat{S}, \hat{I}, \tilde{S}, \tilde{I}) &= \hat{\alpha}(\tilde{S}\hat{I} + \hat{S}\tilde{I}) - \hat{\beta}\tilde{I}, \\
\alpha_3(\hat{S}, \hat{I}, \tilde{S}, \tilde{I}) &= \hat{\beta}\tilde{I}.
\end{aligned} \tag{3.74}$$

Let us calculate the first derivatives of these functions with respect to their variables:

$d\alpha_1/d\tilde{S}|_{(0,0,0)} = -\hat{\alpha}\hat{I}$; $d\alpha_1/d\tilde{I}|_{(0,0,0)} = -\hat{\alpha}\hat{S}$; $d\alpha_1/d\tilde{R}|_{(0,0,0)} = 0$; $d\alpha_2/d\tilde{S}|_{(0,0,0)} = \hat{\alpha}\hat{I}$; $d\alpha_2/d\tilde{I}|_{(0,0,0)} = -\hat{\alpha}\hat{S} - \hat{\beta}$; $d\alpha_2/d\tilde{R}|_{(0,0,0)} = 0$; $d\alpha_3/d\tilde{I}|_{(0,0,0)} = -\hat{\beta}$; $d\alpha_3/d\tilde{S}|_{(0,0,0)} = 0$; $d\alpha_3/d\tilde{R}|_{(0,0,0)} = 0$.

From here, (3.62) and (3.74) we obtain the following system of equations for the mean values $(E\tilde{S}(t), E\tilde{I}(t), E\tilde{R}(t))^T$:

$$\begin{cases} dE\tilde{S}(t) &= -\hat{\alpha}\hat{I}(t)E\tilde{S}(t)dt - \hat{\alpha}\hat{S}(t)E\tilde{I}(t), \\ dE\tilde{I}(t) &= \hat{\alpha}\hat{I}(t)E\tilde{S}(t)dt + (\hat{\alpha}\hat{S}(t) - \hat{\beta})E\tilde{I}(t)dt, \\ dE\tilde{R}(t) &= \beta E\tilde{I}(t). \end{cases} \quad (3.75)$$

We note that error of this approximation is $O(r^2)$, where $r := \sqrt{S^2 + I^2 + R^2}$.

The stability in mean of normal deviated epidemic model is completely defined by the behaviour of the system (3.75), which contains only averaged epidemic model components $(\hat{S}, \hat{I}, \hat{R})$ and its averaged coefficients $\hat{\alpha}$ and $\hat{\beta}$.

We note that since

$$\mathbf{X}^{\epsilon}(t) \approx \hat{\mathbf{X}}(t) + \sqrt{\epsilon}\tilde{\mathbf{X}}(t),$$

where

$$\mathbf{X}^{\epsilon}(t) := (S^{\varepsilon}_{\nu(t/\varepsilon)}, I^{\varepsilon}_{\nu(t/\varepsilon)}, R^{\varepsilon}_{\nu(t/\varepsilon)})^T,$$

(see section 3.6)

$$\hat{\mathbf{X}}(t) := (\hat{S}(t), \hat{I}(t), \hat{R}(t))^T,$$

(see section 3.3) and

$$\tilde{\mathbf{X}}(t) := (\tilde{S}(t), \tilde{I}(t), \tilde{R}(t))^T$$

(see section 3.6, system (3.62)), then the stability in mean of the epidemic model in normal deviations scheme (see section 3.6) is defined by the following relation

$$E\mathbf{X}^{\epsilon}(t) \approx \hat{\mathbf{X}}(t) + \sqrt{\epsilon}E\tilde{\mathbf{X}}(t),$$

where $E\tilde{\mathbf{X}}(t)$ is defined in (3.75).

In this way we finally obtain the following relations with respect to the behaviour of mean values of epidemic model in normal deviations scheme:

$$ES^{\epsilon}(t) \approx \hat{S}(t) + \sqrt{\epsilon}E\tilde{S}(t),$$

$$EI^{\epsilon}(t) \approx \hat{I}(t) + \sqrt{\epsilon}E\tilde{I}(t),$$

$$ER^{\epsilon}(t) \approx \hat{R}(t) + \sqrt{\epsilon}E\tilde{R}(t).$$

CHAPTER 4: GENETIC SELECTION MODELS

4.1. Deterministic Genetic Selection Models.

Genetics is the study of heredity and variation among organisms. It is known that in all living organisms, reproduction involves the passing from one generation to the next a genetic code that determines all physical aspects of the offspring. This code is carried on chromosomes. *Chromosomes* are large molecules in living cells that carry the information for all of its chemical needs. Some human cells (sperm and eggs) have single chromosomes, but most other cells in the body have chromosomes, occurring in matched pairs. Plants can have chromosomes appearing in matched set of three or four or even more. When a single chromosome occurs, the cell is called *a haploid cell.* It is *a diploid cell* if the chromosomes occur in matched pairs, and *a polyploid cell* otherwise. Human beings are diploid organisms having 23 chromosome pairs, and bactgeria are haploid cells having a single chromosome that is arranged in a closed loop.

Genes are segments of a chromosome that code for some specific (identifiable) cell function, such as the production of a protein molecule. The location of a gene is called its *locus.* A gene may appear in several slightly variant forms within a population that are detectable by experiments. These variants are called *alleles.*

Our attention is directed at one locus having two alleles. These are denoted by A and B. If the organism is haploid, then it can be either of type A or type B at that locus. Therefore, a population of these organisms is partitioned by this locus into those of type A and those of type B. If the organism is diploid, then the possible types at the locus are AA, AB, and BB. Note that AB and BA are indistinguishable in the organism, and so are lumped together in the single notation AB. These genetic types in a population are called the *genotypes.*

Cell reproduction occurs either through asexual reproduction (*mitosis*) or sexual reproduction (*miosis*). In sexual reproduction, diploid parents each form haploid cells called *gametes.* These are the sperm (male) and the ova (female). The gametes combine to form a fertile cell called the *zygote.* The zygote is a diploid cell that goes on to reproduce by mitosis. The gametes can be thought of as having chromosomes being one strand each from each parent. If the parents, for example, have genotypes AA and BB, respectively, then the gametes are A and B, respectively, so the offspring must have genotype AB.

Cells having genotypes AA and BB are called *homozygotes* and the AB's are *heterozygotes.* So, mating of homozygotes results in homozyous or heterozygous progeny, depending on whether or not the homozytoes are identical. The type of matings and resulting frequencies of progeny genotypes were observed in 1850 by G. Mendel [57] for randomly mating populations. There are Mendelian trains in diploid organisms in human genetics.

Many types of bacteria have additional genetic material called extrachromosomal elements or *plasmids.* Plasmids are small circular pieces of DNA that also carry genes. However, plasmids can pass from cell to cell, and some genes can "jump" from plasmids to chromosomes.

The cell cycle begins with a newborn daughter. All of the cell's components, including the chromosome, are replicated. The replication is followed by splitting or division of the cell into two daughters, each receiving one replicate.

4.1.1. BACTERIAL GENETICS.

Let us consider *bacterial genetics* first. Bacteria are single-celled organisms that are enclosed within the cell wall; the interior is made up of cytoplasmic material and it contains the various mechanisms needed to cell life and reproduction.

In particular, there is the chromosome, a circular loop of deoxyribomucleic acid (DNA), which carries a code for all cell functions. The chromosome is a double felix like a twisted latter, the rungs correspond to pairs of nucleic acids: that is, pairs of the bases *adenine, theymine, cytosine,* and *guamine.* The only possible pairs are AT and GC. Therefore, the sequence of base pairs can be labeled by one strand. The complement strand is made by replacing each A by T, each G by C, and so on.

Let us first model frameshift mutations on the chromosome, and then describe the replication and distribution of plasmids among daughters.

4.1.1.A. Frameshift Mutation: a Two-State Markov Chain. A segment of DNA is processed in several ways during the cell cycle: it is replicated during reproduction, it is copied during transcription, and so on. Errors can occur in these processes. For example, during replication a mucleotide can be deleted or inserted. This can have a dramatic effect on the message carried by the chromosome. The nucleotides are read in triples: each triple codes for an amino acid. Deletion of one mucleotide from the sequence results in an entirely new sequence past the point of deletion, in this case one that has a transcription fermitration triple appearing prematurely (UAG). Such a change is called a *frameshift mutation.*

Let ε be the probability of a frameshift replication error at a given mucleotide during replication of the chromosome, and p be a probability of repair, that a correct mucleotide is inserted immediately after a given mucleotide or that a given incorrect mucleotide is deleted during replication. Suppose that a gene is made up of a number N of mucleotides. Then the probability that a *frameshift error occurs in the gene* is εN. The probability that such an error is repaired is p =probability of repair of a frameshift error. We write $\mu := \varepsilon N$ for mutation probability for a given gene, and $\nu = p$ for reverse mutation probability.

A population of bacteria can be divided into two groups, *mutants,* and wild types, depending on whether they have or have not suffered a frameshift mutation in a given gene. An example of this is the bacterium *Salmonella typhisnurium* which has genes for the production of the amino acid histidine.

Let $p_{w,n}$ be the proportion of the nth generation of wild type, and $p_{m,n}$ be the proportion of the nth generation of mutant type. Then

$$p_{w,n+1} = (1-\mu)p_{w,n} + \nu p_{m,n},$$

$$p_{m,n+1} = \mu p_{w,n} + (1-\nu)p_{m,n}. \tag{4.1}$$

Putting $\bar{p}_n = (p_{w,n}, p_{m,n})$ and

$$\mathbf{P} = \begin{pmatrix} 1-\mu & \mu \\ \nu & 1-\nu \end{pmatrix}, \tag{4.2}$$

we have

$$\bar{p}_{n+1} = \bar{p}_n \cdot \mathbf{P}. \tag{4.3}$$

That is, we obtain a *two state Markov chain*, because the sequence y_n is equal to cell type of a random sample from the nth generation: it has two possible states w and m.

Equation (4.3), where $\bar{p}_n$ and $\mathbf{P}$ are defined in (4.1)-(4.2), respectively, can be solved by successive back substitutions:

$$\bar{p}_n = \bar{b}_{n-1} \cdot \mathbf{P}^{n-1} = \ldots = \bar{p}_0 \cdot \mathbf{P}^n.$$

Using the spectral decomposition of $\mathbf{P}$

$$\bar{p}_0 \cdot \mathbf{P} = c_1\lambda_1\phi_1 + C_2\lambda_2\phi_2,$$

where $\lambda_1 = 1, \lambda_2 = 1-(\mu+\nu)$ are eigenvalues of $\mathbf{P}$, ϕ_1 and ϕ_2 are the (left) eigenvectors corresponding to λ_1 and λ_2, respectively, and C_1, C_2 are some constants, we obtain

$$\lim_{n\to+\infty} \lambda_1^{-n}\bar{p}_n = C_1\phi_1,$$

as $0 < \mu, \nu < 1, \| \lambda_2 \| < 1$. Now, the left eigenvectgor corresponding to $\lambda_1 = 1$ is $\phi_1 = (\mu/\nu, 1)$. Therefore,

$$C_1\phi_1 = (p_{w,0} + p_{m,0})(\mu/\nu, 1) = (\mu/\nu, 1).$$

In particular, the ratio of wild types to mutants will be ν/μ after many generations. The eigenvector corresponding to the eigenvalue $\lambda = 1$ gives the asymptotic distribution of the population.

4.1.1.B. Plasmid Incompatibility: Hyper-Geometric and Polya Chains. We consider a plasmid P and two homogenic derivatives P' and P''. The population distribution of the classes of cells with various P'/P'' ratios in any generation can be calculated from the distribution in the preceding generation. The plasmid P is assumed to appear with copy number N in all newborn cells and to replicate according to an $N \xrightarrow{replication} 2N \xrightarrow{partitioning} N$ cycle. Let p_i be the proportion of newborn cells having $iP'-$ and $(N-i)$ $P''-$ plasmids. After replication, there will be $2N$ plasmids, $2i$ of them P' and $(2N-2i)$ of them $P''-$ plasmids.

We suppose that each daughter receives N copies. j copies of P' can be selected for one daughter in C_{2i}^{j} different ways. The total number of partitions is C_{2N}^{N}, and the number of ways one daughter can have $jP'-$ and $(N-j)P''-$ plasmids is

$$C_{2i}^{j}C_{2N-2i}^{N-j}$$

if the mother originally had $iP'-$ and $(N-1)$ $P''-$ plasmids, where

$$C_N^k := \begin{cases} \frac{N!}{k!(N-k)!}, & 0 \leq k \leq N \\ 0, & \text{for other} k, \quad N. \end{cases}$$

Therefore,

$$\pi_{ij} := \frac{C_{2i}^{j}C_{2N-2i}^{N-j}}{C_{2N}^{N}}$$

gives the transition probability of one daughter having j P'- and $(N-j)P''$- plasmids.

A Markov chain with the transition probabilities π_{ij} is called a *hypergeometric chain*.

Since the reverse combination is also possible, the probability of an i P'- plasmid mother bearing one daughter with j and one with $(N-j)$ P'- plasmids is $2\pi_{ij}$.

If there are M bacteria in the nth generation, then after replication and division there will be $2M$ of them. If $p_{j,n+1}$ denotes the proportion having jP'' cells among the daughter $[(n+1)st]$ generation, then $2Mp_{j,n+1} = \sum_{j=0}^{N}(Mp_{i,n})(2\pi_{ij})$. Thus

$$p_{j,n+1} = \sum_{j=0}^{N} p_{i,n}\pi_{ij} \text{ for } j = 0, \ldots, N.$$

If we set $\bar{p}_n = (p_{o,n}, p_{1,n}, \ldots, p_{N,n})$ and $\pi = (\pi_{ij})_{i,j=0}^{N}$, then

$$\bar{p}_{n+1} = \bar{p}_n\pi.$$

4.1.1.C. Selective Growth. A cell having at least one copy of each plasmid is referred to as a heteroplasmid cell. The plasmids might govern resistance to antibiotics, say P' gives resistence to streptomycin and P'' resistence to antibiotics. The clone is first grown on a medium that selects for both plasmids (say, containing both streptomycin and ampicillin) and then transferred to a non-selective growth medium with scoring for the number of heteroplasmid cells in each generation.

We now describe cell growth under selective conditions. Starting with a proportion p_i of newborn cells having $iP'-$ plasmids, the proportion of the population after division having $jP'-$ plasmids now is

$$p_{j,n+1} = \frac{\sum_{i=1}^{N} p_{i,k}\pi_{ij}}{\sum_{k=1}^{N-1}\sum_{m=1}^{N-1} p_{k,n}\pi_{km}} \qquad \text{for } j = 1, \ldots, N-1.$$

These frequencies will equilibrate at values $p_i^*, \dots p_{N-1}^*$, which satisfy the equations

$$p_j^* = (1/\lambda^*) \sum_{k=1}^{N-1} p_k^* \cdot \pi_{kj}, \text{ for } j = 1, \dots, N-1,$$

where

$$\lambda^* = \sum_{k=1}^{N-1} \sum_{m=1}^{N-1} p_k^* \pi_{km}.$$

Let

$$\tilde{\pi} := (\pi_{ij})_{ij=1}^{N-1}$$

and

$$\bar{p}^* = (p_1^*, \dots, p_{N-1}^*).$$

Then $\lambda^* \bar{p}^* = \bar{p}^* \tilde{\pi}$. From the Perron-Frobenius theory for positive matrices (see [41]) it follows that λ^* is an eigenvalue of $\tilde{\pi}$ and $\bar{p}^*$ a corresponding eigenvector: λ^* is the largest eigenvalue of $\tilde{\pi}$. The value of λ^* is equal to

$$\lambda^* = 1 - (1/(2N-1)),$$

and it can be used directly to calculate the rate of increase of the heteroplasmid population. A starting population of H_0 heteroplasmid cells give rise to $(2\lambda^*)^n \cdot H_0$ after n generations of growth.

4.1.1.D. Non-selective Growth. We consider the rate at which heteroplasmid cells disappear through segregation during growth under non-selective conditions. Let $p_0^*, \dots, p_n^*$ be the equilivrium frequencies that will be approached asymptotically by the population after long-term growth under non-selective conditions. These must satisfy the $(N+1)$ equations $p_j^* = \sum_{i=0}^{N} p_i^* \pi_{ij}$ for $j = 0, 1, \dots, N$. These equations can be solved in terms of the initial population distribution $\bar{p}^0$. Note that

$$\sum_{j=0}^{N} j\pi_{ij} = i \text{ for } i = 0, ..., N,$$

we have *martingale property.* It follows that

$$\bar{p}_0^* = p_0^0 + \sum_{i=1}^{N} (1 - \frac{i}{N}) p_i^{(0)}, p_N^* = 1 - p_0^*,$$

and $p_j^* = 0$ for $j = 1, \dots, N-1$.

4.1.1.E. Random Replication, Regular Partitioning. This is a special case of the probability system known as *Polya's urn* (See Feller [21, p.118]).

Given copy number N and a newborn cell having i P'-plasmids, let Q_{ij} be the probability of the number of P' plasmids after replication $= k$, condition on number of P' plasmids in newborn $= 1$. Then

$$Q_{ij} = C_N^{k-i} \frac{\eta_{ik} \cdot \eta'_{ik}}{N(N+1) \cdots (2N-1)},$$

where

$$\eta_{ik} = \begin{cases} 1, & \text{if } k = i, \\ i(i+1)\dots(k-1), & \text{if } i < k \le N+i, \\ 0, & \text{otherwise,} \end{cases}$$

and

$$\eta'_{ik} = \begin{cases} 1, & \text{if } 2N-K = N-i \\ (N-1)\cdots(2N-K-1), & \text{if } 2N-K > N-i \\ 0, & \text{otherwise} \end{cases}$$

Let p_i denote the proportion of the population (after cell division) having i P' and $N-i$ P'' plasmids, and let p_j^+ denote the proportion of the daughter cells having j P' and $(N-j)$ P''-plasmids. Then, as before, we have

$$p_j^+ = \sum_{i=0}^{N}\sum_{k=0}^{2N} p_i Q_{ik} C_k^j C_{2N-K}^{N-j} (C_{2N}^N)^{-1}, i = 0, 1, \dots, N,$$

or in the matrix form, we have

$$p_j^+ = \sum_{i=0}^{N} p_i \pi_{ij},$$

where

$$\pi_{ij} = \sum_{k=0}^{2N} Q_{ik} C_k^j C_{2N-K}^{N-j} / C_{2N}^N,$$

and we obtain the *Polya Markov chain*.

4.1.2. HUMAN GENETICS.

As we have seen in the beginning of Section 4.1.1, some human cells have single chromosomes, but most have chromosomes occurring in matefied sets. Here we are interested in Mendelian traits in diploid organisms. The material in this section and their developments are taken from Ewens [25], Crow and Kimura [21], Moran [63], Cavalli-Sforza and Bodmer [14], Ludwig [57], and Feller [26,27].

4.1.2.A. Random Sampling For One-Locus Two-Allele Traits. Creation of one generation by its parents can be viewed as a *random process.* A population of N individuals carries $2N$ genes, and the distribution of their gametes at reproduction time gives the probabilities of various ones being used in reproduction.

The gene pool of the progeny can be viewed as being formed by sampling the adult gamete pool. Since the population is finite, this random sampling can have significant effects on the population's genetic structure, particularly if the population is small or the time scale is large, as in evolutionary studies.

4.1.2.B. Fisher-Wright Model. A one-locus two-allele trait carried by a population of N diploid individuals defines a gene pool of size $2N$. Let the alleles be denoted by A and B, and let α_n denote the number of A genes in the nth generation. Then, $p_n = \alpha_n/2N$ gives the frequency of A genes in the gene pool. The population is assumed to mate at random and be synchronized with nonoverlapping generations. Furthermore, the population size is assumed to remain constant ($= N$) through the generations.

The sequence of random variables $\{\alpha_n\}$ describes the offspring gene pool immediately after reproduction in successive generations. Wright [82] and Fisher [24] studied the dynamic problem with the help of a model described below.

If α_n is known, then α_{n+1} has a binomial distribution with parameter $p_n = \alpha_n/2N$ and index $2N$.

We first consider the case of *no selection and no mutation*. Under these conditions, the adult gamete pool at the end of the nth reproduction period has $\alpha_n \cdot A$ genes and $(2N - \alpha_n)B$ genes. Thus,

$$p_{ij} = \mathcal{P}\{\alpha_{n+1} = j|\alpha_n = i\} = C_{2N}^j p_n^j \cdot (1 - p_n)^{2N-j}$$

$$= C_{2N}^j \cdot (\frac{i}{2N})^j (1 - \frac{i}{2N})^{2N-j}, \qquad \text{for } j = 0, 1, \ldots, 2N.$$

This is the famous *Fisher-Wright model.* Since α_{n+1} is a binomially distributed random variables, it follows that

$$E[\alpha_{n+1}|\alpha_n] = \alpha_n, \text{Var}\,[\alpha_{n+1}|\alpha_n] = 2Np_n \cdot (1 - p_n). \tag{4.4}$$

If $\phi_{m,n} := \mathcal{P}[\alpha_n = m]$, then $\phi_{m,n+1} = \sum_{k=0}^{2N} \phi_{k,n} p_{k,m}$, or in the vector form we have

$$\bar{\phi} = \bar{\phi}\mathbf{P}^m$$

where $\mathbf{P} = (p_{ij})_{i,j=0}^{2N}$.

If $S_0, \ldots, S_{2N}$ correspond to the gene pool having $0, \ldots, 2N$ A-genes, then there are $(2N + 1)$ possible states for the population each generation. The state in time period n is determined by α_n. Through the generation, the population goes from state to state with transition probabilities (p_{ij}). The sequence $S_{\alpha n}$ describes the states through which the population passes. The sequence (α_n) forms *a Markov chain.*

The interior states $S_1, \ldots, S_{2N-1}$ are transient, for if $p_{ij}^{(n)}$ indicates the probability of passing from state S_i to state S_j in n steps, then $p_{ij}(n) \to 0$ as $n \to +\infty$ for $k = 1, 2, \ldots, 2N - 1$. This follows from spectral decomposition of $\mathbf{P}$ and $p_{ij}^{(n)} = (\mathbf{P}^n)_{ij} =$ the (ij)th element of $\mathbf{P}$.

Also, it is true that

$$\lim_{n\to\infty} p_{j,2N}^{(n)} = j/2N, \qquad \lim_{n\to\infty} p_{j0}^{(n)} = 1 - j/2N.$$

That is, a population beginning in state S_j has probability $j/2N$ of being absorbed in state S_{2N}, and probability $(1 - j/2N)$ of being absorbed in the state S_0.

In addition, we have

$$\lim_{n\to+\infty} P^n = \frac{1}{2N}\begin{bmatrix} 2N & 0 & \dots & 0 \\ 2N-1 & 0 & \dots & 1 \\ 1 & 0 & \dots & 2N-1 \\ 0 & 0 & \dots & 2N \end{bmatrix}.$$

Sampling effects will eventually lead to one of the alleles being eliminated from the population, as the above calculation shows. However, if N is large, the approach to fixation of one or the other of the genes is very slow.

These calculations illustrate *the phenomenon of random genetig drict.* We see that if $\alpha_0 = 1$, then the probability is $1/2N$ that the A gene will eventually dominate the gene pool supply as a result of gene sampling at reproduction.

We now consider the case with *selection and mutation.* The dynamics can be described just as in the preceding case, with the exception that the gamete pool will be sampled at the end of the nth reproductive cycle to form the gene pool in the $(n+1)$th period. The expected frequencies of AA, AB and BB genotypes in the next generation are

$$(\alpha_n/2N)^2, \qquad 2\alpha_n \cdot (2N-\alpha_n)/(2N)^2, \qquad [(2N-\alpha_n)/2N]^2,$$

respectively. If these genotypes have relative viabilities r, s, and 1, respectively, then the expected frequency of A genes at the next reproduction time is

$$\frac{r\cdot\alpha_n^2 + S\cdot\alpha_n(2N-\alpha_n)}{r\cdot\alpha_n^2 + 2\cdot S\cdot\alpha_n\cdot(2N-\alpha_n)+(2N-\alpha_n)^2} = (p/\tilde{\alpha}_n).$$

If, in addition to natural selection, A genes mutate to B genes with probability μ and B genes mutate to A genes with probability ν, then the expected frequency of A genes is

$$(p/\alpha_n) = (1-\mu)(p/\tilde{\alpha}_n) + \nu\cdot(1-(p/\tilde{\alpha}_n)).$$

It follows that

$$p_{ij} = \mathcal{P}[\alpha_{n+1} = j|\alpha_n = i] = C_{2N}^j (p/i)^j (1-(p/i))^{2N-j}.$$

Hence, (α_n) is a *Markov chain.* However, the martingale property fails, and the analysis of the gene pool's dynamics becomes more complicated.

4.1.2.C. Fisher-Wright-Haldane model.

Let us denote by g_n the proportion of the gene pool of type A immediately preceding the nth reproduction. The probability of survival to the next reproduction and their fertility will be denoted by r_n, S_n and t_n for the AA, AB and BB geotypes, respectively.

The gene pool is modeled from generataion to generation by the sequence (g_n), which is, in turn, determined from the equation

$$g_{n+1} = \frac{r_n \cdot g_n^2 + S_n \cdot g_n \cdot (1 - g_n)}{r_n \cdot g_n^2 + 2S_n \cdot (1 - g_n) \cdot g_n + t_n \cdot (1 - g_n)^2}. \tag{4.5}$$

This is *called Fisher-Wright-Haldane model*, and will be our main model in the study of slow genetic selection.

Let $r_n = r, S_n = S, t_n = t$, so that all the fitnesses remain constant through the generations. Then the qualitative behavior of this model can be determined by geometric iteration. In this case, we have

$$g_{n+1} = \frac{rg_n^2 + S(1 - g_n)g_n}{rg_n^2 + 2 \cdot S \cdot g_n \cdot (1 - g_n) + t \cdot (1 - g_n)^2}. \tag{4.6}$$

There are four possible cases:

(i) *A dominant*: if $r > s > t$, then $g_n \to 1$, as $n \to +\infty$, so A eventually dominates the gene pool;

(ii) *B dominant*: if $r < s < t$, then $g_n \to 0$, as $n \to +\infty$, so A eventually disappears, and B dominates the gene pool;

(iii) *Polynuphism*: if $s > r$ and $s > t$, then $g_n \to g^*$, and natural selection of genotypes acts to maintain both alleles in the population. The sicke-cell trait is a well documented case of this: $AA's$ have normal red blood cells but are susceptible to malaria; $BB's$ have badly deformed cells that do not carry oxygen well, resulting in anemia; and the heterozygotes AB don't suffer from anemia and enjoy some immunity to malaria;

(iv) *Disruptive selection*: if $s < r$ and $s < t$, and if $g_o < g^*$, then $g_n \to 1$, as $n \to +\infty$.

4.1.2.D. One-Locus Three-allele Traits. There are now three gamete types making up the gamete pool at each reproduction time, say A, A', and B. Let g_n' be the frequency of A' gametes. Then the frequency of the B gametes is given by $1 - g_n - g_n'$. Then

$$\begin{cases} g_{n+1} = \frac{r_1 g_n^2 + r_2 g_n g_n' + r_3 g_n \cdot (1 - g_n - g_n')}{w(g_n, g_n')}, \\ g_{n+1}' = \frac{r_2 \cdot g_n \cdot g_n' + r_5 g_n' \cdot (1 - g_n - g_n') + r_4 (g_n')^2}{w(g_n, g_n')}, \end{cases} \tag{4.7}$$

where the mean fitness w is given by

$$w(g_n, g_n') = r_1 \cdot g_n^2 + 2r_2 g_n g_n' + 2r_3 g_n g_n'' + r_4 (g_n')^2 + 2r_5 g_n' \cdot g_n'' + r_6 (g_n'')^2, \tag{4.8}$$

and

$$g_n'' = 1 - g_n - g_n'. \tag{4.9}$$

Genotype distribution of a one-locus, three-allele genetic trait is usually given by frequencies and fitnesses.

4.1.3. CONTINUOUS GENETIC SELECTION MODELS.

4.1.3.A. Continuous Fisher-Wright-Haldane Model: Multitime Method of Averaging. The method of geometric iteration (4.6) does not work if the fitness changes from generation to generation, namely, if r, s, t depend on n. In this case the method of averaging applies if the sequences $\{r_n, S_n, t_n\}$ in (4.5) are oscillatory. For example, if the generation time is much less than one year, seasonal changes can have an effect on the population's genetic structure. Let us suppose that the fitnesses are almost constant, say with period T. Then

$$r_n = 1 + \varepsilon p_n, S_n = 1 + \varepsilon\sigma_n, t_n = 1 + \varepsilon\tau_n, \tag{4.10}$$

where $p_{n+T} = p_n, \sigma_{n+T} = \sigma_n$, and $\tau_{n+T} = \tau_n$, for all $n \in Z^+$, and where $0 < \varepsilon \ll 1$ is a small parameter. Then model (4.5) becomes

$$g_{n+1} = g_n + \varepsilon[\sigma_n - \tau_n + g_n(p_n - 2\sigma_n + \tau_n)]g_n(1 - g_n) + o(\varepsilon), \tag{4.11}$$

where $o(\varepsilon)$ indicates terms that are multiplied by ε^2 and higher power of $\varepsilon, r_n, S_n, t_n$ in (4.10).

The multitime method proceeds by replacing g_n in (4.11) with $G(n, \varepsilon_n, \varepsilon) = g_n$, where G is a smooth function of its variables. Substituting this into (4.11) gives

$$G(n+1, s+\varepsilon, \varepsilon)$$
$$= G(n, s, \varepsilon) + \varepsilon[\sigma_n - \tau_n + G(n, s, \varepsilon)(p_n - 2\sigma_n + \tau_n)] \times (1 - G) \times G. \tag{4.12}$$

where we have set $s = \varepsilon_n$. The idea is that two time scales are indicated by (4.11): n and a slower time scale $s = \varepsilon_n$ on which selection acts. The solution of (4.12) should depend smoothly on ε, so we write

$$G(n, s, \varepsilon) = G_0(n, s) + \varepsilon G_1(n, s) + o(\varepsilon).$$

The coefficients, $G_0, G_1, \ldots$, can be found by differentiating the equation (4.12) successively with respect to ε. We have

$$\begin{cases} G_0(n+1, s) &= G_0(n, s), \\ G_1(n+1, s) &= \frac{dG_o(n,s)}{ds} = G_1(n+s)+ \\ &+ [\sigma_n - \tau_n + (p_n - 2\sigma_n + \tau_n)]G_0(n, s)(1 - G_0(n, s)). \end{cases} \tag{4.13}$$

The first equation in (34) shows that G_0 is independent of n, so we write $G_0 \equiv G_0(s)$. The second equation in (34) can be solved for G_1 by successive back substitutions as follows

$$\begin{aligned} G_1(n, s) &= G_1(0, s) \\ &+ \textstyle\sum_{k=0}^{n-1}(\sigma_k - \tau_k + (p_k - 2\sigma_k + \tau_k)G_0)(1 - G_0)G_0 - n\frac{dG_0}{ds}. \end{aligned} \tag{4.14}$$

Dividing equation (4.14) by n and passing to the limit $n \to +\infty$, we see that

$$\frac{dG_0}{ds} = (\bar{\sigma} - \bar{\tau} + (\bar{p}_{n-1} - 2\bar{\sigma} + \bar{\tau})G_0)G_0(1 - G_0), \tag{4.15}$$

where

$$\bar{p} = \lim_{n \to +\infty} \frac{1}{n} \sum_{k=0}^{n-1} p_k = \frac{1}{T} \sum_{k=0}^{T-1} p_k, \tag{4.16}$$

$$\bar{\sigma} = \lim_{n \to +\infty} \frac{1}{n} \sum_{k=0}^{n-1} \sigma_k = \frac{1}{T} \sum_{k=0}^{T-1} \sigma_k,$$

$$\bar{\alpha} = \lim_{n \to +\infty} \frac{1}{n} \sum_{k=0}^{n-1} \tau_k = \frac{1}{T} \sum_{k=0}^{T-1} \tau_k.$$

It is shown in Hoppensteadt and Minanker [33] that $g_n = G_0(\varepsilon n) + O(\varepsilon)$ for $0 \leq n \leq O(1/\varepsilon)$. Let

$$\bar{a} := \bar{\sigma} - \bar{\tau} \quad \text{and } \bar{b} := \bar{p} - 2\bar{\sigma} + \bar{\tau}, \tag{4.17}$$

where $\bar{\sigma}, \bar{\tau}, \bar{p}$ are defined in (4.16). Then behaviours of gene-pool frequencies in the multitime method can be described as follows (see [33] for details):

(i) if $\bar{a} > 0$ and $\bar{b} < 0$, then $G_0(+\infty) = G^* := |\bar{a}/\bar{b}|$;
(ii) if $\bar{a} < 0$ and $\bar{b} < 0$, then $G_0(+\infty) = 0$;
(iii) if $\bar{a} < 0$ and $\bar{b} > 0$, then: $G_0(0) < G^*$ implies $G_0(+\infty) = 0$, and $G_0(0) > G^*$ implies $G_0(+\infty) = 1$;
(iv) if $\bar{a} > 0$ and $\bar{b} > 0$, then $G_0(+\infty) = 1$.

Therefore, the mean values of the fitness determine the gene pool's solution.

4.1.3.B. Continuous One-Local Three-Allele Traits Under Slow Selection: The Method of Matched Asymptotic Expansions.

Geometric iteration is not useful in dealing with more complicated genetic structure such as the ABO system of blood types. A method called *the method of matched asymptotic expansions* has been developed to study slow selection in such systems.

We consider system (4.8)-(4.9), and introduce a method that gives a straightforward description of the gamete frequencies in the special case of slow selection.

If the fitness of the various gemotypes are nearly identical, then natural selective is acting on a slow time scale. Let

$$r_i = 1 + \varepsilon p_i; \qquad i = 1, 2, \ldots, 6,$$

where $\varepsilon \ll 1, \varepsilon p_i \ll 1$, and r_i are defined in Section 4.1.2 D.

With this change in the model's parameters, a rearrangement leads to

$$\begin{cases} g_{n+1} & = & \frac{g_n + \varepsilon \cdot (p_1 g_n^2 + p_2 g_n g_n' + p_3 \cdot g_n \cdot (1 - g_n - g_n'))}{1 + \varepsilon \tilde{w}(g_n, g_n')}, \\ g_{n+1}' & = & \frac{g_n^1 + \varepsilon (p_2 g_n \cdot g_n' + p_5 \cdot g_n' \cdot (1 - g_n - g_n') + p_6 (g_n')^2}{1 + \varepsilon \cdot \tilde{w}(g_n, g_n')}, \end{cases} \tag{4.18}$$

where

$$\begin{aligned}\tilde{w}(g_n, g_n') &= p_1^2 \cdot g_n^2 + 2p_2 \cdot g_n g_n' \\ &= 2p_3 \cdot g_n \cdot g_n'' + p_4 \cdot (g_n')^2 + 2p_5 \cdot g_n' \cdot g_n'' + p_6 \cdot (g_n')^2.\end{aligned}$$

Hoppensteadt and Minanker [33] developed a method to deal with the above system, their method shows that these difference equations look like forward Euler approximations to differential equations for smooth functions $G(s)$ and $G'(s)$. We write

$$g_n = G(\varepsilon n) + O(\varepsilon)$$

and

$$g_n' = G'(\varepsilon n) + O(\varepsilon),$$

where the functions G and G' are determined by the differential equations

$$\begin{cases} dG/dt = (p_3 + (p_1 - p_3)G + (p_2 - p_3)G^1)G - \tilde{w}(G, G')G \\ dG'/dt = (p_5 + (p_2 - p_5)G + (p_4 - p_5)G^1)g' - \tilde{w}(G, G')G', \end{cases} \tag{4.19}$$

and where

$$\begin{aligned}\tilde{w} &= G^2(p_1 + p + 6 - 2p_3) + 2GG'(p_2 + p_6 - p_3 - p_5) \\ &+ (g')^2 \cdot (p_4 + p_6 - 2p_5) + 2G(p_3 - p_6) + 2G' \cdot (p_5 - p_6) + p_6.\end{aligned} \tag{4.20}$$

System (4.19) can be analyzed by standard-phase plane techniques.

One-locus three-allele traits arise frequently in blood group studies. The ABO system in human blood (McKusick [55]) and other systems in baboon blood (Jolly and Brett [38]) are two well-known examples.

4.2. Stochastic GSM (GSM in Random Media).

We consider a genetic selection model, introduced in Section 4.1.2.C, in random environment. We first consider a discrete-time genetic selection model (GSM) in Markov random environment, and then a continuous-time GSM in semi-Markov random environment.

4.2.1. DISCRETE GENETIC SELECTION MODEL IN MARKOV RANDOM MEDIA.

Let $(y_n)_{n\in Z^+}$ be a homogeneous Markov chain and $r(y), s(y), t(y)$ be bounded nonnegative measurable function on Y.

A discrete genetic selection model in Markov random environment is defined as

$$g_{n+1} = \frac{r(y_{n+1})g_n^2 + S(y_{n+1}) \cdot (1 - g_n) \cdot g_n}{r(y_{n+1}) \cdot g_n^2 + 2 \cdot S(y_{n+1}) \cdot g_n(1 - g_n) + t(y_{n+1}) \cdot (1 - g_n)^2}. \tag{4.21}$$

As before, g_n is the proportion of the gene pool of type A immediately preceding the nth reproduction. Functions $r(y), s(y)$ and $t(y)$ describe the probability of survival to the next reproduction and their fertility for the AA, AB and BB genotypes, respectively, for each fixed $y \epsilon Y$.

In such a way, we have a GSM in Markov random environment. In this model, the proportion g_n of the gene pool of type A depends on and changes from the one reproduction to another with respect to the states of Markov chain $(y_n)_{n\in Z^+}$ by functions $r(y_n), s(y_n)$ and $t(y_n)$.

Function g_n in (4.21) is a random function.

4.2.2. CONTINUOUS GENETIC SELECTION MODEL IN MARKOV RENEWAL RANDOM MEDIA.

This model is defined by the following expression:

$$g_{\nu(t)+1} = \frac{r_{(\nu(t)+1)} g^2_{\nu(t)} + S(y_{\nu(t)+1}) \cdot (1 - g_{\nu(t)}) \cdot g_{\nu(t)}}{r(y_{\nu(t)+1}) g^2_{\nu(t)} + 2 \cdot S(y_{\nu(t)+1}) g_{\nu(t)} \cdot (1 - g_{\nu(t)}) + t(y_{\nu(t)+1}) \cdot (1 - g^2_{\nu(t)})}, \tag{4.22}$$

where $(y_n; \theta_n)_{n\in Z^+}$ is a Markov renewal process on $Y \times R_+, \nu(t) := \max\{n : \tau_n \leq t\}$ is a counting process.

If $t\epsilon[\tau_n, \tau_{n+1}]$, where $\tau_n := \sum_{k=0}^n \theta_k$, then $g_{\nu(t)+1} = g_{n+1}$ and expression (4.22) coincides with (4.21).

In (4.22), times of reproduction are random, and are given by $(\tau_n)_{n\in Z^+}$. The times of reproducting are $(\theta_n)_{n\in Z^+}$. Function g in (4.22) is a random process.

4.2.3. CONTINUOUS GENETIC SELECTION MODEL IN SEMI-MARKOV RANDOM MEDIA

This model is defined by the following differential equation:

$$\begin{cases} \frac{dg(t)}{dt} &= (a(y(t)) \cdot g(t) + b(y(t)) \cdot (1 - g(t)) \cdot g(t), \\ g(0) &= g. \end{cases} \tag{4.23}$$

where $y(t) := y_{\nu(t)}$ is a semi-Markov process constructed by Markov renewal process $(y_n; \theta_n)_{n\in Z^+}$, $a(y)$ and $b(y)$ are measurable bounded functions on Y. Function $g(t)$ in (4.23) is a random process.

We can now describe the measurability of genetic selection models in random media.

Discrete-time genetic selection model in Markov random environment in (4.21) is $\mathcal{F}_{n+1}/\mathcal{R}$-measurable, $r(y_{n+1}), s(y_{n+1})$, and $t(y_{n+1})$ are $\mathcal{F}^Y_{n+1}/\mathcal{R}$-measurable, and g_n is $\mathcal{F}_n/\mathcal{R}$-measurable and $\mathcal{F}^Y_n \subset \mathcal{F}^Y_{n+1} \subset \mathcal{F}_{n+1}$.

Continuous-time genetic selection model in Markov renewal random environment in (4.22) is $\mathcal{G}_t/\mathcal{R}$-measurable, as such are $r(y_{\nu(t)+1}), s(y_{\nu(t)+1}), t(y_{\nu(t)+1})$ and $g_{\nu(t)}$ are $\mathcal{H}_t/\mathcal{R}$-measurable and $\mathcal{H}_t \subset \mathcal{G}_t$.

Continuous-time genetic selection model in semi-Markov random environment in (4.23) is $\mathcal{H}_t/\mathcal{R}$-measurable, as $a(y(t))$ and $b(y(t))$ in (4.23) are $\mathcal{H}_t/\mathcal{R}$.

4.3. Averaging of Slow Genetic Selection Model in Random Media.

Let p, σ and τ be positive constants and let the function g_n be determined by Mendelean genetics for a single locus genetic trait having two allelie form, say A and B, in a synchronized population. Namely, let g_n denote the proportion of the gene pool that is of type A in the n-th generation. They are (see Subsection 4.2):

$$g_{n+1} = f(g_n),$$

g_0 is given, where

$$f(g_n) = \frac{pg_n^2 + \sigma g_n(1-g_n)}{pg_n^2 + 2\sigma g_n(1-g_n) + \tau \cdot (1-g_n)^2}. \tag{4.24}$$

The parameters p, σ, τ are the relative fitness of the genotypes AA, AB and BB, respectively. When selection is slow relative to reproduction, then these parameters are near 1, so in this case we write $p = 1 + \varepsilon\alpha, \sigma = 1 + \varepsilon\beta, \tau = 1 + \varepsilon\gamma$.

We consider this model in random environment perturbed by semi-Markov process. The perturbed system has the form:

$$g^{\varepsilon}_{\nu(t/\varepsilon)+1} = f(g^{\varepsilon}_{\nu(t/\varepsilon)}, p^{\varepsilon}_{\nu(t/\varepsilon)}, \sigma^{\varepsilon}_{\nu(t/\varepsilon)}, \tau^{\varepsilon}_{\nu(t/\varepsilon)}), \tag{4.25}$$

where $f(g)$ is defined in (4.24), and

$$p^{\varepsilon}_{\nu(t/\varepsilon)} = 1 + \varepsilon\alpha(y_{\nu(t/\varepsilon)}),$$

$$\sigma^{\varepsilon}_{\nu(t/\varepsilon)} = 1 + \varepsilon\beta(y_{\nu(t/\varepsilon)}),$$

$$\tau^{\varepsilon}_{\nu(t/\varepsilon)} = 1 + \varepsilon\gamma(y_{\nu(t/\varepsilon)}),$$

$\nu(t) := \max\{n : \tau_n \le t\}$ is a counting process, $(y_n)_{n\in Z^+}$ is a Markov chain.

The genetic selection model (4.25) is *continuous-time GSM in Markov renewal random environment* in series scheme.

We suppose that the functions $\alpha(y), \beta(y)$ and $\gamma(y)$ are bounded measurable function maping Y to R and $y_{\nu(t/\varepsilon)}$ is an ergodic semi-Markov process.

The iteration (4.25) may be written as

$$g^{\varepsilon}_{\nu(t/\varepsilon)+1} - g^{\varepsilon}_{\nu(t/\varepsilon)} = \frac{\varepsilon Q(g^{\varepsilon}_{\nu(t/\varepsilon)})}{1 + \varepsilon P(g^{\varepsilon}_{\nu(t/\varepsilon)})} g^{\varepsilon}_{\nu(t/\varepsilon)}, \tag{4.26}$$

where

$$g^{\varepsilon}_0 = g_0, \qquad P(g) = \alpha g^2 + 2\beta g(1-g) + \gamma(1-g)^2 \tag{4.27}$$

and

$$Q(g) = (1-g)(ag+b),$$

$$a \equiv a(y) = \alpha(y) + \gamma(Y) - 2\beta(y) \text{ and } b \equiv b(y) = \alpha(y) - \gamma(y).$$

The model (4.26)-(4.27) becomes:

$$\begin{aligned} g & {}^{\varepsilon}_{\nu(t/\varepsilon)+1} - g^{\varepsilon}_{\nu(t/\varepsilon)} \\ &= \varepsilon[b(y_{\nu(t/\varepsilon)} + g^{\varepsilon}_{\nu(t/\varepsilon)}a(y_{\nu(t/\varepsilon)})]g^{\varepsilon}_{\nu(t/\varepsilon)}(1 - g^{\varepsilon}_{\nu(t/\varepsilon)}) + o(\varepsilon^2), \end{aligned} \tag{4.28}$$

where $o(\varepsilon^2)$ indicates terms that are multiplied by ε^2 and higher powers of ε. We note that if $t\epsilon[\varepsilon\tau_n, \varepsilon\tau_{n+1})$, then (4.28) becomes the *discrete slow genetic selection model in Markov random environment:*

$$g_{n+1} - g^{\varepsilon}_n = \varepsilon[b(y_n) + g^{\varepsilon}_n a(y_n)]g^{\varepsilon}_n(1 - g^{\varepsilon}_n) + o(\varepsilon^2);$$

Replacing g^{ε}_n with $g^{\varepsilon}_n = G(n, \varepsilon n, \varepsilon)$, where G is a smooth function of its variables and substituting this into (4.28) gives:

$$G(n+1, t+\varepsilon, \varepsilon) = G(n, t, \varepsilon) + \varepsilon[b(y)n) + G(n, t, \varepsilon)a(y)n)](1 - G)G + o(\varepsilon), \tag{4.29}$$

where $t = \varepsilon\nu(t/\varepsilon)$, as $t\epsilon[\varepsilon\tau_n, \varepsilon\tau_{n+1})$ and $\nu(t/\varepsilon) = n$ (namely, for such t we have $t = \varepsilon \cdot n$). The solution of the above equation can be written as

$$G(n, t, \varepsilon) = G(n, t) + \varepsilon G_1(n, t) + o(\varepsilon).$$

The coefficients $G_0, G_1, \ldots$ can be found by differentiating the equation successively with respect to ε. The results are:

$$G_0(n+1, t) = G_0(n, t),$$

$$\begin{aligned} G_1(n+1, t) + \tfrac{dG_0}{dt}(n, t) &= \\ G_1(n, t) & \\ &+ [b(y_n) + a(y_n) \cdot G_0(n, t)]G_0(n, t)(1 - G_0(n, t)). \end{aligned} \tag{4.30}$$

The first equation shows that G_0 is independent of n, so

$$G_0(n, t) \equiv G_0 \equiv G_0(t).$$

The second equation can be solved for G_1 by successive back substitutions:

$$G_1(n, t) = G_1(0, t) + \sum_{k=0}^{n-1}(g(y_k) + a(y_k) \cdot G_0)(1 - G_0)G_0 - n \cdot \frac{dG_0}{dt}. \tag{4.31}$$

If the expression is valid, then G_1 is bounded for large n. Assuming this to be the case, dividing this equation by n and passing to the limit $n \to +\infty$, we get

$$\frac{dG_0(t)}{dt} = (\bar{b} + \bar{a}G_0)(1 - G_0)G_0, \tag{4.32}$$

where

$$\bar{b} := \lim_{n\to\infty} \frac{1}{n}\sum_{k=0}^{n-1}(\alpha(y_k) - \gamma(y_k)) =: \bar{\alpha} - \bar{\gamma},$$

$$\bar{a} := \lim_{n\to\infty} \frac{1}{n}\sum_{k=0}^{n-1}(\alpha(y_k) + \gamma(y_k) - 2\beta(y_k)) =: \bar{\alpha} + \bar{\gamma} - 2\bar{\beta}. \tag{4.33}$$

In such a way, we have proved that

$$g^{\varepsilon}_{\nu(t/\varepsilon)} \to G_0(t)$$

as $\epsilon \to 0$, where $g^{\varepsilon}_{\nu(t/\varepsilon)}$ and $G_0(t)$ are defined in (4.26) and (4.32), respectively.

$$\begin{aligned} \bar{\alpha} &:= \textstyle\int_Y p(dy)\alpha(y), \\ \bar{\beta} &:= \textstyle\int_Y p(dy)\beta(y), \\ \bar{\gamma} &:= \textstyle\int_Y p(dy)\gamma(y), \end{aligned}$$

as $(y_n)_{n\in Z_+}$ is an ergodic Markov chain.

If we take $n = \nu(t/\varepsilon)$ in (4.31), we obtain from (4.32)-(4.33) that:

$$\begin{aligned} G_1(\nu(t/\varepsilon), t) &= G_1(0, t) \\ &+ \textstyle\sum_{k=0}^{\nu(t/\varepsilon)+1}[b(y_k) + a(y_k)G_0]G_0 \cdot (1 - G_0) - -\nu(t/\varepsilon)\frac{dG_0(t)}{dt}. \end{aligned} \tag{4.34}$$

Dividing this equation by $\nu(t/\varepsilon)$ and passing to the limit $\varepsilon \to 0$ we get

$$\frac{dG_0(t)}{dt} = [\hat{b} + \hat{a} \cdot G_0]G_0 \cdot (1 - G_0), \tag{4.35}$$

where $\hat{b} = \bar{b}/m$, $\hat{a} = \hat{b}/m$, since

$$\frac{1}{\nu(t/\varepsilon)} \sum_{k=0}^{\nu(t/\varepsilon)-1} b(y_k) \to_{\epsilon\to 0} \bar{b}/m, \tag{4.36}$$

$$m = \int_Y m(y)p(dy).$$

We note that

$$\nu(t/\varepsilon) \sim \frac{t}{\varepsilon m},$$

and $\nu(t/\varepsilon) \to +\infty$ as $\varepsilon \to 0, \forall t \epsilon R_+$.

Remark. We can also obtain the above result on averaging using Theorem 2 (Chapter 2, Subsection 2.2).

We note, that the result (4.32) on averaging of discrete-time slow genetic selection model in Markov random environment can be obtained from Theorem 1 (Subsection

2.2, Chapter 2). Also, the result (4.35) on averaging of continuous-time slow genetic selection model in Markov renewal random environment follows from Theorem 2 (Subsection 2.2, Chapter 2).

As in previous sections, the averaged models (4.32)-(4.33) and (4.35) differ by scaling factor m in (4.35). Here, m^{-1} is the mean stationary intensity of the jumps of random environment.

Let us consider now *continuous genetic selection model in semi-Markov* random environment:

$$\begin{cases} \frac{dg^\varepsilon(t)}{dt} &= (a(y(t/\varepsilon))g^\varepsilon(t) + b(y(t/\varepsilon))(1-g^\varepsilon(t))g^\varepsilon(t) \\ \tilde{g}^\epsilon(0) &= g_0, \end{cases}$$

where $a(y), b(y)$ are defined in (4.27). From the results of Chapter 2 it follows that $g^\epsilon(t)$ converges weakly as $\epsilon \to 0$ to the function $\tilde{g}(t)$:

$$\begin{cases} \frac{d\tilde{g}(t)}{dt} &= (\tilde{a}\tilde{g}(t) + \tilde{b})(1-\tilde{g}(t))\tilde{g}(t), \\ \tilde{g}(0) &= g_0, \end{cases}$$

where

$$\begin{aligned} \tilde{a} &:= \textstyle\int_Y p(dy)m(y)a(y)/m, \\ \tilde{b} &:= \textstyle\int_Y p(dy)m(y)b(y)/m, \\ m(y) &:= \textstyle\int_0^\infty tG_y(dt), \\ m &:= \textstyle\int_Y p(dy)m(y). \end{aligned}$$

Here $m(y)$ is the lifetime in state y of the continuous-time slow genetic selection model, and m is the mean stationary lifetime in the states of the model.

Remark. The same result on averaging of discrete-time slow genetic selection model in Markov random environment was obtained in [36] using another method.

4.4. Merging of Slow Genetic Selection Model in Random Media.

We now consider the following genetic selection model:

$$g^\varepsilon_{(\nu(t/\varepsilon)+1)} = f(g^\varepsilon_{\nu(t/\varepsilon)}, p^\varepsilon_{\nu(t/\varepsilon)}, \sigma^\varepsilon_{\nu(t/\varepsilon)}, \tau^\varepsilon_{\nu(t/\varepsilon)}), \tag{4.37}$$

where

$$f(g^\varepsilon, p^\varepsilon, \sigma^\varepsilon\tau^\varepsilon) := \frac{p^\varepsilon g^\varepsilon_n + \sigma^\varepsilon g^\varepsilon_n\cdot(1-g^\varepsilon_n)}{p^\varepsilon\cdot(g^\varepsilon_n)^2 + 2\sigma^\varepsilon g^\varepsilon_n\cdot(1-g^\varepsilon_n) + \tau^\varepsilon\cdot(1-g^\varepsilon_n)^2}. \tag{4.38}$$

When selection is slow relative to reproduction, then the parameters $p^\varepsilon, \sigma^\varepsilon$ and τ^ε are close to 1, in this case we can write

$$
\begin{aligned}
p^{\varepsilon}_{\nu(t/\varepsilon)} &= 1+\varepsilon\cdot\alpha(y^{\varepsilon}_{\nu(t/\varepsilon)}),\\
\sigma^{\varepsilon}_{\nu(t/\varepsilon)} &= 1+\varepsilon\beta(y^{\varepsilon}_{\nu(t/\varepsilon)}),\\
\tau^{\varepsilon}_{\nu(t/\varepsilon)} &= 1+\varepsilon\cdot\gamma(y^{\varepsilon}_{\nu(t/\varepsilon)}),
\end{aligned}
\tag{4.39}
$$

here $y^{\varepsilon}_{\nu(t/\varepsilon)}$ is a perturbed semi-Markov process with stochastic kernel $Q_{\varepsilon}(y, A, t) = P_{\varepsilon}(y, A)G_y(t)$.

Using (4.39), the iteration (4.37) with function g in (4.38) may be written as

$$
\begin{cases}
g^{\varepsilon}_{\nu(t/\varepsilon)+1} - g^{\varepsilon}(\nu(t/\varepsilon)) = \frac{\varepsilon\cdot Q(g^{\varepsilon}(\nu(t/\varepsilon)))}{1+\varepsilon\cdot P(g^{\varepsilon}(\nu(t/\varepsilon)))} g^{\varepsilon}(\nu(t/\varepsilon)),\\
g^{\varepsilon}(0) = g_0,
\end{cases}
\tag{4.40}
$$

where

$$
\begin{aligned}
&Q(g) := (1-g)(ag+b),\, P(g) := \alpha g^2 + 2\beta g(1-g) + \gamma(1-g)^2,\\
&a := a(y) := \alpha(y) + \gamma(y) - 2\beta(y) \text{ and } b := b(y) := \alpha(y) - \gamma(y).
\end{aligned}
\tag{4.41}
$$

We note that functions $\alpha(y), \beta(y)$ and $\gamma(y)$ are bounded and continuous on Y.

Using the expansion of $(1+\varepsilon P(g^{\varepsilon}))^{-1}$ with respect to ε we can rewrite (4.40) as

$$
g^{\varepsilon}(\nu(t/\varepsilon)+1) - g^{\varepsilon}(\nu(t/\varepsilon)) = \varepsilon\cdot Q(g^{\varepsilon}(\nu(t/\varepsilon)))g^{\varepsilon}(\nu(t/\varepsilon)) + o(\varepsilon), \tag{4.42}
$$

where $Q(g^{\varepsilon})$ is defined in (4.41), and

$$
|o(\varepsilon)|/\epsilon \to_{\varepsilon\to 0} 0.
$$

Applying the Theorem 6 (Chapter 2, Subsection 2.5) to the equation (4.42) with the function

$$
g(x, y) = (ax+b)(1-x)x,
$$

we obtain that $g^{\varepsilon}(\nu(t/\varepsilon))$ converges weakly as $\varepsilon \to 0$ to the process $\tilde{g}(t)$ which satisfies the following equation:

$$
\begin{cases}
\frac{d\tilde{g}(t)}{dt} = -((\tilde{\alpha}(\hat{y}(t)) + \tilde{\gamma}(\hat{y}(t)) - 2\tilde{\beta}(\hat{y}(t)))\tilde{g}(t) + \tilde{\alpha}(\hat{y}(t))\\
\qquad - \tilde{\gamma}(\hat{y}(t))(1-\tilde{g}(t))\tilde{g}(t),\\
\tilde{g}(0) = g_0,
\end{cases}
\tag{4.43}
$$

where $\hat{y}(t)$ is the merged Markov process in phase space $(V, \mathcal{V})$,

$$
\tilde{\alpha}(v) := \int_{Y_v} p_v(dy)\alpha(y)/m_v;\, \tilde{\beta}(v) := \int_{Y_v} p_v(dy)\beta(y)/m_v,
$$

$$\tilde{\gamma}(v) := \int_{Y_v} p_v(dy)\gamma(y)/m_v. \tag{4.44}$$

In what follows, we show now that the result (4.43) can be obtained by another method. We note that if $t\epsilon[\varepsilon\tau_n, \varepsilon\tau_{n+1})$, then we have from (4.42) that

$$g^\varepsilon(n+1) - g^\varepsilon(n) = \varepsilon[Q(g^\varepsilon(n)g^\varepsilon(n)] + o(\varepsilon), \tag{4.45}$$

where $Q(g^\varepsilon)$ is defined in (4.41).

Using the same arguments as in (4.29)-(4.32) (Section 4.3) we obtain

$$g(n, t/\varepsilon) = g(n,t) + \varepsilon g_1(n,t) + O(\varepsilon), \tag{4.46}$$

where $g_n^\varepsilon = g(n, \varepsilon n, \varepsilon)$, $t = \varepsilon n$, and

$$g(n+1, t) = g(n,t),$$

$$\begin{aligned} g\quad & {}_1(n+1,t) + \tfrac{dg}{dt}(n,t) \\ = \quad & g_1(n,t) + [b(y^\varepsilon_{(n)}) + a(y^\varepsilon_{(n)})g(n,t)]g(n,t)(1-g(n,t)). \end{aligned} \tag{4.47}$$

The first equation shows that $g(n,t)$ is independent of n, so

$$g(n,t) \equiv g = \tilde{g}(t).$$

The second equation can be solved for $g_1(n,t)$ by successive back substitutions:

$$g_1(n,t) = g_1(0,t) + \sum_{n=1}^{n-1}[b(y_k^\varepsilon) + a(y_k^\varepsilon)\tilde{g}(t)]\tilde{g}(t)(1-\tilde{g}(t)) - n\cdot\frac{d\tilde{g}(t)}{dt}. \tag{4.48}$$

Function $g_1(n,t)$ is bounded for large $n, \forall t\epsilon R_+$. If we take in (4.48) $n = \nu(t/\varepsilon)$ and dividing (4.47) by $\nu(t/\varepsilon)$ and then pass to the limit $\varepsilon \to 0$, then we obtain

$$\begin{cases} \frac{d\tilde{g}(t)}{dt} &= (\tilde{b}_t + \tilde{a}_t\cdot\tilde{g}(t))\cdot(1-\tilde{g}(t))\tilde{g}(t), \\ \tilde{g}(0) &= g_0, \end{cases} \tag{4.49}$$

where

$$\tilde{a}_t := \lim_{\varepsilon\to 0}\frac{1}{\nu(t/\varepsilon)}\sum_{k=1}^{\nu(t/\varepsilon)-1}(\alpha(y_k^\varepsilon) + \gamma(y_n^\varepsilon) - 2\beta(y_k^\varepsilon)),$$

$$\tilde{b}_t := \lim_{\varepsilon\to 0}\frac{1}{\nu(t/\varepsilon)}\sum_{k=1}^{\nu(t/\varepsilon)-1}(\alpha(y_k^\varepsilon) - \gamma(y_k^\varepsilon)). \tag{4.50}$$

Taking into account the following convergence (see (3.54), Subsection 3.4, Chapter 3)

$$\frac{1}{\nu(t/\varepsilon)}\sum_{k=1}^{\nu(t/\varepsilon)-1}(\alpha(y_k^\varepsilon) \to_{\varepsilon\to 0} \int_0^t \tilde{\alpha}(\hat{y}(s))ds, \tag{4.51}$$

where

$$\tilde{\alpha}(v) := \int_{Y_v} p_v(dy)\alpha(y)/m_v,$$

we obtain from (4.50) and (4.51) that

$$\tilde{a}_t = \tilde{\alpha}(\hat{y}(t)) \text{ and } \tilde{b}_t = \tilde{b}(\hat{y}(t)),$$

where

$$\tilde{a}(v) := \tilde{\alpha}(v) + \tilde{\gamma}(v) - 2\tilde{\beta}(v), \tag{4.52}$$

and

$$\tilde{a}(v) := \tilde{\alpha}(v) - \tilde{\gamma}(v);$$

functions $\tilde{\alpha}(v), \tilde{\gamma}(v)$ and $\tilde{\beta}(v)$ are defined in (4.44).

Taking into account (4.49)-(4.52) we obtain that $\tilde{g}(t)$ in (4.49) satisfies the equation (4.43).

Equation (4.43) is a continuous-time genetic selection model in merged Markov random environment $\hat{y}(t)$ in phase space $(V, \mathcal{V})$.

4.5. Diffusion Approximation of Slow Genetic Selection in Random Media.

Let us suppose that coefficients $\bar{\alpha}, \bar{\gamma}, \bar{\beta}$ are equal to zero: $\bar{\alpha} = \bar{\gamma} = \bar{\beta} = 0$ (in (4.33)). Then $\bar{a} = \bar{b} = 0$, and *the balance condition is fulfilled* for the *averaged equations* (4.32) and (4.35).

We consider our difference equation in slow genetic selection model in scale of time t/ε^2, using the representation (4.26)

$$g^{\varepsilon}_{\nu(t/\varepsilon^2)+1} - g^{\varepsilon}_{\nu(t/\varepsilon^2)} = \frac{\varepsilon \cdot Q^{\varepsilon}(g^{\varepsilon}_{\nu(t/\varepsilon^2)})}{1 + \varepsilon \cdot P(g^{\varepsilon}_{\nu(t/\varepsilon^2)})} \cdot g^{\varepsilon}_{\nu(t/\varepsilon^2)}, \tag{4.53}$$

where functions $Q(g^{\varepsilon})$ and $P(g^{\varepsilon})$ are defined in (4.27).

Using the representation (4.26), we obtain

$$\begin{aligned} g^{\varepsilon}_{\nu(t/\varepsilon^2)+1} \quad - \quad & g^{\varepsilon}_{\nu(t/\varepsilon^2)} = \varepsilon \cdot [b(y_{\nu(t/\varepsilon^2)}) + a(y_{\nu(t/\varepsilon^2)}) g^{\varepsilon}_{\nu(t/\varepsilon^2)}] g^{\varepsilon}_{\nu(t/\varepsilon^2)} \\ \times \quad & (1 - g^{\varepsilon}_{\nu(t/\varepsilon^2)}) + \varepsilon^2 \cdot Q(g^{\varepsilon}) \cdot P(g^{\varepsilon}) g^{\varepsilon}_{\nu(t/\varepsilon^2)} + o(\varepsilon^2). \end{aligned} \tag{4.54}$$

To see that, we note the following

$$\begin{aligned} \frac{\varepsilon Q(g^{\varepsilon})}{1+\varepsilon \cdot P(g^{\varepsilon})} &= \varepsilon Q(g^{\varepsilon} \cdot g^{\varepsilon} \cdot (1 + \varepsilon P(g^{\varepsilon}) + \varepsilon^2 \cdot P(g^{\varepsilon}) + 0(\varepsilon^2)) \\ &= \varepsilon Q(g^{\varepsilon}) g^{\varepsilon} + \varepsilon^2 Q(g^{\varepsilon}) P(g^{\varepsilon}) \cdot g^{\varepsilon} + \varepsilon^3 Q(g^{\varepsilon}) P^2 g^{\varepsilon} g^{\varepsilon} + \ldots \\ &= \varepsilon [b(y) + a(y) g^{\varepsilon}] \cdot g^{\varepsilon} \cdot (1 - g^{\varepsilon}) + \varepsilon^2 Q(g^{\varepsilon}) P(g^{\varepsilon}) g^{\varepsilon} + O(\varepsilon^3), \end{aligned}$$

and (4.54) follows. Let $\tilde{g}(t)$ be the limiting process for g^{ε} as $\varepsilon \to 0$ in (4.54).

From the theory of RE and Theorem 4 (Chapter 2, Subsection 2.2) it follows that the process $g^{\varepsilon}_{\nu(t/\varepsilon^2)}$ converges weakly as $\varepsilon \to 0$ to the diffusion process $\tilde{g}(t)$ with

$$d\tilde{g}(t) = \tilde{\alpha}(\tilde{g}(t))dt + \tilde{\beta}(\tilde{g}(t))dw(t), \tag{4.55}$$

where

$$\begin{aligned} \tilde{\alpha}(g) &:= \textstyle\int_Y p(dy)[G(y,g)\mathbf{R}_0 \cdot G_g(y,g) \\ &+ 1/2G(y,g)G_g(y,g) + gG(y,g)P(g)]/m, \\ \tilde{\beta}^2(g) &:= 2\textstyle\int_Y p(dy)[G(y,g)\mathbf{R}_0 G(y,g) + 1/2G^2(y,g)]/m, \end{aligned} \tag{4.56}$$

where

$$G(y,g) := [b(y) + a(y)\cdot g]g(1-g),$$

$P(g)$ is defined in (4.27).

4.6. Normal Deviations of Slow Genetic Selection Model in Random Media.

Let g^{ε}_n be a sequence defined by

$$\begin{cases} g^{\varepsilon}_{n+1} - g^{\varepsilon}_n &= \frac{\varepsilon Q(g^{\varepsilon}_n)}{1+\varepsilon\cdot P(g^{\varepsilon}_n}g^{\varepsilon}_n, \\ g^{\varepsilon}_0 &= g_0, \end{cases} \tag{4.57}$$

where

$$P(g) := \alpha g^2 + 2\beta g(1-g) + \gamma(1-g)^2, \tag{4.58}$$

$$Q(g) := (1-g)(ag+b),$$

$\alpha \equiv a(y) := \alpha(y) + \gamma(y) - 2\beta(y)$ and $b \equiv b(y) := \alpha(y) - \gamma(y)$. We define the sequence $\bar{g}^{\varepsilon}$ by

$$\begin{cases} \bar{g}^{\varepsilon}_{n+1} - \bar{g}^{\varepsilon}_n &= \varepsilon(\bar{b} + \bar{a}\bar{g}^{\varepsilon}_n)(1-\bar{g}^{\varepsilon}_n)\bar{g}^{\varepsilon}_n, \\ \bar{g}^{\varepsilon}_0 &= g_0, \end{cases} \tag{4.59}$$

where

$$\begin{aligned} \bar{b} &:= \lim_{n\to\infty} \tfrac{1}{n}\textstyle\sum_{k=0}^{n-1}(\alpha(y_k) - \gamma(y_k)) =: \bar{\alpha} - \bar{\gamma}, \\ \bar{a} &:= \lim_{n\to\infty} \tfrac{1}{n}\textstyle\sum_{k=0}^{n-1}(\alpha(y_k) + \gamma(y_k)) - 2\beta(y(k)) \\ &:= \bar{\alpha} + \bar{\gamma} - 2\bar{\beta}. \end{aligned} \tag{4.60}$$

where

$$Z_n^\varepsilon := [g_n^\varepsilon - \tilde{g}_n^\varepsilon]/\sqrt{\varepsilon} \tag{4.62}$$

is the normalized and deviated process.

Applying the normal deviation theorem from Section 2.4 (Chapter 2) to the process $Z^\varepsilon(t)$ in (4.61), we obtain that the process $Z^\varepsilon(t)$ converges weakly as $\varepsilon \to 0$ to the process $\tilde{Z}(t)$ given by

$$\tilde{Z}(t) = \int_0^t \tilde{G}_g(G_0(s))\tilde{Z}(s)ds + \int_0^t \sigma(G_0(s))dw(s),$$

where

$$\begin{aligned}
\tilde{G}(g) &:= \textstyle\int_Y p(dy)G(y,g)/m, \\
\sigma^2(g) &:= 2\textstyle\int_Y p(dy)[(G(y,g) - \tilde{G}(g))\mathbf{R}_0(G(y,g) - \tilde{G}(g)) \\
&+ (G(y,g) - \tilde{G}(g))^2/2]/m, \\
G(y,g) &:= [b(y) + a(y)g]g(1-g),
\end{aligned}$$

$b(y)$ and $a(y)$ are defined in (4.58), and $G_0(t)$ satisfies the following equation:

$$\begin{cases} \frac{dG_0(t)}{dt} = [\hat{b} + \hat{a}G_0(t)]G_0(t)(1 - G_0(t)) \\ G_0(0) = g_0, \end{cases}$$

where

$$\hat{b} := \bar{b}/m, \hat{a} := \bar{a}/m, m := \int_Y p(dy)m(y),$$

$\mathbf{R}_0$ is a potential of Markov chain $(y_n)_{n\in Z^+}$, and $w(t)$ is a standard Wiener process.

4.7. Stochastic Stability of Slow Genetic Selection Model.

In this section, we study stochastic stability of slow genetic selection model in series scheme in averaging and diffusion approximation schemes, using the results from Chapter 2.

4.7.1. STABILITY OF SLOW GENETIC SELECTION MODEL IN AVERAGING SCHEME.

Let us consider slow genetic selection model in series scheme (see Chapter 4, Subsection 4.3):

$$\begin{aligned}
g^\varepsilon_{\nu(t/\varepsilon)+1} - g^\varepsilon_{\nu(t/\varepsilon)} &= \varepsilon[b(y_{\nu(t/\varepsilon)}) + g^\varepsilon_{\nu(t/\varepsilon)} \cdot a(y_{\nu(t/\varepsilon)})]g^\varepsilon_{\nu(t/\varepsilon)} \\
&\quad 1 - g^\varepsilon_{\nu(t/\varepsilon)}) + o(\varepsilon),
\end{aligned} \tag{4.63}$$

where $a(y) := \alpha(y) + \gamma(y) - 2\beta(y)$ and $b(y) := \alpha(y) - \gamma(y), g_n^2$ are the proportion of the gene pool that is of type A in the n-th generation for a single locus genetic traits having two allelie form, A and B (see (4.25)).

We have stated in Subsection 4.3 that under averaging conditions the process $g^{\varepsilon}_{\nu(t/\varepsilon)}$ in (4.26) converges weakly as $\varepsilon \to 0$ to the averaged process $\hat{g}(t)$ given by

$$\begin{cases} \frac{d\hat{g}(t)}{dt} &= (\hat{b} + \hat{a}\hat{g}(t))\hat{g}(t)(1 - \hat{g}(t)), \\ \hat{g}(0) &= g_0, \end{cases} \tag{4.64}$$

where

$$\hat{b} := \int_Y p(dy)b(y)/m, \hat{a} := \int_Y p(dy)a(y)/m.$$

We note that the right hand side of (4.64) is equal to zero if $\hat{g}(t) = 0$. It means that condition (iii) for function $g(x, y) == [b(y) + xa(y)]x(1 - x)$ is satisfied (see Theorem 7, Chapter 2) and we can therefore study the stability of zero state of the process $g^{\varepsilon}_{\nu(t/\varepsilon)}$.

If there exists a smooth function $V(x)$ satisfying the following condition:

$$[\hat{b} + \hat{a}x]x(1 - x)V_x(x) \leq -\beta V(x), \tag{4.65}$$

for some $\beta > 0$, and condition (i) of Theorem 7, in Subsection 2.7.

Then by applying this Theorem we conclude that the process $g^{\varepsilon}_{\nu(t/\varepsilon)}$ in (4.63) is stochastically exponentially stable, and is asymptotically stocastically stable. In particular,

$$P_{g_{0,y}}\{\lim_{t \to +\infty} g^{\varepsilon}_{\nu(t/\varepsilon)} = 0\} = 1. \tag{4.66}$$

From the result (4.66) it follows that the proportion $g^{\varepsilon}_{\nu(t/\varepsilon)}$ of the gene pool that is of type A in the $\nu(t/\varepsilon)$-th generation for a single locus genetic traits having two allelie form, A and B, tends to zero as $t \to +\infty$ for small $\varepsilon > 0 : 0 \leq \varepsilon \leq \varepsilon_0$. Unfortunately, constructing such a function of V is not a trivial task.

4.7.2. STABILITY OF SLOW GENETIC MODEL IN DIFFUSION APPROXIMATION SCHEME.

Let us consider the slow genetic selection model (4.54) in series scheme with the balance condition

$$\hat{a} = \hat{b} = 0. \tag{4.67}$$

Namely,

$$\begin{aligned} g^{\varepsilon}_{\nu(t/\varepsilon^2)+1} - g_{\nu(t/\varepsilon^2)} &= \varepsilon[b(y_{\nu(t/\varepsilon^2)}) + a(y_{\nu(t/\varepsilon^2)})g^{\varepsilon}_{\nu(t/\varepsilon^2)}] \\ &\quad g^{\varepsilon}_{\nu(t/\varepsilon^2)} \times (1 - g^{\varepsilon}_{\nu(t/\varepsilon^2)}) + \varepsilon^2 Q(g^{\varepsilon}) \cdot P(g^{\varepsilon})g^{\varepsilon}_{\nu(t/\varepsilon^2)} + o(\varepsilon^2), \end{aligned} \tag{4.68}$$

where

$$
\begin{aligned}
Q(g) &:= (1-g)(a(y)g+b(y)),\\
P(g) &:= \alpha(y)g^2+2\beta(y)g(1-g)+\gamma(y)(1-g),\\
a(y) &:= \alpha(y)+\gamma(y)-2\beta(y),\\
b(y) &:= \alpha(y)-\gamma(y).
\end{aligned}
\tag{4.69}
$$

Under condition (4.67) and the diffusion approximation condition (Section 4.5), we have obtained that the process $g^{\varepsilon}_{\nu(t/\varepsilon^2)}$ in (4.68) converges weakly as $\varepsilon \to 0$ to the process $\tilde{g}(t)$ given by

$$
d\tilde{g}(t) := \tilde{\alpha}(\tilde{g}(t))dt + \tilde{\beta}(\tilde{g}(t))dw(t), \tag{4.70}
$$

where

$$
\begin{aligned}
\tilde{\alpha}(y) &:= \textstyle\int_Y p(dy)[G(y,g)\mathbf{R}_0 G^1_y(y,g)+\\
&+ \ 1/2G(y,g)\cdot G^1_g(y,g)+g\cdot G(y,g)P(g)]/m,\\
\tilde{\beta}^2 &:= 2\cdot \textstyle\int_y p(dy)[G(y,g)\mathbf{R}_0 G(y,g)+1/2G^2(y,g)]/m,\\
G(y,g) &:= [b(y)+a(y)g]\cdot g\cdot(1-g),
\end{aligned}
\tag{4.71}
$$

$P(g), a(y)$ and $b(y)$ are defined in (4.69).

We note that the right hand side of (4.70) is equal to zero if $\tilde{g}(t) = 0$. Since $\tilde{\alpha}(0) = \tilde{\beta}(0) = 0$, condition (iii) (see Theorem 8 in Subsection 2.6.2) is satisfied. We now study the stability of the zero state of the process $g^{\varepsilon}_{\nu(t/\varepsilon^2)}$ in (4.68). Assume there exists a smooth function $W(x)$ which satisfies the following condition:

$$
\tilde{\alpha}(x)W_x(x) + 1/2\tilde{\beta}^2(x)W_{xx}(x) \le -\gamma W(x), \tag{4.72}
$$

for some $\beta > 0, \tilde{\alpha}(y)$ and $\tilde{\beta}(x)$ are defined in (4.71), and the condition (i), in Theorem 8, subsection 2.6.2, then from this Theorem 8 we obtain that the process $g^{\varepsilon}_{\nu(t/\varepsilon^2)}$ in (4.68) is stochastically exponentially stable and is asymptotically stochastically stable:

$$
P_{g_{0},y}\{\lim_{t\to+\infty} g^{\varepsilon}_{\nu(t/\varepsilon^2)} = 0\} = 1. \tag{4.73}
$$

From the result (4.73) we obtain that under balance condition (4.67) and condition (4.72) the proportion $g^{\varepsilon}_{\nu(t/\varepsilon^2)}$ of the gene pool that is of type A in the $\nu(t/\varepsilon^2)$-th generation for a single locus genetic traits, having two allelie form, A and B, tends to zero as $t \to +\infty$ for small $\varepsilon > 0$.

CHAPTER 5: BRANCHING MODELS

5.1. Branching Models with Deterministic Generating Function.

Another point of view can be taken toward the Fisher-Wright model, considered in Section 4.1.2.B. This approach is in the spirit of *branching processes.*

5.1.1. THE GALTON-WATSON-FISHER MODEL.

Suppose that each gene leaves a random number of offspring genes with probability of 0,1,2,..., being given by the numbers $p_0, p_1, p_2, \ldots$, respectively. The sequence $\{p_j\}$ then gives the probability distribution of offsprings. The generating function for this distribution is defined by the formula

$$b(u) = \sum_{j=0}^{+\infty} p_j u^j, |u| \leq 1. \tag{5.1}$$

Again let α_n denote the number of A genes in the nth generation. We now derive a formula for the *generating function of the* $(n+1)$*th generation.*

First, we note that

$$P\{\alpha_{n+1} = k | \alpha_n = i\} = \sum_{j_1+\ldots+j_i=k} p_{j_1} \cdots p_{j_i}. \tag{5.2}$$

The conditional generating function is given by

$$\begin{aligned} F_{n+1,j}(u) &= \sum_{k=0}^{+\infty} \mathcal{P}\{\alpha_{n+1} = k | \alpha_n = i\} u^k \\ &= \sum_{k=0}^{+\infty} \sum_{j_1+\ldots+j_i=k} p_{j_1} \cdots p_{j_i} u^{j_1} \ldots u^{j_i} = [b(u)]^i. \end{aligned} \tag{5.3}$$

Next, we note

$$P\{\alpha_{n+1} = k\} = \sum_{i=0}^{+\infty} \mathcal{P}\{\alpha_{n+1} = k | \alpha_n = i\} \mathcal{P}\{\alpha_n = i\}.$$

The generating function of the $(n+1)$th generation is given by

$$\begin{aligned} F_{n+1}(u) &= \sum_{k=0}^{+\infty} \mathcal{P}\{\alpha_{n+1} = k\} u^k \\ &= \sum_{k=0}^{+\infty} \sum_{i=0}^{+\infty} \mathcal{P}\{\alpha_{n+1} = k | \alpha_n = i] \mathcal{P}\{\alpha_n = i\} u^k \\ &= \sum_{i=0}^{+\infty} \mathcal{P}\{\alpha_n = i\} [b(u)]^i = F_n[b(u)]. \end{aligned} \tag{5.4}$$

This is the generating function of the $(n+1)$th generation for Galton-Watson-Fisher process.

Let us now differentiate this recursion equation (5.3) with respect to u to get

$$F_n'(u) = F_{n-1}'[b(u)]b'(u) = \ldots = F_0'(u)[b'(u)]^n. \tag{5.5}$$

In particular,

$$F_n'(1) = F_0'(1)[b'(1)]^n. \tag{5.6}$$

Note that

$$b'(1) = \sum_{j=0}^{+\infty} jp_j$$

gives the *expected number of offspring of A genes* and $F_n'(1)$ gives the *expected number* of A genes *in the nth generation.* If $b'(1) < 1$, then the expected number of A genes approaches zero; if $b'(1) = 1$, this number remains constant; and if $b' > 1$, the expected number of A genes grows exponentially.

Suppose that a particular locus has only one gene, a. When a single mutation $a \to A$ occurs, the population consists of all aa's except for one mutant having the Aa genotype. If the fitness of the mutant is s (that is, the expected number of progeny from Aa genotypes in the next generation is s), then

$$b'(1) = \sum_{i=0}^{+\infty} ip_i = s.$$

Thus, $b'(1)$ gives the mutant's fitness.

If a mutation $(a \to A)$ occurs at time $n = 0$, then $\alpha_0 = 1$. Let ε be the *probability of eventual extinction* of the A gene. If the mutant leave j offspring, then the probability of eventual extinction of all j progeny is εj. Therefore, if

$$\varepsilon = \mathcal{P}\{\text{actual extinction}\},$$

then

$$\varepsilon = \sum_{j=0}^{+\infty} \mathcal{P}\{j \text{ offspring}\}\mathcal{P}\{\text{all } j \text{become extinct}\} = \sum_{j=0}^{\infty} p_j \cdot \varepsilon^j = b(\varepsilon),$$

and the equation

$$\varepsilon = b(\varepsilon) \tag{5.7}$$

determines *the extinction probability.*

The function $b(u)$ and its derivatives are non-negative, so in particular, the function is convex. If $b'(u) \leq 1$, there is only one static state in the unit interval: $u = 1$. In this case $\varepsilon = 1$, so extinction is certain. If $b'(1) > 1$, there are two static states: $u = \varepsilon < 1$ and $u = 1$. If $b(1) \leq 1$, then from (5.6) we obtain

$$F_n(u) = b^n(u) > 1,$$

whereas if $b'(1) > 1, F_n(u) \to \varepsilon$ as $n \to +\infty$. Note that in any case

$$F_1(0) = \mathcal{P}\{\alpha_1 = 0\} = \mathcal{P}\{\text{extinction by first generation}\},$$

$$\begin{aligned} F_2(0) &= \textstyle\sum_{j=0}^{\infty} p_j(b(0))^j = \sum_{j=0}^{+\infty}\sum_{i=0}^{+\infty} \mathcal{P}\{\alpha_2 = j|\alpha_n = i\}\mathcal{P}\{\alpha_1 = i\}u^j|_{u=0} \\ &= \mathcal{P}\{\text{extinction by second generation}\}. \end{aligned}$$

In general,

$$F_n(0) = \mathcal{P}\{\text{extinction by thenth generation}\}.$$

If u_n is defined by $u_{n+1} = b(u_n), u_0 = 0$, we see that $u_n = F_n(0)$. As before, $u_n \to \varepsilon$.

Next, we consider *the probability of fixation* $\phi := 1 - \varepsilon$. Here ϕ satisfies

$$\begin{aligned} \phi &= 1 - b(1-\phi) = 1 - b(1) + f^1(1)\phi - b''(1)\phi^2/2 + 0(\phi^3) \\ &\sim s\phi + (\sigma^2 + s^2)\phi^2/2, \end{aligned}$$

where σ^2 is the variance of the distribution of offspring genes p_j. If $\phi \ll 1$, we get

$$\phi \sim 2 \cdot (s-1)/(\sigma^2 + s^2),$$

provided $s > 1$ (recall that $s \leq 1$ implies that $\phi = 0$).

Suppose then that the offspring have a *Poisson distribution.* If a mutant has fitness s, then $f(u) = e^{s\cdot(u-1)}$. The probability of eventual extinction, ε, is determined from the equation $\varepsilon = e^{s\cdot(\varepsilon-1)}$. Moreover, the probability ϕ that the A gene is eventually fixed in the population if $\phi = 1-\varepsilon$. This is determined by $\phi == 1-e^{-s\cdot\phi}$, so $\phi \sim 2 \cdot (s-1)s^2$ if $s \sim 1$.

In such a way, this recursion equation (5.4) for the generating function of the Galton-Watson process is very useful.

5.1.2. BELLMAN-HARRIS BRANCHING PROCESS.

Branching process $\{\xi_t\}_{t\in R^+}$ with a single type of particles is defined by the homogeneous Markov branching Bellman-Harris process with the intensity $a > 0$ of the exponential distribution of the lifetimes of particles and with the generating function

$$b(u) = \sum_{k=0}^{\infty} p_k u^k, |u| \leq 1, \tag{5.8}$$

of the number of direct descendants of one particle.

The generating function of the branching process

$$\Phi(t,u) = Eu^{\xi_t} = \sum_{k=0}^{\infty} \mathcal{P}\{\xi_t = k|\xi_0 = 1\}u^k \tag{5.9}$$

satisfies the ordinary differential equation:

$$\begin{cases} \frac{d\Phi(t,u)}{dt} = g(\Phi(t,u)), \\ \Phi(0,u) = u, \end{cases} \tag{5.10}$$

Here

$$g(u) := a[b(u) - u]. \tag{5.11}$$

The solution of equation (5.10) is

$$\Phi(t,u) = \psi(t + \int_0^u \frac{dv}{g(v)}), \tag{5.12}$$

where $\psi(t)$ is the inverse function of

$$t = y(u) := \int_0^u \frac{dv}{g(v)}.$$

Let $\alpha := \sum_{k=1}^{\infty} k \cdot p_k = b'(1)$, where $b(u)$ is defined in (5.1). As $\alpha \leq 1$, then extinction probability q is equal to 1; if $\alpha > 0$ then $q < 1$ and the value q is given by solving the equation $b(u) = u$.

In the Galton-Watson process, the lifetime of each particle was one unit of time. A natural generalization is to allow these lifetime to be random variables. Instead of the discrete time Markov chain $(\alpha_n)_{n \in Z_+}$, as in the case of the Galton-Watson-Fisher model, we consider the process $(\xi(t))_{t \in R^+}$, where $\xi(t)$ = number of particles at time t, and so, we have Bellman-Harris branching process.

The equation (5.10) is a *backward equation.* The corresponding *forward equation* for $p(t,u)$ is

$$\begin{cases} \frac{d\Phi(t,u)}{dt} &= g(u)\frac{d}{du}\Phi(t,u), \\ \Phi(0,u) &= u. \end{cases} \tag{5.13}$$

An interesting question then arises: when it is possible for the process to produce infinitely many particles in a finite time (by having infinitely many transitions in $(0,t)$, i.e., to explode). We suppose here the *non-explosion hypothesis*: for every $\varepsilon > 0$

$$\int_{1-\varepsilon}^{1} \frac{du}{b(u) - u} + \infty. \tag{5.14}$$

This condition (5.14) is necessary and sufficient for $\mathcal{P}\{\xi(t) < +\infty\} \equiv 1$, where $\xi(t)$ is the Bellman-Harris process [29,81].

We note that from Chapman-Kolmogorov equation it follows

$$\Phi(t+s,u) = \Phi(\Phi(t,u),s), |u| \leq 1, \qquad s \geq 0, t \geq 0. \tag{5.15}$$

This is the analog of the functional iteration formula (5.4) for $F_n(u)$ in the Galton-Watson process.

Let us study the *extinction probability* for Bellman-Harris process. Let

$$B := \{w : \xi(t,w) \to 0 \quad as\ t \to +\infty\}.$$

The set B is called *the extinction set*, and its probability the *extinction probability*. We note that

$$q(t) \equiv \mathcal{P}\{\xi(t) = 0 | \xi(0) = 1\} = \Phi(t,0)$$

is a non-decreasing function of t. Furthermore, from the backward equation (5.10) we see that

$$\frac{dq(t)}{dt} = g(q(t)). \tag{5.16}$$

Integrating (5.16), and using $q(0) = 0$, we get

$$q(t) = \int_0^t g(q(y))dy. \tag{5.17}$$

Let q^* be the smallest root in [0,1] of the equation $b(u) = u$, or equivalently $g(u) = 0$. Since $q(t)$ is non-decreasing and (5.16) holds, we have $g(q(t)) \geq 0$, and since $g(0) \geq 0$ we have $q(t) \leq q^*$ for all $t \geq 0$. But $q := \lim_{t\to+\infty} q(t)$. Thus $q \leq q^*$.

Suppose $g(q) > 0$. Then by the continuity of $g(y)$, we have

$$\lim_{t\to+\infty} \int_0^t g(q(y))dy = +\infty, \tag{5.18}$$

while $g(t)$ lies in [0,1]. This contradicts (58). Hence, we must have $g(q) = 0$. Therefore, by definition of q^*, we obtain $q = q^*$. Summarizing the discussions, we conclude that the *extinction probability* q is the smallest root in [0,1] of the equation $g(u) = 0$.

5.2. Branching Models in Random Media.

Here, we are mainly interested in *Bellman-Harris branching processes* (introduced in section 5.1) in random environment.

Although, our methods fit to other models of branching process, for example, Galton-Watson, but we cannot concern it here. Discrete-time and continuous-time branching processes are studied in Markov and semi-Markov random environments.

5.2.1. BELLMAN-HARRIS BRANCHING PROCESS IN MARKOV RANDOM MEDIA.

Let $a(y)$, $y\epsilon Y$ intensities of lifetimes of particles and $p_k(y), y\epsilon Y$ the probability distributions of the number of direct decendants, be a bounded and measurable function on $Y, k \geq 0$.

Also, define

$$b(u,y) := \sum_{k=0}^{+\infty} p_k(y)u^k, |u| \leq 1, \tag{5.19}$$

and

$$g(u,y) := a(x)[b(u,y) - u]. \tag{5.20}$$

Let's define the *Bellman-Harris branching process in Markov random environment* by the difference equation for its generating function $\Phi_n(a)$:

$$\Phi_{n+1}(u) - \Phi_n(u) = g(\Phi_n(u), y_{n+1}), \tag{5.21}$$

where $(y_n)_{n\in Z^+}$ is a homogeneous Markov chain.

It means that $\Phi_n(u)$ is a random process.

5.2.2. BELLMAN-HARRIS BRANCHING PROCESS IN MARKOV RENEWAL RM.

This process is defined by the difference equation for its generating function:

$$\Phi_{\nu(t)+1}(u) - \Phi_{\nu(t)}(u) = g(\Phi_{\nu(t)}(u), y_{\nu(t)+1}), \tag{5.22}$$

where $\nu(t) := \max\{n : \tau_n \le t\}$.

5.2.3. BELLMAN-HARRIS BRANCHING PROCESS IN SEMI-MARKOV RANDOM MEDIA.

This process is defined by its generating function as follows:

$$\begin{cases} \frac{d\Phi(t,u)}{dt} &= g(\Phi(t,u), y(t)) \\ \Phi(0,u) &= u, \end{cases} \tag{5.23}$$

where $y(t) := y_{\nu(t)}$ is a semi-Markov process, and function $g(u, y)$ is defined in (5.20). Hence, function $\Phi(t, u)$ in (5.23) is a random process.

Measurability of Branching Processes in Random Media. *Bellman-Harris branching process in Markov random environment* in (5.21) is $\mathcal{F}_{n+1}/\mathcal{R}$-measurable, as $a(y_{n+1})$ and $p_k(y_{n+1})$ are $\mathcal{F}^Y_{n+1}/\mathcal{R}$- measurable, hence function $g(u, y_{n+1})$ in (5.19)-(5.20) is $\mathcal{F}_{n+1}/\mathcal{R}$-measurable.

Bellman-Harris branching process in Markov renewal random environment in (5.22) is $\mathcal{G}_t/\mathcal{R}$-measurable, and the same process in semi-Markov random environment in (5.22) is $\mathcal{H}_t/\mathcal{R}$-measurable.

5.3. Averaging of Branching Models in Random Media.

Let us consider Bellman-Harris homogeneous Markov proces (see subsection 5.1, Chapter 5) in random environment (see subsection 5.2, Chapter 2) with single type of particles.

This process is given by the generating function $Q^\varepsilon_{\nu(t/\varepsilon)}$ which satisfies the following difference equation:

$$Q^\varepsilon_{\nu(t/\varepsilon)+1} - Q^\varepsilon_{\nu(t/\varepsilon)} = \varepsilon \cdot g(Q^\varepsilon_{\nu(t/\varepsilon)}, y_{\nu(t/\varepsilon)+1}), \quad Q^\varepsilon_0 = u. \tag{5.24}$$

where $\nu(t) := \max\{n : \tau_n \le t\}$ is a counting process, $y_{\nu(t)}$ is a semi-Markov process, $(y_n)_{n\in Z_+}$ is a Markov chain, and

$$g(u,y) := a(y) \cdot [b(u,y) - u], \tag{5.25}$$

$$b(u,y) := \sum_{k=0}^{+\infty} p_k(y) \cdot u^k, |u| \leq 1. \tag{5.26}$$

Function $g(u,y)$ satisfies all the conditions of Theorem 2 (Chapter 2) and we can apply this result to the process $Q^\varepsilon_{\nu(t/\varepsilon)}$.

Then process $Q^\varepsilon_{\nu(t/\varepsilon)}$ converges weakly as $\varepsilon \to 0$ to the process $\tilde{Q}_t$ such that:

$$\frac{d\tilde{Q}_t}{dt} = \tilde{g}(\tilde{Q}_t), \tilde{Q}_0 = u, \tag{5.27}$$

$$\tilde{g}(u) := \int_Y p(dy) g(u,y)/m.$$

Let's consider the following difference equation for generating function $Q^\varepsilon_{\nu[t/\varepsilon]}$:

$$Q^\varepsilon_{\nu[t/\varepsilon]+1} - Q^\varepsilon_{\nu[t/\varepsilon]} = \varepsilon g(Q^\varepsilon_{\nu[t/\varepsilon]}, y_{\nu[t/\varepsilon]+1}), Q^\varepsilon_0 = u. \tag{5.28}$$

Then from Theorem 1 it follows (see Chapter 2, subsection 2.2) that $Q^\varepsilon_{[t/\varepsilon]}$ converges weakly as $\varepsilon \to 0$ to the process $\hat{Q}_t$ such that:

$$\frac{d\hat{Q}_t}{dt} = \hat{g}(\hat{Q}_t), \quad \hat{Q}_0 = u, \tag{5.29}$$

where

$$\hat{g}(u) := \int_Y p(dy) g(u,y). \tag{5.30}$$

Equation (5.27) differs on equation (5.29) only by factor $1/m$, where

$$m = \int_Y m(y) p(y), m(y) := \int_0^\infty t \; G_y(dt).$$

It means that equation (5.27) is averaged by mean value of jumps up to the moment of time t.

Let us consider continuous-time Bellman-Harris process in semi-Markov random environment:

$$\begin{cases} \frac{dQ^\varepsilon(t)}{dt} &= g(Q^\varepsilon(t), y(t/\varepsilon)) \\ Q^\varepsilon(0) &= u. \end{cases} \tag{5.31}$$

From the theory of random evolution (see Chapter 2, subsection 2.1) it follows that $Q^\varepsilon(t)$ converges weakly as $\varepsilon \to 0$ to the process $\bar{Q}(t)$ such that:

$$\begin{cases} \frac{d\bar{Q}(t)}{dt} &= \bar{g}(\bar{Q}(t)) \\ \bar{Q}(0) &= u, \end{cases} \tag{5.32}$$

where

$$\bar{g}(u) := \int_Y p(dy)g(u,y)m(y)/m,$$

$$m(y) := \int_0^{\infty} tG_y(dt), m := \int_Y p(dy)m(y).$$

Here, $m(y)$ is a lifetime in state y of the continuous-time Bellman-Harris branching process, and m is a mean stationary lifetime in the states of this model.

5.4. Merging of Branching Model in Random Media.

Let us consider the following difference equation for generating function of Bellman-Harris branching process in perturbed semi-Markov random media $y^{\varepsilon}(\nu(t/\varepsilon))$:

$$\begin{cases} Q^{\varepsilon}(\nu(t/\varepsilon)+1) - Q^{\varepsilon}(\nu(t/\varepsilon)) &= \varepsilon \cdot g(Q^{\varepsilon}(\nu(t/\varepsilon)), y^{\varepsilon}(\nu(t/\varepsilon)+1)) \\ Q^{\varepsilon}(0) &= u, \end{cases} \tag{5.33}$$

Let the merging conditions be satisfied (see subsection 1.9.4, Chapter 1 and subsection 2.5, Chapter 2). Then, applying Theorem 6 (subsection 2.5, Chapter 2), we obtain that process $Q^{\varepsilon}(\nu(t/\varepsilon))$ converges weakly as $\varepsilon \to 0$ to the process $\tilde{Q}(t)$, which satisfies the equation:

$$\begin{cases} \frac{d\tilde{Q}(t)}{dt} &= \tilde{g}(\tilde{Q}(t), \hat{y}(t)) \\ \tilde{Q}(0) &= u, \end{cases} \tag{5.34}$$

where $\tilde{g}(u,v) := \int_{Y_v} p_v(dy)g(u,y)/m_v$. We note, that equation (5.34) looks like the equation for generating function of Bellman-Harris branching process in semi-Markov random media (see subsection 5.2, Chapter 5), but in place of semi-Markov process in (5.34) we have merged Markov process $\hat{y}(t)$.

5.5. Diffusion Approximation of Branching Process in Random Media.

Let us consider Bellman-Harris homogeneous Markov process in RM (Chapter 5, subsection 5.2) which describes by the generating function $Q^{\varepsilon}_{\nu(t/\varepsilon^2)}$ in series scheme t/ε^2, satisfying the following difference equation:

$$Q^{\varepsilon}_{\nu(t/\varepsilon^2)+1} - Q^{\varepsilon}_{\nu(t/\varepsilon^2)} = \varepsilon g(Q^{\varepsilon}_{\nu(t/\varepsilon^2)}, y^{\varepsilon}_{\nu(t/\varepsilon^2)+1}), \tag{5.35}$$

where $\nu(t) := \max\{u : \tau_u \le t\}$ is a counting process, $y_{\nu(t)}$ is a semi-Markov process,

$$g(u,y) := a(y)[b(u,y) - u],$$

$$b(u,y) := \sum_{k=0}^{+\infty} p_K(y)u^k, \quad |u| \le 1. \tag{5.36}$$

We suppose here that *balance condition* for branching process is fulfilled:

$$\tilde{g}(u) := \int_Y p(dy)g(u,y) = 0, \quad \forall |u| \le 1, \tag{5.37}$$

where $g(u,y)$ is defined in (5.36).

Under conditions of Theorem 4 (Chapter 2, subsection 2.2) for function $g(u,y)$ in place of function $g(x,y)$ and with *balance condition* (5.37) we obtain that process $Q^{\varepsilon}_{\nu(t/\varepsilon^2)}$ in (5.35) converges weakly as $\varepsilon \to 0$ to the diffusion process $\tilde{Q}(t)$ which satisfies the following stochastic differential equation:

$$d\tilde{Q}(t) = \tilde{\alpha}(\tilde{Q}(t))dt + \tilde{\beta}(\tilde{Q}(t))dw(t), \tag{5.38}$$

where

$$\tilde{\alpha}(u) := \int_y p(dy)[g(u,y)\mathbf{R}_0 g_u(u,y)]/m, \tag{5.39}$$

$$\tilde{\beta}^2(y) := 2\int_Y p(dy)[g(u,y)\mathbf{R}_0 g(u,y) + 1/2g^2(u,y)]/m,$$

$w(t)$ is a standard Wiener process, $\mathbf{R}_0$ is a potential of Markov chain $(y_n)_{n\in Z_+}$, $g_u(u,y)$ is a derivative of g by u.

Let us consider the following branching model (see subsection 5.2, Chapter 5):

$$\begin{cases} \frac{dQ^{\varepsilon}(t)}{dt} & = & \frac{1}{\varepsilon}g(Q^{\varepsilon}(t), y(t)\varepsilon^2), \\ Q^{\varepsilon}(0) & = & u, \end{cases}$$

where $y_{\nu(t)}$ is a semi-Markov process, $|u| \le 1$, and

$$g(u,y) := a(y)[b(u,y) - u].$$

Then, using the result on diffusion approximation of RE with application to branching model, we obtain, that $Q^{\varepsilon}(t)$ converges weakly as $\varepsilon \to 0$, to the diffusion process $\hat{Q}(t)$ such that:

$$d\hat{Q}(t) = \hat{\alpha}(\hat{Q}(t))dt + \hat{\beta}(\hat{Q}(t))dw(t),$$

where

$$\begin{aligned} \hat{\alpha}(u) &:= \textstyle\int_Y p(dy)[m(y)g(u,y)\mathbf{R}_0 m(y)g_u(u,y)]/m, \\ \hat{\beta}^2(u) &:= 2\textstyle\int_Y p(dy)[m(y)g(u,y)\mathbf{R}_0 m(y)g(u,y) \\ &+ 1/2m_2(y)g^2(u,y)]/m, \end{aligned}$$

where $m(y) := \int_0^\infty tG_y(dt)$.

5.6. Normal Deviations of Branching Process in Random Media.

Let us consider the eqution for generating function of Bellman-Harris homogeneous Markov process (see subsection 5.3, Chapter 5) in random media:

$$\begin{cases} Q^\varepsilon_{n+1} - Q^\varepsilon_n &= \varepsilon \cdot g(Q^\varepsilon_n, y_{n+1}) \\ Q^\varepsilon_0 &= u, \end{cases} \tag{5.40}$$

where

$$g(u,y) := a(y) \cdot [b(u,y) - u], \tag{5.41}$$

$$b(u,y) := \sum_{k=0}^{+\infty} p_k(y) \cdot u^k, \quad |u| \le 1.$$

Let the sequence $\tilde{Q}^\varepsilon_u$ is defined by the following relation:

$$\tilde{Q}^\varepsilon_{n+1} - \tilde{Q}^\varepsilon_u = \varepsilon \cdot \tilde{g}(Q^\varepsilon_u), \tag{5.42}$$

where $\tilde{g}(u)$ is defined as follows:

$$\tilde{g}(u) := \int_Y p(dy) g(u,y)/m. \tag{5.43}$$

Let us define the deviation process

$$Q^\varepsilon(t) := \sum_{n=0}^{+\infty} Q^\varepsilon(n) \mathbf{1}\{\tau_n \le t/\varepsilon < \tau_{n+1}\}, \tag{5.44}$$

where

$$Q^\varepsilon(n) := [Q^\varepsilon_n - \tilde{Q}^\varepsilon_n]/\sqrt{\varepsilon}, \tag{5.45}$$

and $Q^\varepsilon_n, \tilde{Q}^\varepsilon_n$ are defined in (5.40) and (5.42), respectively.

Applying the normal deviation theorem from subsection 2.4 (Chapter 2) to the process $Q^\varepsilon(t)$ in (5.44) we obtain that $Q^\varepsilon(t)$ converges weakly as $\varepsilon \to 0$ to the process $\tilde{Q}(t)$ such that:

$$\tilde{Q}(t) = \int_0^t \tilde{g}^1_u(\tilde{Q}(s))\tilde{Q}(s)ds + \int_o^t \sigma(\tilde{Q}(s))dw(s),$$

where

$$\sigma^2(y) := \int_Y p(dy)[(g(u,y) - \tilde{g}(u))\mathbf{R}_0(g(u,y) - \tilde{g}(u)) +$$
$$+(g(u,y) - \tilde{g}(u))^2/2]/m.$$

5.7. Stochastic Stability of Branching Model in Averaging and Diffusion Approximation Schemes.

5.7.1. STABILITY OF BRANCHING MODEL IN AVERAGING SCHEME.

We would like to study here the stability of Bellman-Harris process $\xi_{t/\varepsilon}$, namely, the stability of zero state of $\xi_{t/\varepsilon}$.

It means, that we are interested in the behavior $\xi_{t/\varepsilon}$ as $\varepsilon \to +0$.

Since

$$\Phi_{\nu(t/\varepsilon)} := Eu^{\xi_{\nu(t/\varepsilon}}, \tag{5.46}$$

then, to study $\xi_{\nu(t/\varepsilon} \to_{\varepsilon\to 0} 0$ is equivalent to study

$$\Phi_{\nu(t/\varepsilon)} \to_{\varepsilon\to+0} 1, \quad \forall t \epsilon R^+.$$

It is the same as to study

$$\ln \Phi_{\nu(t/\varepsilon)} \to_{\varepsilon\to+0} 0.$$

In this way, we will consider the equation for $\ln \Phi_{\nu(t/\varepsilon)}$ rather than for $\Phi_{\nu(t/\varepsilon)}$.

Let us consider the difference equation for logarithm of generating function of Bellman-Harris branching process $\xi_{\nu(t/\varepsilon)}$:

$$\ln \Phi^{\varepsilon}_{\nu(t/\varepsilon)+1} - \ln \Phi^{\varepsilon}_0 = \varepsilon \sum_{k=0}^{\nu(t/\varepsilon)} g(\Phi^{\varepsilon}_k, y_{k+1}), \tag{5.47}$$

that follows from the equation (5.24), subsection 5.3, for $\Phi^{\varepsilon}_{\nu(t/\varepsilon)}$, where

$$g(u,y) := a(y)[b(u,y) - u],$$

$$b(u,y) := \sum_{k=0}^{+\infty} p_k(y)u^k, \quad |u| \le 1, \tag{5.48}$$

$a(y)$ is bounded and measurable function.

Since $g(1,y) = 0$ and $\ln \Phi|_{Q=1} = 0$, then the state 1 is stable for equation (5.47).

We note that by renewal theorem:

$$\ln(\Phi^{\varepsilon}_{\nu(t/\varepsilon)+1}/\Phi^{\varepsilon}_0) = \varepsilon \sum_{k=0}^{\nu(t/\varepsilon)} g(\Phi^{\varepsilon}_k, y_{k+1}) \longrightarrow$$

$$\to_{\varepsilon\to 0} t \cdot \int_Y p(dy) g(u,y)/m, \tag{5.49}$$

where

$$\Phi^{\varepsilon}_0 = u.$$

For $\hat{g}(u) := \int_Y p(dy)g(u,y)$ to be zero it needs $\hat{g}(u) = 0$ or

$$\int_Y p(dy)a(y)b(u,y) = \hat{a}u,$$

$\forall y \epsilon Y$, or $\tilde{b}(u) = u$, where

$$\tilde{b} := \int p(dy)a(y)b(u,y),$$

$$\tilde{a} := \int_Y p(dy)a(y).$$

We note that $b(0,y) = p_0(y)$ and $b(1,y) = 1, \forall y \epsilon Y$, that follows from definition of $b(u,y)$ in (5.48). Moreover, for each fixed $y \epsilon Y$ function $b(u,y)$ is strictly convex and increasing in $[0,1]$; for first moment of ξ_t we have:

$$E(\xi_1 | y(0) = y) = \sum p_k(y) k = b_u(1,y) := m(y), \tag{5.50}$$

and:

1) if $m(y) \leq 1$, then $b(u,y) > u$ for $t \epsilon [0,1]$;
2) if $m(y) > 1$, then $b(u,y) = u$ has a unique root in $[0,1]$.

Let $\hat{g}$ be the smallest root of $\hat{b}(u) = u$ for $u \epsilon [0,1]$. Then from above it follows that if $m(y) \leq 1$, then $g(y) = 1$, and if $m(y) > 1$, then $\hat{g} < 1$. In this way, we have the following result, using the reasonings (5.46)-(5.50): if $b(u) = u$, then

$$\ln \Phi^{\varepsilon}_{\nu(t/\varepsilon} / \Phi^{\varepsilon}_0 \to_{\varepsilon \to +\infty} 0,$$

and

$$\Phi^{\varepsilon}_{\nu(t/\varepsilon} \to_{\varepsilon \to 0} 1, \quad \forall t \epsilon R^+.$$

It means that

$$\xi_{\nu(t/\varepsilon)} \to_{\varepsilon \to 0} 0, \quad \forall t \epsilon R^+.$$

Let $A := \{w : \xi_t(w) \to +\infty, t \to +\infty\}$ and $B := \{w : \xi_t(w) \to 0, t \to +\infty\}$, then $Q(A \bigcup B) = 1$.

The set B is called the *extinction set*, and its probability the *extinction probability*. From the basic property of branching process it is clear that

$$\mathcal{P}(B | \xi_0(w) = k, y(0) = y) = [\mathcal{P}(B | \xi_0(w) = 1, y(0) = y]^k, \tag{5.51}$$

and hence it suffices to find

$$g(y) := \mathcal{P}(B | \xi_0(w) = 1, y(0) = y).$$

From the definition of $\Phi(t,w) := E u^{\xi_t}$ we note that

$$q(t,y) := \mathcal{P}(\xi_t(w) = 0 | \xi_0(w) = 1, y(0) = y) = \Phi(t,0) = P_{10}(t), \tag{5.52}$$

is a nondecreasing function of t.

From the equation for $\Phi(t,u)$ we see that $q(t,y)$ is differentiable in t and satisfies

$$\frac{dq(t,y)}{dt} = g(q(t,y),y), q(0,y) = 0, \tag{5.53}$$

or

$$q(t,y) = \int_0^t g(q(s,y))ds.$$

Let $q^*(y)$ be the smallest root in $[0,1]$ of the equation $b(u,y) = u$, or equivalently, $g(u,y) = 0$. Since $q(t,y)$ is nondecreasing and (5.53) holds, we have $g(g,(t),y) \geq 0, \forall y \epsilon Y$, and since $g(0,y) \leq 0$ it follows that $q(t,y) \leq q^*(y)$ for all $t \geq 0$ and $y \epsilon Y$. But $q(y) = \lim_{t \to +\infty} q(t,(y)$. Thus $q(y) \leq q^*(y), \forall y \epsilon Y$.

Now suppose that $g(q, y) > 0$. Then by continuity of g we have

$$\lim_{t\to+\infty} \int_0^t g(q(s, y), y)ds = +\infty, \text{ while } q(t, y) \in [0, 1]. \qquad (5.54)$$

This contradicts (5.53). Hence, we must have $g(q, y) = 0$, and by definition of $q^*(y)$, get $q(y) = q^*(y)$. In this way, we have the following result.

Theorem 1. The extinction probability $q(y)$ is the smallest root in $[0, 1]$ of the equation $g(u, y) = 0$ or $b(u, y) = u$.

Let us consider the process $\xi_{\nu(t/\varepsilon)}$ and set

$$A^\varepsilon := \{w : \xi_{\nu(t/\varepsilon)} \to +\infty, t \to +\infty, 0 \le \varepsilon \le \varepsilon_0\},$$

$$B^\varepsilon := \{w : \xi_{\nu(t/\varepsilon)} \to +0, t \to +\infty, 0 \le \varepsilon \le \varepsilon_0\},$$

then

$$\mathcal{P}\{A^\varepsilon \bigcap B^\varepsilon\} = 0, \text{ for } 0 \le \varepsilon \le \varepsilon_0.$$

The set B^ε is called the *extinction set* for branching process $\xi_{\nu(t/\varepsilon)}$, and its probability the *extinction probability*. From (5.51) we obtain:

$$\mathcal{P}\{B^\varepsilon|\xi_0(w) = k, y(0) = y\} = [\mathcal{P}\{B^\varepsilon|\xi_0(w) = 1, y(0) = y]^k,$$

and hence it sufficies to find

$$q^\varepsilon(y) := \mathcal{P}\{B^\varepsilon|\xi_0(w) = 1, y(0) = y\}.$$

From the definition of $\Phi_{\nu(t/\varepsilon)}(u) := Eu^{\xi_{\nu(t/\varepsilon)}}$ we note that

$$q_{\nu(t/\varepsilon)}(y) := \mathcal{P}\{\xi_{\nu(t/\varepsilon)} = 0/\xi_0(w) = 1, y(0) = y\} =$$

$$= \Phi_{\nu(t/\varepsilon)}(0) = P_{10}(\nu(t/\varepsilon)),$$

is a nondecreasing function of t.

From the difference equation for $\Phi_{\nu(t/\varepsilon)}(u)$ (see subsection 5.3) we have:

$$q_{\nu(t/\varepsilon)+1}(y) - q_{\nu(t/\varepsilon)}(y) = g(q_{\nu(t/\varepsilon)}(y), y_{\nu(t/\varepsilon)+1}), \quad q(0, y) = 0. \qquad (5.55)$$

From averaging theorem for branching process we obtain that $q_{\nu(t/\varepsilon)}(y)$ converges weakly as $\varepsilon \to 0$ to the process $\hat{q}(t)$:

$$\frac{d\hat{q}(t)}{dt} = \hat{g}(\hat{q}(t)), \hat{q}(0) = 0, \qquad (5.56)$$

where

$$\hat{g}(u) := \int_Y p(dy)a(y)[b(u, y) - u]/m.$$

The rate of convergence has the following form [72]:

$$|E[q_{\nu(t/\varepsilon)}(y) - \hat{q}(t)]| \le \varepsilon \cdot C_T, \quad \forall t \epsilon [0, T].$$

Let $\hat{q}^*$ be the smallest root in $[0,1]$ of the equation $\tilde{b}(u) = u$, where

$$\tilde{b}(u) := \int_Y p(dy)a(y)b(u,y)/\tilde{a}, \tilde{a} := \int_Y p(dy)a(y).$$

Since $q_{\nu(t/\varepsilon)}(y)$ in nondecreasing and (5.53) holds, we have $\hat{g}(\hat{g}(t)) \geq 0$ and since $\hat{g}(0) \geq 0$ it follows that $\hat{g}(t) \leq \hat{q}^*$ for all $t \geq 0$.

Now suppose that $\hat{g}(\hat{g}(t)) > 0$. Then by continuity of $\hat{g}$ we have from (5.53)

$$\lim_{t\to+\infty} \int_0^t \hat{g}(q(s)ds = +\infty,$$

while $\hat{q}(t)\epsilon[0,1]$. This contradicts (5.53) that

$$\hat{q}(t) = \int_0^t \hat{g}(\hat{q}(s))ds.$$

Hence, we must have $\hat{g}(q) = 0$ and by definition of $\hat{q}^*$ get $\hat{q} = \hat{q}^*$.

In this way, we have the following result.

Theorem 2. The extinction probability $\hat{q}$ is the smallest root in $[0,1]$ of the equation $\hat{g}(u) = 0$ or $\tilde{b}(u) = u$.

Remark. We can not apply directly the stability theorem in averaging scheme (see subsection 4.7), since the condition $g(0,y) = 0$ fails:

$$g(0,y) = a(y)[b(0,y) - 0] =$$
$$= a(y)p_0 > 0,$$

and state 0 is not the solution of the initial equation.

5.7.2. STABILITY OF BRANCHING MODEL IN DIFFUSION APPROXIMATION SCHEME.

Let us consider the following branching model in scale of time t/ε^2 :

$$\begin{cases} \frac{d\Phi^\varepsilon(t)}{dt} &= \frac{1}{\varepsilon} g(\Phi^\varepsilon(t), y(t/\varepsilon^2)) \\ \Phi^\varepsilon(t) &= \Phi^\varepsilon(t/\epsilon^2), \\ \Phi^\varepsilon(0) &= u. \end{cases} \tag{5.57}$$

Let us study asymptotic behavior of extinction probability $P_{10}^\varepsilon(t)$ for time interval $(0, t/\varepsilon^2)$ as $t \to +\infty, \varepsilon\epsilon[0,\varepsilon_0]$.

Let $P_{10}^\varepsilon(t) := q^\varepsilon(t)$ and $p^\varepsilon(t) := 1 - q^\varepsilon(t)$.

Since $P_{10}^\varepsilon(t) = \Phi^\varepsilon(t,0)$ then from the equation (5.57) we obtain that function $p^\varepsilon(t)$ satisfies equation:

$$\begin{cases} \frac{dp^\varepsilon(t)}{dt} &= \frac{1}{\varepsilon} g(1 - p^\varepsilon(t), y(t/\varepsilon^2)) \\ p^\varepsilon(0) &= 1, \end{cases} \tag{5.58}$$

By formula of finite differences we obtain:

$$\frac{dp^\varepsilon(t)}{dt} = -\frac{1}{\varepsilon}g(1, y(t/\varepsilon^2) + \frac{p^\varepsilon(t)}{\varepsilon}g_u(\theta^\varepsilon, y(t/\varepsilon^2)),$$

where $\theta^\varepsilon \equiv \theta_t^\varepsilon \epsilon(P_{10}^\varepsilon(t), 1)$. Since $g(u, y)$ is monotarically increasing function, $\theta_t^\varepsilon \to 1$ as $t \to +\infty$, then

$$g_u(\theta^\varepsilon y) = g_u(1, y) - \gamma^\varepsilon(t), \text{ where } \gamma^\varepsilon(t) > 0$$

and

$$\lim_{t\to+\infty} \gamma^\varepsilon(t) = 0, \forall y \epsilon Y.$$

In this way,

$$\frac{dp^\varepsilon(t)}{dt} = \frac{p^\varepsilon(t)}{\varepsilon} \cdot (\alpha(y(t/\varepsilon^2)) - \gamma^\varepsilon(t)), \tag{5.59}$$

where

$$\alpha(y) := g_u(1, y),$$

and from here we obtain:

$$p^\varepsilon(t) = \exp \{\frac{1}{\varepsilon}\int_0^t \alpha(y(s/\varepsilon^2))ds - \int_0^t \gamma^\varepsilon(s)ds\}. \tag{5.60}$$

We note, that ($\theta^\varepsilon < \theta_\varepsilon^1 < 1$)

$$0 < \gamma^\varepsilon(t) = g_u(1, y) - g_u(\theta^\varepsilon, y) = g_{uu}(\theta_\varepsilon^1, g) \cdot (1 - \theta_\varepsilon^1) \leq$$

$$\leq g_{uu}(1, y)(1 - P_{10}(t/\varepsilon^2)) \leq g_{uu}(1, y(t/\varepsilon^2))e^{\frac{1}{\varepsilon}\int_0^t \alpha(y(s/\varepsilon^2))ds} \tag{5.61}$$

Let's find the limit for $\frac{1}{\varepsilon}\int_0^t \alpha(y)s/(\varepsilon^2)ds$ as $\varepsilon \to 0$. We suppose that

$$g_u(1, y) = 0 \text{ or } b_u(1, y) = 1, \forall y \epsilon Y.$$

then

$$\int_Y p(dy)m(y)a_u(1, y) = 0,$$

and *balance condition* is fulfilled for the integral functional

$$\frac{1}{\varepsilon}\int_0^t \alpha(y(s)\varepsilon^2))ds = \varepsilon \int_0^{t/\varepsilon^2} \alpha(y(s))ds = \tag{5.62}$$

$$= \varepsilon \int_0^{t/\varepsilon^2} g_u(1, y)(s))ds.$$

We can apply the result on diffusion approximation of integral functional in (5.62). We obtain that:

$$\varepsilon \int_0^{t/\varepsilon^2} g_u(1, y(s))ds \to_{\varepsilon\to 0} \sigma \cdot w(t), \tag{5.63}$$

where

$$\sigma^2 := 2\int_Y p(dy)[m(y)g_u(1, y)\mathbf{R}_0 g_u(1, y) + m_2(y)g_u(1, y))^2/2]/m,$$

$m_2(y) := \int_Y t^2 G_y(dt), w(t)$ is a standard Wiener process, $\sigma > 0$. From (5.63) we have:

$$0 < \gamma^\varepsilon(t) \le \hat{g} e^{\sigma w(t)} \text{ with } P = 1,$$

where

$$\hat{a} := \sup a(y), \text{ and } \varepsilon \epsilon [0, \varepsilon_0].$$

Hence,

$$\exp\{-\int_0^t \gamma^\varepsilon(s) ds\} \le 1, \forall t \epsilon R_+.$$

From (5.62) and (5.63) we obtain:

$$p^\varepsilon(t) \le \exp\{\frac{1}{\varepsilon}\int_0^t \alpha(y(s/\varepsilon^2))ds\} \to_{\varepsilon\to 0} e^{\sigma w(t)},$$

and

$$E(p^\varepsilon(t))^m \le e^{\frac{\sigma^2 \cdot tm^2}{2}}.$$

In particular, if $m = 2$, then:

$$E(p^\varepsilon(t))^2 \le e^{2\sigma^2 t}.$$

It means, that extinction probability $p^\varepsilon(t)$ do not go to zero as $t \to +\infty$ and $\varepsilon \in [0, \varepsilon_0]$.

CHAPTER 6: DEMOGRAPHIC MODELS

Census data are usually collected at five-, or-ten-year intervals, and these data give a profile of the population's age structure. There are several methods such as renewal theory and reproduction matrix theory, developed to study age distribution and its effects on population changes. Here, we are mainly interested in Fibonacci's sequence to illustrate the method of reproduction matrix. The models described here are linear and are extensions of the Malthus model [58] to account for age structure.

6.1. Deterministic Demographic Model.

6.1.1. FIBONACCI SEQUENCE.

Fibonacci [28] described a hypothetical rabbit population that starts with a single pair, one male and one female. The pair will reproduce twice, and at each reproduction produce a new pair, one male and one female. These will go on to reproduce twice, and so on.

Let B_n denote the number of newborn females at the nth reproduction time. Then for $n = 2, 3, 4, \ldots$,

$$B_{n+1} = B_n + B_{n-1}, B_0 = 1, B_1 = 1.$$

This is the *Fibonacci renewal equation.* This equation can be solved by setting $B_n = r^n$. It follows that r satisfies $r^2 - r - 1 = 0$, whose roots are $r_1 = \frac{1+\sqrt{5}}{2}$ and $r_2 = \frac{1-\sqrt{5}}{2}$. Since $B_0 = 1$ and $B_1 = 1$, we have that

$$B_n = (\frac{\sqrt{5}+1}{2\sqrt{5}})r_1^n + (\frac{\sqrt{5}-1}{2\sqrt{5})})r_2^n.$$

This formula gives the number of females born at each reproduction time. Note that since $r_2/r_1 < 1$, we have that

$$\lim_{n\to\infty} r_1^{-n} \cdot B_n = \frac{\sqrt{5}+1}{1\sqrt{5}}.$$

It means that

$$B_n \simeq r_1^n \cdot [\frac{\sqrt{5}+1}{2\sqrt{5}}]$$

for large n, and the percent error made in this approximation goes to zero:

$$100 \times \left|\frac{B_n - r_1^n(\frac{\sqrt{5}+1}{2\sqrt{5}})}{B_n}\right| = 100 \times \frac{1}{1 + (\frac{\sqrt{5}+1}{\sqrt{5}-1})(\frac{r_1}{r_2})^n} \to 0,$$

as $n \to +\infty$. Hence, the numbers of reproducing pairs at any reproduction time can be easily determined directly from the sequence B_n.

6.1.2. FIBONACCI REPRODUCTION MATRIX.

Let $v_{0,n}$ be the number of newborn females at the nth reproduction, and $v_{1,n}$ be the number of females of age one at the nth reproduction. Let

$$\vec{v}= \begin{pmatrix} v_{0,n} \\ \vdots \\ v_{1,n} \end{pmatrix}$$

Since

$$v_{0,n+1} = v_{0,n} + v_{1,n}, \qquad v_{1,n+1} = v_{0,n},$$

we can define matrix

$$\mathbf{M} := \begin{pmatrix} 1 & 1 \\ 1 & 0 \end{pmatrix}$$

to simplify the notation and get a vector equation

$$\vec{v}_{n+1}= \mathbf{M}\ \vec{v}_n\ .$$

This is a *Fibonacci reproduction model* and $\mathbf{M}$ is called the *Fibonacci reproduction matrix.* The solution can be found by successive back substitutions of the Fibonacci reproduction model

$$\vec{v}_{n+1}= \mathbf{M}\ \vec{v}_n= \mathbf{M}(\mathbf{M}\ \vec{v}_{n-1}) = \ldots = \mathbf{M}^n \cdot \vec{v}_1,$$

where $\vec{v}_1$is the initial distribution vector

$$\vec{v}_1= \begin{pmatrix} 1 \\ 1 \end{pmatrix}.$$

The eigenvalues of $\mathbf{M}$ play an important role in studying $\mathbf{M}^n$. These are defined as the solutions of the algebraic equation:

$$det(\mathbf{M} - \lambda\ I) = 0,$$

where det is the determinant and $I = \begin{pmatrix} 1 & 0 \\ 0 & 1 \end{pmatrix}$. We obtain $\lambda^2 - \lambda - 1 = 0$, and hence,

$$\lambda_1 = r_1 = \frac{1+\sqrt{5}}{2}, \qquad \lambda_2 = r_2 = \frac{1-\sqrt{5}}{2}.$$

The matrix $\mathbf{M}$ can be written in the form of its spectral decomposition

$$\mathbf{M} = \lambda_1 P_1 + \lambda_2 P_2,$$

where

$$P_1 = \frac{1}{\lambda_2 + 2}\begin{pmatrix} \lambda_1^2 \, \lambda_1 \\ \lambda_1 \, 1 \end{pmatrix} \qquad P_2 = \frac{1}{\lambda_2 + 2}\begin{pmatrix} \lambda_2^2 \, \lambda_2 \\ \lambda_2 \, 1 \end{pmatrix}.$$

It follows easily that

$$P_1 P_1 = P_1, \qquad P_2 P_2 = P_2, \qquad P_1 P_2 = \begin{pmatrix} 0 \, 0 \\ 0 \, 0 \end{pmatrix}.$$

Matrices P_1 and P_2 satisfying these relations are called *projection matrices.* Note that $M^2 = \lambda_1^2 P_1 + p_2^2 P_2$. For any integer n, we have

$$\mathbf{M}^n = \lambda_1^n \cdot P_1 + \lambda_2^n \cdot P_2.$$

The age distribution vector is given by

$$\vec{v}_{n+1} = \mathbf{M}^n \cdot \vec{v}_n = \lambda_1^n \cdot P_1 \, \vec{V}_1 + \lambda_2^n \cdot P_2 \cdot \vec{v}_1,$$

which gives an easily calculated formula for $\vec{v}_n$. We then obtain:

$$\lambda_1^{-n} \cdot \vec{v}_{n+1} = P_1 \, \vec{v}_1 + \frac{\lambda_2}{\lambda_1}^n P_2 \, \vec{v}_2 \, .$$

Since $\lambda_2/\lambda_1 < 1$, we get

$$\lim_{n \to \infty} \lambda_1^{-n} \cdot \vec{v}_{n+1} = P_1 \, \vec{v}_1,$$

and

$$P_1 \cdot \vec{v}_1 = \frac{1}{\lambda_2 + 2}\begin{pmatrix} 2\lambda_1 + 1 \\ \lambda_2 + 1 \end{pmatrix}.$$

The above vector $P_1 \, \vec{v}_1$ is called the *stable age distribution* of the population.

For large n, we have

$$\lambda_1^n \cdot P_1 \, \vec{v}_1 \simeq \vec{v}_{n+1},$$

where the percent error in this approximation is negligible. Therefore, the ratio

$$v_{0,n}/v_{1,n} = (2\lambda_1 + 1)(\lambda_1 + 1) = 1.618$$

which is independent of n. Thus, the ratio of the age classes remains constant even though the numbers in these classes grow geometrically.

In Fibonacci case, since $\lambda_1 = \frac{1+\sqrt{5}}{2}$ we have

$$(2\lambda_2 + 1)/(\lambda_1 + 1) \simeq 1.618.$$

Therefore, the Fibonacci sequence, as a sequence of integers, is approximately a geometric sequence with common ratio 1.618. From the renewal equation, we know that $B_{n+1}/B_n \simeq r_1$.

As a final remark of this subsection, we note that *Renewal Theory* is also used to analyze human population data, described in terms of the birth rate or in terms of the age distribution vector. This analysis was made by Feller [14] in general cases.

We will not get into details, but refer to [8, 14] for same discussions of the renewal theory.

6.1.3. LESLIE REPRODUCTION MATRIX.

The age distribution of a population can be determined by census, and the age profile at the nth census is now described by the numbers $v_{0,n}, v_{1,n}, \ldots, v_{A-1,n}$. The age classes beyond the reproductive one are ignored. This information is summarized in the vector

$$\vec{v}_n = \begin{pmatrix} v_{0,n} \\ \vdots \\ v_{A-1,n} \end{pmatrix}. \tag{6.1}$$

For example, a census might be taken every 5 years, females counted, and data kept up through the end of reproductive ages, say age 60. Then 12 age classes would be monitored.

Let $\sigma_i, i = 1, ..., A$, be the proportion of the births in the nth census that survive to the next census, i.e., $\sigma_i \in [0, 1]$ are the *survival probabilities* of the various age groups to the next census. Let b_m be the *fertilities* of the various age groups, $b_m \geq 0, m = 1, ..., A$. The age profile at the next census is determined from the equations

$$\begin{cases} v_{0,n+1} & = & \sigma_1 b_1 v_{0,n} + \sigma_2 b_2 v_{1,n} + \ldots + \sigma_A b_a v_{A-1,n}, \\ v_{1,n+1} & = & \sigma_1 v_{0,n}, \\ v_{2,n+1} & = & \sigma_2 v_{1,n}, \\ & \ldots & \\ v_{A-1,n+1} & = & \sigma_{A-1} v_{A-1,n}, \end{cases} \tag{6.2}$$

or, in vector notation,

$$\vec{v}_{n+1} = \mathbf{M}\, \vec{v}_n,$$

where the matrix $\mathbf{M}$, called *Leslie reproduction matrix*, has the form

$$\mathbf{M} = \begin{pmatrix} \sigma_1 b_1 & \sigma_2 b_2 & \sigma_3 b_3 & \dots & \sigma_A b_A \\ \sigma_1 & 0 & 0 & \dots & 0 \\ 0 & \sigma_2 & 0 & \dots & 0 \\ \vdots & \vdots & \vdots & \ddots & \vdots \\ 0 & 0 & \dots & \sigma_{A-1} & 0 \end{pmatrix}. \tag{6.3}$$

The profile in the nth census is therefore given by the *Leslie model* [47]

$$\vec{v}_n = \mathbf{M}^n \cdot \vec{v}_0 \, . \tag{6.4}$$

We now use the spectral decomposition of $\mathbf{M}$ and apply the Perron-Frobenius theorem to unravel the matrix $\mathbf{M}$.

Suppose that we know the A eigenvalues of $\mathbf{M}$ and they are distinct, say $\lambda_1, \dots, \lambda_A$. Then there are A vectors $\phi_1, \dots, \phi_A$, such that $\mathbf{M}\phi_i = \lambda_i \phi_i, i = 1, \dots, A$. The eigenvectors $\phi_i, i = 1, ..., A$, span the space in the sense that given a vector $\vec{v}_0$, there are constants $C_1, \dots, C_A$ such that $\vec{v}_0 = C_1\phi_1 + \dots + C_A\phi_A$. Applying $\mathbf{M}$ to $\vec{v}_0$ gives

$$\mathbf{M}\, \vec{v}_0 = \mathbf{M}(C_1\phi_1 + \dots + C_A\phi_A) = C_1\lambda_1\phi_1 + \dots + C_A\lambda_A\phi_A,$$

and

$$\mathbf{M}^2\, \vec{v}_0 = \mathbf{M}(\mathbf{M}\, \vec{v}_0) = C_1\lambda_1^2 \cdot \phi_1 + \dots + C_A\lambda_A^2\phi_A,$$

and so on. In general, we have

$$\mathbf{M}^n\, \vec{v}_0 = C_1\lambda_1^k\phi_1 + \dots + C_A\lambda_A^n\phi_A.$$

Suppose that the λ_i are ordered so that $|\lambda_1| > |\lambda_2| > \dots > |\lambda_A|$. Then $\lambda_1^{-n} \cdot M^n\, \vec{V}_0 \rightarrow C_1\phi_1$.

In the general case, where the eigenvalues $\lambda_i, i = 1, ..., A$, are not necessarily distinct, the Perron-Frobenius theorem gives useful information. From this theory it follows that:

(i) $\mathbf{M}$ has a unique positive eigenvalue λ_*;
(ii) λ_* is an eigenvalue for $\mathbf{M}^*$, the adjoint of $\mathbf{M}$;
(iii) the corresponding eigenvectors, say C_* and C_*^*, of $\mathbf{M}$ and $\mathbf{M}^*$, respectively, have non-negative components;
(iv) if λ is another eigenvalue $(\lambda \neq \lambda_*)$ of either $\mathbf{M}$ or $\mathbf{M}^*$, then $|\lambda| < \lambda_*$;
(v) the power of $\mathbf{M}$ can be calculated asymptotically

$$\mathbf{M}^n\, \vec{v}_0 = C_*\lambda_*^n\phi_* + E_n,$$

where $|E_n|/\lambda_*^n \rightarrow \mathbf{0}$ as $n \rightarrow +\infty$. It follows that

$$\lambda_*^{-n} \cdot \mathbf{M}^n\, \vec{v}_0 \rightarrow C_* \cdot \phi_* \qquad \text{as } n \rightarrow +\infty.$$

These results show that the eigenvalues of the reproduction matrix play a critical role in describing the population's age structure after many generations. In particular,

$$\mathbf{M}^n \vec{v}_0 \simeq \lambda_*^n \cdot C_* \phi_*.$$

It follows that the components of the vector $\mathbf{M}^n \vec{v}_0$ eventually have ratios that are independent of n. The distribution of the population among the age classes remains constant even as the population grows (if $\lambda_* > 1$) or dies out ($\lambda_* < 1$). For this reason the vector ϕ_* is called the *stable age distribution.*

We note that the reproduction model $\vec{v}_{n+1} = \mathbf{M}\ \vec{v}_n$ is quite similar in spirit to the Malthus [58] model $p_{n+1} = r\ p_n$, where r is the intrinsic growth rate and p_n is the number of adults at the nth reproduction time. The largest real eigenvalue corresponds to the r.

Finally, we note that Bernardelli [10] pointed out that if fertility is concentrated in one age group, then the birth rate could be a periodic function of time. This is an extreme example of a population wave. The birth rate will appear to oscillate if several characteristic roots are imaginary but have the same modules as the dominant real root. This can occur if $\mathbf{M}$ does not satisfy the conditions of the Perron-Frobenius theorem. This phenomena is called *Bernardelli population waves* (i.e., a periodic age structure). Population waves are observed in human populations [64].

6.2. Stochastic Demographic Models (Demographic Models in Random Media).

The demographic model, introduced in Section 6.1, is considered here in random environment. Discrete-time and continuous-time demographic models are studied in Markov and semi-Markov random environments. We note that in this case we will have a random Leslie reproduction matrix.

6.2.1. DISCRETE DEMOGRAPHIC MODEL IN MARKOV RANDOM MEDIA.

Let $\mathbf{M}(y)$ denote a matrix of the same form as $\mathbf{M}$ in (6.3), but having coefficients $\beta_i(y)$ and $\sigma_i(y)$ instead of β_i and σ_i, respectively, where b_i and σ_i are defined in Section 6.1, $i = 1, ..., \bar{A}$.

The functions $b_i(y)$ and $\sigma_i(y)$ are assumed to be bounded and measurable functions mapping Y to R.

A discrete demographic model in Markov random enviroment is defined by

$$\vec{v}_{n+1} = \mathbf{M}(y_{n+1}) \cdot \vec{v}_n, \tag{6.5}$$

where $(y_n)_{n \in Z^+}$ is a homogeneous Markov chain, and $\vec{v}$ is defined in Section 6.1 (where instead of σ_i and b_i we should take $\sigma_i(y)$ and $b_i(y), i = 1, ..., \bar{A}$).

In this way, the vector $\vec{v}_n$ of age class at the nth census is randome one, since the survival probabilities $\sigma_i(y_{n+1})$ of the various age groups to the next census and the fertilities of the various age groups $b_i(y_{n+1})$ in the nth census depend on the state of Markov chain $(y_n)_{n \in Z^+}$.

6.2.2. CONTINUOUS DEMOGRAPHIC MODEL IN MARKOV RENEWAL RANDOM MEDIA.

This model is defined as

$$\vec{v}_{\nu(t)+1} = \mathbf{M}(y_{\nu(t)+1})\ \vec{v}_{\nu(t)}, \tag{6.6}$$

where $\nu(t) := \max\{n : \tau_n \leq t\}$, $(y_n)_{n \in Z^+}$ is a homogeneous Markov chain.

We note that if $t \epsilon [\tau_n, \tau_{n+1}]$, then $\nu(t) = n$ and model (6.6) reduce to the model (6.5). This means that the census occurred in random time τ_n, and the number of such census for time t is defined by the counting process $\nu(t)$.

6.2.3. DEMOGRAPHIC MODELS ON INFINITE TIME INTERVAL.

We now consider the behavior of $\vec{v}$ in (6.5).

Suppose that $(y_n)_{n \in Z^+}$ is a stationary ergodic Markov chain with ergodic distribution $p(dy)$, i.e., for any function $g : Y \to R$ for which $\int |g(y)| p(dy) < +\infty$, we have

$$P\{\lim_{n \to +\infty} \frac{1}{n} \sum_{k=1}^{n} g(y_k) = \int_Y g(y) p(dy)\} = 1.$$

We also suppose that

$$\int_Y p(dy) |\ln \mathbf{M}(y)| < +\infty. \tag{6.7}$$

If (6.7) is fulfilled, then

$$\lim_{t \to +\infty} \frac{1}{n} \sum_{k=0}^{n-1} \ln \mathbf{M}(y_{k+1}) = \int_Y \ln \mathbf{M}(y) p(dy) \tag{6.8}$$

with probability 1. From (6.5) we obtain

$$\vec{v}_{n+1} = \prod_{k=1}^{w-1} \mathbf{M}(y_{k+1})\ \vec{v}_0\ . \tag{6.9}$$

Then, after taking a ln on both sides of (6.9), we get

$$\ln \vec{v}_{n+1} = \sum_{k=0}^{n-1} \ln \mathbf{M}(y_{k+1}) \cdot \vec{v}_0\ . \tag{6.10}$$

Here, we note that for a matrix A, we have

$$\ln A = \sum_{k=1}^{\infty} \frac{(-1)^k}{k} (A - I)^k,$$

where I is the identity matrix, provided $|\lambda_k - 1| < 1$, for each eigenvalue λ_k of A.

From (6.10) and (6.8), we obtain

$$\begin{aligned} \frac{\ln \vec{v}_{n+1}}{n} &= \frac{1}{n} \textstyle\sum_{k=0}^{n-1} \ln \mathbf{M}(y_{n+1})\ \vec{v}_0 \\ &\to_{n \to +\infty} \textstyle\int_Y p(dy) \ln \mathbf{M}(y)\ \vec{v} \qquad \text{with} \quad \mathcal{P} = 1. \end{aligned} \tag{6.11}$$

Hence, we obtain the following

Theorem 1. Under condition (6.7), we have

$$\frac{\ln \vec{v}_{n+1}}{n} \to_{n\to+\infty} \int_Y p(dy) \ln \mathbf{M}(y)\, \vec{v}_0 \tag{6.12}$$

with probability 1. Therefore, for large n, $\vec{v}_n$ behaves like

$$\vec{v}_n \simeq e^{n\cdot \int_Y p(dy) \ln \mathbf{M}(y)}\, \vec{v}_0\,. \tag{6.13}$$

We now obtain an analogous result for the model (6.6). From (6.6), we get

$$\ln \vec{v}_{\nu(t)+1} = \sum_{k=0}^{\nu(t)-1} \ln \mathbf{M}(y_{k+1})\, \vec{v}_0,$$

and hence,

$$\frac{\ln \vec{v}_{\nu(t)+1}}{t} = \frac{1}{t}\frac{\nu(t)}{\nu(t)} \cdot \sum_{k=0}^{\nu(t)-1} \ln \mathbf{M}(y_{k+1})\, \vec{v}_0\,. \tag{6.14}$$

We note that

$$\frac{\nu(t)}{t} \to_{t\to+\infty} \frac{1}{m}, \tag{6.15}$$

That follows from the renewal theorem in Chapter 1.

From (6.12) in Theorem 1 it follows that

$$\frac{1}{\nu(t)} \sum_{k=0}^{\nu(t)-1} \ln \mathbf{M}(y_{n+1})\, \vec{v}_0 \to_{t\to\infty} \int_Y p(dy) \ln \mathbf{M}(y)\, \vec{v}_0, \tag{6.16}$$

since $\nu(t) \to_{t\to+\infty} +\infty$.

From (6.14)-(6.16), we obtain that

$$\frac{\ln \vec{v}_{\nu(t)+1}}{t} \to_{t\to+\infty} \int_Y p(dy) \ln \mathbf{M}(y_0)\, \vec{v}_0 \,/m \qquad \text{with} \quad \mathcal{P} = 1,$$

where

$$m := \int_Y p(dy) m(y),$$

$$m(y) := \int_0^\infty t G_y(dt).$$

In summary, we obtain

Theorem 2. Under condition (6.7), we have

$$\frac{\ln \vec{v}_{\nu(t)+1}}{t} \to_{t\to+\infty} \int_Y p(dy) \ln \mathbf{M}(y)\, \vec{v}_0 \,/m \tag{6.17}$$

with probability 1. Therefore, for large t, $\vec{v}_{\nu(t)}$ behaves like

$$\vec{v}_{\nu(t)} \simeq e^{t\cdot\int_Y p(dy)\ln \mathbf{M}(y)/m}\ \vec{v}_0\ . \tag{6.18}$$

We now describe measurability of demographic models in random media.

Discrete-time demographic model in Markov random environment in (6.5) is $\mathcal{F}_{n+1}/\mathcal{R}^A$-measurable, as $\sigma_i(y_{n+1})$ and $b_i(y_{n+1})$, $i = 1, ..., A$, are $\mathcal{F}^Y_{n+1}/\mathcal{R}^A$-measurable, $\mathcal{R}^A$ is the σ-algebra generated by Borel sets in R^A, and $\mathcal{F}^Y_{n+1} \subset \mathcal{F}_{n+1}$.

Continuous-time demographic model in Markov renewal random environment in (6.6) is $\mathcal{G}_t/\mathcal{R}^A$-measurable, as $\sigma_i(y_{\nu(t)+1})$ and $b_i(y_{\nu(t)+1})$ are $\mathcal{G}_t/\mathcal{R}^A$-measurable, $i = 1, ..., A$.

6.3. Averaging of Demographic Models in Random Media.

Here we consider the demographic model in random environment in series scheme. We are mainly concerned with the *continuous demographic model* in *Markov renewal random environment* in series scheme of the following form:

$$\begin{cases} \vec{v}^{\,\varepsilon}_{\nu(t/\varepsilon+1)} &= \mathbf{M}(y_{\nu(t/\varepsilon+1)})\ \vec{v}^{\,\varepsilon}_{\nu(t/\varepsilon+1)} \\ \vec{v}^{\,\varepsilon}_0 &= \vec{v}_0, \end{cases} \tag{6.19}$$

where $\mathbf{M}(y)$ is a Leslie matrix of the same form as $\mathbf{M}$ in (6.3), but having coefficients $b_i(y)$ and $\sigma_i(y)$ instead of b_i and σ_i, respectively, where b_i and σ_i are defined in Subsection 6.1.3, $i = 1, ..., A$, and these functions $b_i(y)$ and $\sigma_i(y)$ are assumed to be bounded and measurable functions mapping $Y \to R$.

From (6.19), we obtain that

$$\vec{v}^{\,\varepsilon}_{\nu(t/\varepsilon)+1} = \prod_{k=0}^{\nu(t/\varepsilon)} \mathbf{M}(y_{k+1})\cdot\ \vec{v}_0,$$

and

$$\ln \vec{v}^{\,\varepsilon}_{\nu(t/\varepsilon)+1} = \sum_{k=0}^{\nu(t/\varepsilon)} \ln \mathbf{M}(y_{k+1})\cdot\ \vec{v}_0, \tag{6.20}$$

where

$$\ln \mathbf{M}(y) := \sum_{k=1}^{\infty} \frac{(I - \mathbf{M}(y))^k}{k} = \sum_{k=1}^{\infty} \frac{(-1)^{k-1}}{k}(\mathbf{M}(y) - I)^k,$$

provided that $|\lambda_k - 1| < 1$, for each $\lambda_k \in \sigma(\mathbf{M}(y))$, the spectrum of matrix $\mathbf{M}(y)$. Suppose that

$$\int_Y p(dy)|\ln \mathbf{M}(y)| < +\infty.$$

From renewal theorem and the ergodicity of Markov the chain $(y_n)_{n\in Z_+}$, we obtain

$$\varepsilon\cdot \sum_{k=0}^{\nu(t/\varepsilon)} \ln \mathbf{M}(y_{k+1})\ \vec{v}_0 \to_{\varepsilon\to 0} t\cdot \int_Y p(dy)\ln \mathbf{M}(y)\cdot\ \vec{v}_0\ /m. \tag{6.21}$$

Therefore, by (6.20) and (6.21), we get

$$\varepsilon \ln \overrightarrow{v}^{\varepsilon}_{\nu(t/\varepsilon)+1} \rightarrow_{\varepsilon\to 0} t \cdot \int_Y p(dy) \ln \mathbf{M}(y) \cdot \overrightarrow{v}_0 /m.$$

Hence, vector $\overrightarrow{v}^{\varepsilon}_{\nu(t/\varepsilon)}$ behaves like

$$\overrightarrow{v}^{\varepsilon}_{\nu(t/\varepsilon)} \simeq e^{\frac{t}{\varepsilon}\int_Y p(dy)\ln \mathbf{M}(y)/m}\, \overrightarrow{v}_0,$$

with small $\varepsilon > 0$ and for all $t \in R_+$.

We now consider the following *discrete demographic model in Markov random environment* in series scheme:

$$\begin{aligned} \overrightarrow{v}^{\varepsilon}_{\nu[t/\varepsilon]+1} &= \mathbf{M}(y_{\nu[t/\varepsilon]+1})\, \overrightarrow{v}_{\nu[t/\varepsilon]+1}, \\ \overrightarrow{v}^{\varepsilon}_0 &= \overrightarrow{v}_0 . \end{aligned}$$

By a similar argument, we can see that for small $\varepsilon > 0$ $\overrightarrow{v}^{\varepsilon}_{\nu[t/\varepsilon]}$ behaves like

$$\overrightarrow{v}^{\varepsilon}_{[t/\varepsilon]} \simeq e^{[t/\varepsilon]\cdot\int_Y p(dy)\ln \mathbf{M}(y)}\, \overrightarrow{v}_0 .$$

The scaling factor m^{-1} the above formula for the limiting process characterizes the mean value of stationary intensity of the jumps.

Remark. We should mention that averaging of discrete-time demographic model in Markov random environment was also considered in [41].

6.4. Merging of Demographic Model.

We consider the following demographic model in Markov random environment $y^{\varepsilon}(\nu(t/\varepsilon))$

$$\begin{cases} \overrightarrow{v}^{\varepsilon}(\nu(t/\varepsilon)+1) = \mathbf{M}(y^{\varepsilon}(\nu(t/\varepsilon)+1))\, \overrightarrow{v}^{\varepsilon}(\nu(t/\varepsilon)), \\ \overrightarrow{v}^{\varepsilon}(0) = \overrightarrow{v}_0, \end{cases} \tag{6.22}$$

where $\mathbf{M}(y)$ is a Leslie reproduction matrix defined in (6.19), and $y^{\varepsilon}(\nu(t/\varepsilon))$ is a perturbed semi-Markov process.

From (6.22) we obtain that

$$\overrightarrow{v}^{\varepsilon}(\nu(t/\varepsilon)+1) = \prod_{k=0}^{\nu(t/\varepsilon)} \mathbf{M}((y^{\varepsilon}_{k+1})\, \overrightarrow{v}_0,$$

and

$$\ln \overrightarrow{v}^{\varepsilon}(\nu(t/\varepsilon)+1) = \sum_{k=0}^{\nu(t/\varepsilon)} \ln \mathbf{M}(y^{\varepsilon}_{k+1})\, \overrightarrow{v}_0, \tag{6.23}$$

where $\ln \mathbf{M}(y)$ is defined in (6.20).

Suppose that

$$\int_{Y_v} p_v(dy)|\ln \mathbf{M}(y)| < +\infty, \forall v \epsilon V. \tag{6.24}$$

From the merging theorem for perturbed Markov chain (see Chapter 1, Subsection 1.9) we obtain

$$\varepsilon \cdot \sum_{k=0}^{\nu(t/\varepsilon)} \ln \mathbf{M}(y^{\varepsilon}_{k+1}) \overrightarrow{v}_0 \to_{\varepsilon\to 0} \int_0^t \tilde{\mathbf{M}}(\hat{y}(s)) \overrightarrow{v}_0 \, ds, \tag{6.25}$$

where

$$\tilde{\mathbf{M}}(v) := \int_{Y_v} p_v(dy) \ln \mathbf{M}(y)/m_v. \tag{6.26}$$

Therefore,

$$\varepsilon \cdot \ln \overrightarrow{v}^{\varepsilon} (\nu(t/\varepsilon) + 1) \to_{\varepsilon\to 0} \int_0^t \tilde{\mathbf{M}}(\hat{y}(s)) \overrightarrow{v}_0 \, ds. \tag{6.27}$$

In this way, vector $\overrightarrow{v}^{\varepsilon} (\nu(t/\varepsilon))$ behaves like

$$\overrightarrow{v}^{\varepsilon} (\nu(t/\varepsilon)) \simeq \exp(\frac{1}{\varepsilon} \int_0^t \tilde{\mathbf{M}}(\hat{y}(s)) \overrightarrow{v}_0 \, ds.$$

for small $\varepsilon > 0$ and $\forall t \in R^+$, where $\hat{y}(t)$ is a merged Markov process.

6.5. Diffusion Approximation of Demographic Model.

We suppose that *balance condition* for the demographic model is fulfilled:

$$\int_Y p(dy) \ln \mathbf{M}(y) \overrightarrow{v_o} = 0, \tag{6.28}$$

where $\mathbf{M}(y)$ is the Leslie representation matrix in RM, $\ln \mathbf{M}$ is defined in (6.20). We consider here the following demographic model in scale of time t/ε^2 :

$$\overrightarrow{v}^{\varepsilon}_{\nu(t/\varepsilon^2)+1} = \mathbf{M}(y_{\nu(t/\varepsilon^2)+1}) \overrightarrow{v}^{\varepsilon}_{\nu(t/\varepsilon^2)}, \overrightarrow{v}^{\varepsilon}_0 = \overrightarrow{v}_0 . \tag{6.29}$$

From (6.29) we obtain that

$$\overrightarrow{v}^{\varepsilon}_{\nu(t/\varepsilon^2)+1} = \prod_{k=0}^{\nu(t/\varepsilon^2)} \mathbf{M}(y_{k+1}) \overrightarrow{v}_0$$

and

$$\ln \overrightarrow{v}^{\varepsilon}_{\nu(t/\varepsilon^2)+1} = \sum_{k=0}^{\nu(t/\varepsilon^2)} \ln \mathbf{M}(y_{k+1}) \overrightarrow{v}_0 . \tag{6.30}$$

From Central Limit Theorem for additive functionals of Markov chains (see Chapter 1), we obtain that $\varepsilon \ln \overrightarrow{v}^{\varepsilon}_{\nu(t/\varepsilon^2)+1}$ converges weakly as $\varepsilon \to 0$ to the Wiener process with variance

$$\tilde{\beta}^2 := 2 \int_Y p(dy)[\ln \mathbf{M}(y) \mathbf{R}_0 \ln \mathbf{M}(y) + 1/2(\ln \mathbf{M}(y))^2]/m,$$

where $\mathbf{R}_0$ is a potential of the Markov chain $(y_n)_{n\in Z^+}$.

6.6. Normal Deviations of Demographic Models in Random Media.

We consider the following continuous-time demographic model in Markov random environment in series scheme:

$$\begin{cases} \overrightarrow{v}^{\varepsilon}_{\nu(t/\varepsilon)+1} &= \mathbf{M}(y_{\nu(t/\varepsilon)+1})\ \overrightarrow{v}^{\varepsilon}_{\nu(t/\varepsilon)}, \\ \overrightarrow{v}^{\varepsilon}_{0} &= \overrightarrow{v}_{0}, \end{cases} \tag{6.31}$$

where $\mathbf{M}(y)$ is a Leslie matrix of the same form as $\mathbf{M}$ in (6.3), but having coefficients $b_i(y)$ and $\sigma_i(y)$ instead of b_i and σ_i, respectively, where b_i and σ_i are defined in subsection 6.1.3, $i = 1, ..., \bar{A}$.

From the results of Subsection 6.3 it follows that

$$\varepsilon \cdot \ln \overrightarrow{v}^{\varepsilon}_{\nu(t/\varepsilon)+1} \rightarrow_{\varepsilon\rightarrow 0} t \int_Y p(dy) \ln \mathbf{M}(y)\ \overrightarrow{v}_0 /m, \tag{6.32}$$

where $\overrightarrow{v}_{\nu(t/\varepsilon)}$ is defined in (6.31).

We consider the following deviations of the process $\varepsilon \ln \overrightarrow{v}^{\varepsilon}_{\nu(t/\varepsilon)}$ from the averaged process $t \cdot \hat{\mathbf{M}}$:

$$v^{\varepsilon}(t) := [\varepsilon \ln \overrightarrow{v}^{\varepsilon}_{\nu(t/\varepsilon)} - t \cdot \hat{\mathbf{M}}]/\sqrt{\varepsilon}, \tag{6.33}$$

where

$$\hat{\mathbf{M}} := \int_Y p(dy) \ln \mathbf{M}(y)\ \overrightarrow{v}_0 /m.$$

Applying the normal deviation theorem from Subsection 2.4 (of Chapter 2) to the process $v^{\varepsilon}(t)$ in (6.33), we obtain that $v^{\varepsilon}(t)$ converges weakly as $\varepsilon \rightarrow 0$ to the process $\hat{v}(t)$:

$$\tilde{v}(t) = \sigma w(t),$$

where

$$\sigma^2 := \int_Y p(dy)(\ln \mathbf{M}(y) - \hat{\mathbf{M}})\mathbf{R}_0(\ln \mathbf{M}(y) - \hat{\mathbf{M}}) + (\ln \mathbf{M}(y) - \hat{\mathbf{M}})^2/2]m,$$

and $w(t)$ is a standard Wiener process.

6.7. Stochastic Stability of Demographic Model in Averaging and Diffusion Approximation Schemes.

6.7.1. STABILITY OF DEMOGRAPHIC MODEL IN AVERAGING SCHEME.

Let us consider the following demographic model in series scheme:

$$\begin{cases} \overrightarrow{v}^{\varepsilon}_{\nu(t/\varepsilon)+1} &= \mathbf{M}(y_{\nu(t/\varepsilon)+1})\ \overrightarrow{v}^{\varepsilon}_{\nu(t/\varepsilon)}, \\ \overrightarrow{v}^{\varepsilon}_{0} &= \overrightarrow{v}_{0}, \end{cases} \tag{6.34}$$

where $\mathbf{M}(y)$ is a Leslie matrix.

As shown in Subsection 6.3, we have

$$\varepsilon \ln \overrightarrow{v}^{\varepsilon}_{\nu(t/\varepsilon)} \to_{\varepsilon\to 0} t \cdot \int_Y p(dy) \ln M(y) \overrightarrow{v}_0 /m \tag{6.35}$$

under condition $\int_Y p(dy) |\ln \mathbf{M}(y)| < +\infty$.

This means that vector $\overrightarrow{v}^{\varepsilon}_{\nu(t/\varepsilon)}$ behaves like:

$$\overrightarrow{v}^{\varepsilon}_{\nu(t/\varepsilon)} \simeq e^{t/\varepsilon \int_Y p(dy) \ln \mathbf{M}(y)/m} \overrightarrow{v}_0$$

for small $\varepsilon > 0$.

We are interested in the behavior of $\overrightarrow{v}^{\varepsilon}_{\nu(t/\varepsilon)}$ as $t \to +\infty$ and $0 \le \varepsilon \le \varepsilon_0$ for small enough $\varepsilon_0 > 0$.

Let $\lambda_k(y) \epsilon \sigma(\mathbf{M}(y))$. We need $|\lambda_k(y) - 1| < 1, \quad \forall y \epsilon Y$. Namely, $0 \le \lambda_k(y) \le 2$.

Let us calculate $\ln \mathbf{M}(y) \overrightarrow{v}_0$, where

$$\mathbf{M}(y) = \begin{pmatrix} \sigma_1(y)b_1(y) & \sigma_2(y)b_2(y) & \dots & \sigma_A(y)b_A(y) \\ \sigma_1(y) & 0 & \dots & 0 \\ \dots & \sigma_2(y) & \dots & \dots \\ 0 & 0 & \dots & \sigma_{A-1}(y) \end{pmatrix}$$

and

$$\overrightarrow{v}_0 = (v_{0,0}, v_{0,1}, \dots, \overrightarrow{v}_{0,A-1}).$$

Then

$$\mathbf{M}(y) \cdot \overrightarrow{v}_0 = (\sum_{k=0}^{A-1} \sigma_k(y) b_k(y) \cdot v_{0,K}, \sigma_1(y) \cdot v_{0,1}, \sigma_2(y) v_{0,2}, \dots, \sigma_{A-1}(y) v_{0,A-1}),$$

and

$$\int_Y p(dy) \ln \mathbf{M}(y) \overrightarrow{v}_0 = (\ln \sum_{K=0}^{A-1} \hat{b}_{k+1}, \ln \hat{\sigma}_1, \ln \hat{\sigma}_2, \dots, \ln \hat{\sigma}_{A-1}),$$

where

$$\begin{aligned} \hat{b}_{k+1} &:= \textstyle\int_Y p(dy) \sigma_{k+1}(y) b_{k+1}(y) v_{0,k}, k = \overline{0, A-1}, \\ \hat{\sigma}_k &:= \textstyle\int_Y p(dy) \sigma_k(y) v_{0,k}, \qquad k = 1, \dots, A-1. \end{aligned}$$

Therefore, if

$$0 < \sum_{k=0}^{A-1} \hat{b}_{k+1} < 1, \quad 0 < \hat{\sigma}_k < 1, \quad k = 1, \dots, A-1,$$

then

$$\int_Y p(dy) \ln \mathbf{M}(y) \overrightarrow{v}_0 < 0$$

and

$$e^{\frac{t}{\varepsilon} \int_Y p(dy) \ln \mathbf{M}(y)} \overrightarrow{v}_0 \to_{t\to+\infty} 0$$

for $0 \leq \varepsilon \leq \varepsilon_0$.

If $\sum_{k=0}^{A-1} \hat{b}_{k+1} > 1, \quad \hat{\sigma}_k > 1, \quad k = 1, ..., A-1$, then

$$\int_Y p(dy) \ln \mathbf{M}(y) \overrightarrow{v}_0 > 0$$

and

$$e^{\frac{t}{\varepsilon} \int_Y p(dy) \ln \mathbf{M}(y)} \overrightarrow{v}_0 \rightarrow_{t \to +\infty} +\infty$$

for $0 \leq \varepsilon \leq \varepsilon_0$.

Suppose we know that the A eigenvalues of $\mathbf{M}(y)$ and assume they are distinct, and denoted by $\lambda_1(y), \lambda_2(y), \ldots, \lambda_A(y)$. Then there are A column vectors $\phi_1, \ldots, \phi_A$ such that

$$\mathbf{M}(y)\phi_i = \lambda_i(y)\phi_i, i = 1, A.$$

The eigenvectors $\phi_i, \quad i = 1, ..., A$, span the space in the sense that given a vector $\overrightarrow{v}_0$, there are constants $C_1, \ldots, C_A$ such that $\overrightarrow{v}_0 = C_1\phi_1 + \cdots + C_A\phi_A$.

Applying $\mathbf{M}(y)$ to $\overrightarrow{v}_0$ gives

$$\begin{aligned} \mathbf{M}(y) \overrightarrow{v}_0 &= \mathbf{M}(y)(C_1\phi_1 + \cdots + C_A\phi_A) \\ &- C_1\lambda_1(y)\phi_1 + \cdots + C_A\lambda_A(y)\phi_A, \end{aligned}$$

and

$$\mathbf{M}^2(y) \overrightarrow{v}_0 = \mathbf{M}(y)(\mathbf{M}(y) \overrightarrow{v}_0) = C_1\lambda_1^2(y)\phi_A,$$

and, in general,

$$\prod_{k=0}^{\nu(t)} \mathbf{M}(y_{k+1}) \overrightarrow{v}_0 = C_1 \prod_{k=0}^{\nu(t)} \lambda_1(y_{k+1}) \cdot \phi_1 + \cdots + C_A \cdot \prod_{k=0}^{\nu(t)} \lambda_A(y_{k+1})\phi_A.$$

Suppose that $\lambda_i(y)$ are ordered so that

$$|\lambda_1(y)| > |\lambda_2(y)| > \cdots > |\lambda_A(y)|. \tag{6.36}$$

Then

$$\left|\frac{\prod_{k=0}^{\nu(t)} \lambda_A(y_{n+1}}{\prod_{k=0}^{\nu(t)} \lambda_1(y_{n+1}}\right| \leq \prod_{k=0}^{\nu(t)} \left|\frac{\lambda_A(y_{k+1})}{\lambda_1(y_{k+1})}\right| \rightarrow_{t \to \infty} 0,$$

since $\nu(t) \rightarrow_{t \to +\infty} +\infty$.

In this way, we have

$$\frac{\prod_{k=0}^{\nu(t)} \mathbf{M}(y_{n+1}) \overrightarrow{v}_0}{\prod_{k=0}^{\nu(t)} \lambda_1(y_{k+1})} \rightarrow_{t \to +\infty} C_1\phi_1.$$

We now consider the demographic model in series scheme in averaging. From previous results it follows that

$$\begin{aligned} \overrightarrow{v}_{\nu(t/\varepsilon)} &= \prod_{k=0}^{\nu(t/\varepsilon)} \mathbf{M}(y_{k+1}) \overrightarrow{v}_0 \\ &= C_1 \cdot \prod_{k=0}^{\nu(t/\varepsilon)} \lambda_1(y_{k+1})\phi_* + \cdots + C_A \cdot \prod_{k=0}^{\nu(t/\varepsilon)} \lambda_A(y_{k+1})\phi_A. \end{aligned} \tag{6.37}$$

Suppose that $\lambda_i(y)$ are ordered in the way as in (6.35), $i = 1, ..., A$. Then

$$\frac{\overrightarrow{v}_{\nu(t/\varepsilon)}}{\prod_{k=0}^{\nu(t/\varepsilon)} \lambda_1(y_{k+1})} \to_{\varepsilon\to 0} C_1\phi_1. \tag{6.38}$$

It means that $\overrightarrow{v}_{\nu(t/\varepsilon)}$ behaves as follows:

$$\overrightarrow{v}_{\nu(t/\varepsilon)} \simeq C_1 \cdot \prod_{k=0}^{\nu(t/\varepsilon)} \lambda_1(y_{k+1}) \cdot \phi_1.$$

Since

$$\begin{aligned}\prod_{k=0}^{\nu(t/\varepsilon)} \lambda_1(y_{k+1}) &= e^{\sum_{k=0}^{\nu(t/\varepsilon)} \ln \lambda_1(y_{k+1})} \\ &\simeq e^{\frac{t}{\varepsilon}\int_Y p(dy)\ln\lambda_1(y)/m}\end{aligned}$$

which follows from renewal theorem (see Chapter 1), then we get

$$\overrightarrow{v}_{\nu(t/\varepsilon)} \simeq C_1 \cdot \phi_1 \cdot e^{\frac{t}{\varepsilon}\hat{\lambda}_1}, \tag{6.39}$$

where $\hat{\lambda}_1 := \int_Y p(dy) \ln \lambda_1(y)/m$. Consequently, the population grows if $\hat{\lambda}_1 > 0$, and dies out if $\hat{\lambda}_1 < 0$. This , of course, corresponds to $\lambda_1(y) > 1$ or $\lambda_1(y) < 1$, respectively.

In the general case, where the eigenvalues $\lambda_i(y), i = 1, ..., A$, are not necessarily distinct, the Perron-Frobenius theorem gives useful information. From this theory it follows that:

(i) $\mathbf{M}(y)$ has a unique positive eigenvalue $\lambda_*(y)$;
(ii) $\lambda_*(y)$ is an eigenvalue for $M^*(y)$, the adjoint of $\mathbf{M}(y)$;
(iii) the corresponding eigenvectors, say C_* and C_*^*, of $\mathbf{M}(y)$ and $\mathbf{M}^*(y)$, respectively, have nonnegative components;
(iv) if $\lambda(y)$ is another eigenvalue $(\lambda(y)/\neq \lambda_*(y))$ of either $\mathbf{M}(y)$ or $\mathbf{M}^*(y)$, then $|\lambda| < \lambda_*$;
(v) $\mathbf{M}^n$ can be calculated by

$$\prod_{k=0}^{\nu(t)} \mathbf{M}(y_{k+1}) \overrightarrow{v}_0 = C_* \cdot \prod_{k=0}^{\nu(t)} \lambda_*(y_{k+1}) \cdot \phi_* + E_{\nu(t)}, \tag{6.40}$$

where

$$|E_{\nu(t)}| / \prod_{k=0}^{\nu(t)} \lambda_*(y_{k+1}) \to_{t\to+\infty} 0.$$

It follows that

$$\frac{\prod_{k=0}^{\nu(t)} \mathbf{M}(y_{k+1}) \overrightarrow{v}_0}{\prod_{k=0}^{\nu(t)} \lambda_*(y_{k+1})} \to_{t\to+\infty} C_* \cdot \phi_*.$$

In particular,

$$\overrightarrow{v}_{\nu(t)} = \prod_{k=0}^{\nu(t)} \mathbf{M}(y_{k+1}) \overrightarrow{v}_0 \simeq C_* \prod_{k=0}^{\nu(t)} \lambda_*(y_{k+1}) \cdot \phi_*. \tag{6.41}$$

The distribution of the population among the age classes remains constant even if the population grows (if $\lambda_*(y) > 1, \forall y \epsilon Y$) or dies out ($\lambda_*(y) < 1, \forall y \epsilon Y$). For this reason, the vector ϕ_* is called the *stable age distribution.*

In general case, for demographic model in series scheme in averaging, where the eigenvalues $\lambda_i(y), i = 1, ..., A$, are not necessarily distinct, the Perron-Frobenius theorems also gives useful information. In particular, from (6.36)-(6.41) and properties (i)-(iv) we get

$$\overrightarrow{v}_{\nu(t/\varepsilon)} = \prod_{k=0}^{\nu(t/\varepsilon)} \mathbf{M}(y_{k+1}) \overrightarrow{v}_0 = C_* \prod_{k=0}^{\nu(t/\varepsilon)} \lambda_*(y_{k+1})\phi_* + E_{\nu(t/\varepsilon)}, \tag{6.42}$$

where

$$|E_{\nu(t/\varepsilon)}| / \prod_{k=0}^{\nu(t/\varepsilon)} \lambda_*(y_{k+1}) \to_{\varepsilon\to 0} .$$

Therefore,

$$\overrightarrow{v}_{\nu(t/\varepsilon)} = \prod_{k=0}^{\nu(t/\varepsilon)} \mathbf{M}(y_{k+1}) \overrightarrow{v}_0 \simeq C_* \prod_{k=0}^{\nu(t/\varepsilon)} \lambda_*(y_{k+1})\phi_*. \tag{6.43}$$

From (6.41)-(6.43), we obtain

$$\overrightarrow{v}_{\nu(t/\varepsilon)} \simeq C_*\phi_* \cdot e^{\frac{t}{\varepsilon}\hat{\lambda}_*},$$

where

$$\lambda_* := \int_Y p(dy) \ln \lambda_*(y)/m.$$

Therefore, we conclude that the population grows if $\hat{\lambda}_* > 0$, and dies out if $\hat{\lambda}_* < 0$.

We can study demographic model in averaging in another way. Namely, we can rewrite the relation as

$$\overrightarrow{v}_{\nu(t/\varepsilon)+1} - \overrightarrow{v}_{\nu(t/\varepsilon)} = (\mathbf{M}(y_{\nu(t/\varepsilon)+1})) - I) \overrightarrow{v}_{\nu(t/\varepsilon)} . \tag{6.44}$$

It follows that

$$\overrightarrow{v}_{\nu(t/\varepsilon)+1)} = \overrightarrow{v}_0 + \sum_{k+0}^{\nu(t/\varepsilon)} [\mathbf{M}(y_{k+1}) - I] \overrightarrow{v}_k$$

We can then apply the renewal theorem to the above relation to obtain same results as obtained earlier based on (6.43).

6.7.2. STABILITY OF DEMOGRAPHIC MODEL IN DIFFUSION APPROXIMATION SCHEME.

Let us consider the demographic model in series scheme in scale of time t/ε^2 with balance condition (6.28)

$$\overrightarrow{v}_{\nu(t/\varepsilon^2)+1} = \mathbf{M}(y_{\nu(t/\varepsilon^2)+1}) \overrightarrow{v}_{\nu(t/\varepsilon^2)}, \overrightarrow{v}^{\varepsilon}_0 = \overrightarrow{v}_0 .$$

From the result of Subsection 6.5 it follows that $\varepsilon \ln \overrightarrow{v}_{\nu(t/\varepsilon^2)}$ converges weakly as $\varepsilon \to 0$ to the Wiener process with variance.

$$\tilde{\beta}^2 := 2 \int_Y p(dy)[\ln \mathbf{M}(y)\mathbf{R}_0 \ln \mathbf{M}(y) + 1/2(\ln \mathbf{M}(y))^2]/m.$$

Therefore, for small $\varepsilon \epsilon [0, \varepsilon_0]$, where ε_0 is fixed, we have

$$\overrightarrow{v}_{\nu(t/\varepsilon^2)} \simeq e^{\frac{1}{\varepsilon}\tilde{\beta} w(t)}, \tag{6.45}$$

where $w(t)$ is a standard Wiener process.

To study the behavior of the right hand side (6.45) when $t \to +\infty$, we need to calculate

$$E(\overrightarrow{v}^{\varepsilon}_{\nu(t/\varepsilon^2)})^m$$

since there is no limit for $e^{\frac{1}{\varepsilon}\tilde{\beta} w(t)}$ as $t \to +\infty$. Since

$$E(\overrightarrow{v}^{\varepsilon}_{\nu(t/\varepsilon^2)})^m \simeq e^{\frac{\tilde{\beta}^2 \cdot tm^2}{\varepsilon^2}},$$

we can see that

$$e^{\frac{\tilde{\beta}^2 \cdot tm^2}{\varepsilon^2}} \longrightarrow_{t \to +\infty} +\infty,$$

using $\tilde{\beta}^2 > 0, \quad m^2 > 0, \quad \varepsilon^2 > 0, \quad \forall m = 1, 2, \ldots$.

In conclusion, we observe that population in demographic model growths exponentially in m-mean sense, and hence, by probability.

CHAPTER 7: LOGISTIC GROWTH MODELS

7.1. Deterministic Logistic Growth Model.

Differential equation models, whether ordinary, delay, partial or stochastic, imply a continuous overlap of generations. Many species leave no overlap between successive generations and so population growth is in discrete steps. For primitive organisms, these discrete steps can be quite short, and hence a continuous (in time) model may be a reasonable approximation. Here, we consider both discrete and continuous logistic growth model (LGM).

7.1.1. DISCRETE LOGISTIC GROWTH MODEL.

Consider a model that relates the population at time $n+1$, denoted by N_{n+1}, in terms of the population N_n at time n. This leads to the following difference equation

$$N_{n+1} = N_n F(N_n) = f(N_n), \tag{7.1}$$

where $f(N_n)$ is in general a non-linear function of N_n. The first form is often used to emphasize the existence of a zero steady state. Whatever the form of $f(N)$, we shall only be interested in non-negative populations. If function F is a constant $r > 0$, that is the population one step later is simply proportional to the current population, and from (7.1) we get

$$N_{n+1} = rN_n \qquad \text{and} \quad N_n = r^n N_0. \tag{7.2}$$

So the population grows or decays geometrically according to whether $r > 1$ or $r < 1$, respectively. This particularly simple model is not very realistic for most populations though it has been used with some justification for the early stages of growth of certain bacteria.

A slight modification to bring in crowding effects could be $N_{n+1} = rN_s, N_s = N_n^{1-b}, b$ is a constant, where N_s is the population that survives to breed. There must be restrictions on b so that $N_s \leq N_n$ otherwise those surviving to breed would be more than the population of which they form a part. Generally, because of crowding and self-regulation, we expect $f(N_n)$ in (7.1) to have some maximum, say at N_m, as a function of N_t with f decreasing for $N_t > N_m$. A variety of $f(N_n)$ have been used in practical biological situations (see, for example, May and Oster [61]).

One such model, sometimes referred to as the *Verhulst process*, is

$$N_{n+1} = r \cdot N_n \cdot (1 - \frac{N_n}{K}), r > 0, K > 0, \tag{7.3}$$

which is a *discrete analogue* of the continuous *logistic growth model*, to be considered in the next subsection. An obvious drawback of this specific model is that if $N_n > K$, then $N_{n+1} < 0$. A more realistic model is such that for large N_n there should be a reduction in the growth rate but N_{n+1} should remain non-negative.

One such frequently used model is

$$N_{n+1} = N_n \cdot \exp\left[r \cdot (1 - \frac{N_n}{K})\right], r > 0, K > 0,$$

which we can be regarded as a modification of (7.3) where there is a mortality factor $\exp(-rN_n/K)$ which is more severe if N_n is large. Here $N_n > 0$ for all of n if $N_0 > 0$.

Let us rescale the non-linear logistic model (7.3) by writing $u_n = N_n/K$ so that the carrying capacity K is now 1. Then

$$u_{n+1} = ru_n(1 - u_n), r > 0, \tag{7.4}$$

where we assume $0 < u_0 < 1$ and we are interested in solutions $u_n \geq 0$. The steady states and corresponding eigenvalues λ are

$$u^* = 0, \lambda = f'(0) = r; u^* = \frac{r-1}{r}, \lambda = f'(u^*) = 2 - r.$$

As r increases from zero to 1 the only realistic equilibrium is $u^* = 0$ which is stable. It is also clear from equation (7.4) that $u_1 < u_0 < 1$ and $u_{n+1} < u_n$ for all n, which implies that $u_n \to 0$ as $n \to +\infty$.

Other properties of equation (7.3), such as chaos, stability, periodic solutions and bifurcations, may be found in Murray [64].

7.1.2. CONTINUOUS LOGISTIC GROWTH MODEL.

Let $N(t)$ be the population of the species at time t, then the rate of change

$$\frac{dN}{dt} = \text{births - deaths + migration}$$

is a *conservation equation* for the population.

For the simplest model, we have no migration and the birth and death terms are proportional to N. Then

$$\frac{dN}{dt} = bN - dN.$$

Therefore, we get

$$N(t) = N_0 \cdot e^{(b-d)/t},$$

where b, d are positive constants and the initial population $N(0) = N_0$. Thus, if $b > d$ the population grows exponentially while if $b < d$ it dies out. This model, due to Malthus [58] but actually suggested earlier by Euler, is pretty unrealistic.

In the long run, of course, there must be some adjustment to such an exponential growth. Verhulst [86] proposed that a self-limiting process should operate when a population becomes too large. He suggested

$$\frac{dN(t)}{dt} = r \cdot N \cdot (1 - N/K), \tag{7.5}$$

where r and K are positive constants. This is now called a *logistic growth* model.

In this model the per capita birth rate is $r(1 - N/K)$, that is, it is dependent on N. The constant K is the *carrying capacity* of the environment, which is usually determined by the available sustaining resources.

If $N(0) = N_0 > 0$, then the solution of (7.5) is

$$N(t) = \frac{N_0 \cdot K \cdot e^{rt}}{[K + N_0 \cdot (e^{rt} - 1)]} \to K \text{as } t \to +\infty. \tag{7.6}$$

From (7.6), if $N_0 < K, N(t)$ simply increases monotonically to K while if $N_0 > K$ it decreases monotonically to K. In the former case, there is a qualitative difference depending on whether $N_0 > K/2$ or $N_0 < K/2$. There are two *steady states* or *equilibrium states* of (7.5), namely, $N = 0$ and $N = K$, that is where $dN/dt = 0$. $N = 0$ is unstable since linearization about it gives $dN/dt \approx rN$, and so N grows exponentially from zero. The other equilibrium $N = K$ is stable: linearization about it gives $d(N - K)/dt \approx -r \cdot (N - K)$ and so $N \to K$ as $t \to +\infty$. The carrying capacity K determines the size of the stable steady state population while r is a measure of the rate at which it is reached, that is, it is a measure of the dynamics: we could incorporate it in the time by a transformation from t to rt. Thus $1/r$ is a representative time scale of the response of the model to any change in population.

7.2. Stochastic Logistic Growth Model (LGM in RM).

We consider *discrete-time and continuous-time logistic growth models*, introduced in Section 7.1, in random environment, Markov and semi-Markov ones.

7.2.1. DISCRETE LOGISTIC GROWTH MODEL IN MARKOV RANDOM MEDIA

Let $r(y)$ and $K(y)$ be bounded and measurable finitious on Y, and $(y_n)_{n \in Z^+}$ be a homogeneous Markov chain in Y.

Then the discrete-time logistic growth model in Markov random environment is defined as follows:

$$N_{n+1} = r(y_{n+1}) \cdot N_n(1 - \frac{N_n}{K(y_{n+1})}). \tag{7.7}$$

We note that N_n in (7.7) is a random process.

The carrying capacity function $K(y)$ of the environment really depends on the state of random environment $y \epsilon Y$, and is usually determined by the available sustaining resources. The rate of growth of the population also depends on the state of random environment $y \epsilon Y$.

7.2.2. CONTINUOUS LOGISTIC GROWTH MODEL IN MARKOV RENEWAL RANDOM MEDIA

This model is defined by the following relation:

$$N_{\nu(t)+1} = r(y_{\nu(t)+1}) N_{\nu(t)} \cdot (1 - \frac{N_{\nu(t)}}{K(y_{\nu(t)+1})}), \tag{7.8}$$

where $\nu(t) := \max\{n; \tau_{n+1}\}$.

If $t\epsilon[\tau_n, \tau_{n+1}]$, then $\nu(t) = n$ and model (7.8) coincides with the model (7.7).

Functions $r(y)$ and $K(y)$ are defined in Section 7.2.1. Of course, function $N_{\nu(t)}$ in (7.8) is also a random process.

7.2.3. CONTINUOUS LOGISTIC GROWTH MODEL IN SEMI-MARKOV RANDOM MEDIA

This model is defined by the following expression:

$$\frac{dN(t)}{dt} = r(y(t)) \cdot N(t) \cdot (1 - N(t)/K(y(t)), \tag{7.9}$$

where $y(t) := y_{\nu(t)}$ is a semi-Markov process, $r(y)$ and $K(y)$ are functions defined above.

The process (7.9) is a random one.

Measurability of Logistic Growth Models in Random Media.

Discrete-time LGM in Markov random environment in (see (7.7)) is $\mathcal{F}_{n+1}/\mathcal{R}$-measurable, since $r(y_{n+1})$ are, and $K(y_{n+1})$ and N_n is $\mathcal{F}^Y_{n+1}/\mathcal{R}$- measurable, $\mathcal{F}^Y_n \subset \mathcal{F}_n \subset \mathcal{F}_{n+1}$.

Continuous-time LGM in Markov renewal random environment in (7.8) is $\mathcal{G}_t/\mathcal{R}$-measurable, since $r(y_{\nu(t)+1})$, and $K(y_{\nu(t)+1})$, and $N_{\nu(t)}$ is $\mathcal{H}_t/\mathcal{R}$-measurable and $\mathcal{H}_t \subset \mathcal{G}_t$.

Continuous-time LGM in semi-Markov random environment in (7.9) is $\mathcal{H}_t/\mathcal{R}$-measurable, since $r(y(t)), K(y(t))$ and $y(t) := y_{\nu(t)}$ are.

7.3. Averaging of Logistic Growth Model in Random Media

Let us consider the discrete logistic growth model (LGM) in Markov random environment in series scheme

$$N^\varepsilon_{\nu(t/\varepsilon)+1} - N^\varepsilon_{\nu(t/\varepsilon)} = \varepsilon \cdot r(y_{\nu(t/\varepsilon)+1}) \cdot N^\varepsilon_{\nu(t/\varepsilon)} \cdot (1 - N^\varepsilon_{\nu(t/\varepsilon)}/K(y_{\nu(t/\varepsilon)+1})), \tag{7.10}$$

where $r(y)$ and $K(y)$ are positive bounded functions on Y, ε is a small parameter, $\varepsilon > 0$.

Function $g(x, y)$ used in (Chapter 2, Subsection 2.2) has the following form:

$$g(x, y) = r(y)x(1 - x/K(y)).$$

Applying Theorem 2 in (Chapter 2, Subsection 2.2) to the equation (7.10), we obtain that $N^\varepsilon_{\nu(t/\varepsilon)} \to_{\varepsilon\to 0} \tilde{N}_t$ with

$$\begin{cases} \frac{d\tilde{N}_t}{dt} = \tilde{R}(\tilde{N}_t), \\ \tilde{N}_0 = N_0, \end{cases} \tag{7.11}$$

where

$$\tilde{R}(x) := \int_Y p(dy) \cdot \frac{r(y)}{K(y)} x(K(y) - x)/m, \tag{7.12}$$

$$m := \int_Y m(y)p(dy).$$

For the discrete LGM in Markov Random environment in series scheme

$$\begin{cases} N^{\varepsilon}_{[t/\varepsilon]+1} - N^{\varepsilon}_{[t/\varepsilon]} = \varepsilon N^{\varepsilon}_{[t/\varepsilon]} \cdot r(y_{[t/\varepsilon]+1})(1 - N^{\varepsilon}_{[t/\varepsilon]+1}/K(y_{[t/\varepsilon]+1})), \\ N^{\varepsilon}_0 = N_0, \end{cases} \tag{7.13}$$

applying Theorem 1 in (Chapter 2, subsection 2.2) we obtain that $N^{\varepsilon}_{[t/\varepsilon]} \to_{\varepsilon\to 0} \hat{N}_t$, with

$$\begin{cases} \frac{d\hat{N}_t}{dt} = \hat{R}(\bar{N}_t), \\ \hat{N}_0 = N_0, \end{cases} \tag{7.14}$$

where

$$\hat{R}(x) := \tilde{R}(x) \cdot m,$$

and $\tilde{R}(s)$ is defined in (7.12).

Factor m^{-1} in (7.12) characterizes the mean stationary intensity of the jumps of random environment in LGM.

We now consider the continuous-time LGM in semi-Markov random environment in series scheme:

$$\begin{cases} \frac{dN^{\varepsilon}(t)}{dt} = r(y(t/\varepsilon)) \cdot N^{\varepsilon}(t) \cdot (1 - N^{\varepsilon}(t))/K(y(t/\varepsilon))), \\ N^{\varepsilon}(0) = N_0, \end{cases} \tag{7.15}$$

where $y(t)$ is a semi-Markov process.

From the general theory of random evolution (see Chapter 2, Subsection 2.1) it follows that $N^{\varepsilon}(t) \to_{\varepsilon\to 0} \bar{N}(t)$ with

$$\begin{cases} \frac{d\bar{N}(t)}{dt} = \bar{R}(\bar{N}(t)), \\ \bar{N}(0) = N_0, \end{cases} \tag{7.16}$$

where

$$\bar{R}(x) := \int_Y p(dy)m(y)\frac{r(y)}{K(y)}x(K(y) - x)/m. \tag{7.17}$$

The value $m(y)$ in (7.14) is a lifetime in the state y of the continuous-time LGM in semi-Markov random environment. Models (7.11),(7.14) and (7.16) are called *averaged LG Verhulst models.*

7.4. Merging of Logistic Growth Model in Random Media

Let us consider the following difference equation, describing the logistic growth model (see Subsection 7.1) in perturbed semi-Markov random environment $y^\varepsilon(\nu(t/\varepsilon))$ (Subsection 7.3):

$$\begin{cases} N^\varepsilon(\nu(t/\varepsilon)+1) - N^\varepsilon(\nu(t/\varepsilon) &= \\ & \varepsilon \quad r(y^\varepsilon(\nu(t/\varepsilon)+1))N^\varepsilon(1 - N^\varepsilon/K(y^\varepsilon(\nu(t/\varepsilon)+1)), \\ N^\varepsilon(0) &= N_0, \end{cases} \tag{7.18}$$

where $r(y)$ and $K(y)$ are bounded continuous functions on Y.

Let the merging conditions be satisfied (see Subsection 1.9.4 of Chapter 1) and conditions of Theorem 6 (in Subsection 2.5 of Chapter 2) be fulfilled with the function $g(x,y) = r(y)x(1 - x/K(y))$. Then the process $N^\varepsilon(\nu(t/\varepsilon))$ in (7.18) converges weakly as $\varepsilon \to 0$ to the process $\tilde{N}(t)$, which satisfies the equation

$$\left\{ \frac{d\tilde{N}(t)}{dt} = \tilde{R}(\tilde{N}(t), \hat{y}(t))\tilde{N}(0) = N_0, \right. \tag{7.19}$$

where

$$\tilde{R}(u,v) := \int_{Y_v} p_v(dy)\frac{r(y)}{K(y)}u \cdot (K(y) - u)/m_v.$$

Equation (7.19) looks like the equation for the continuous-time LGM in semi-Markov random environment, but in place of semi-Markov process $y(\nu(t/\varepsilon))$ we have the merged Markov process $\hat{y}(t)$ in phase space $(V, \mathcal{V})$ with infinitesimal operator $\hat{Q}$, generated by the stochastic kernel

$$\hat{Q}, (y, A, t) := \hat{P}(y, A) \cdot (1 - e^{-\hat{q}(v)\cdot t}),$$

where $\hat{P}$ and $\hat{q}(v)$ are defined in Subsection 1.9.6 of Chapter 1.

7.5. Diffusion Approximation of Logistic Growth Model in Random Media.

Let the following condition be satisfied for LGM

$$\tilde{N}_0 = N_0 = -\frac{\int_Y p(dy)r(y)}{\int_Y p(dy)\frac{r(y)}{K(y)}}. \tag{7.20}$$

Condition (7.20) is the *balance condition* for LGM.

Consider the following difference equation in scale of time t/ε^2 :

$$N^\varepsilon_{\nu(t/\varepsilon^2)+1} - N^\varepsilon_{\nu(t/\varepsilon^2)} = \varepsilon \cdot r(y_{\nu(t/\varepsilon^2)+1})N^\varepsilon_{\nu(t/\varepsilon^2)} \cdot (1 - \frac{N^\varepsilon_{\nu(t/\varepsilon^2)}}{K(y^\varepsilon_{\nu(t/\varepsilon^2)+1})}), \tag{7.21}$$

where functions $r(y)$ and $K(y)$ are bounded and continuous on Y.

The conditions of Theorem 4 in (Chapter 2, Subsection 2.2) are satisfied with balance condition (7.20) and function

$$g(x, y) = r(y) \cdot x \cdot (1 - x/K(y)). \tag{7.22}$$

Applying this theorem to the process N^ε in (7.22) we obtain that $N^\varepsilon_{\nu(t/\varepsilon^2)}$ converges weakly as $\varepsilon \to 0$ to the diffusion process $\tilde{N}(t)$, which satisfies the following stochastic differential equation:

$$d\tilde{N}(t) = \tilde{\alpha}(\tilde{N}(t)dt + \tilde{\beta}(\tilde{N}(t))dw(t), \tag{7.23}$$

where

$$\begin{aligned} \tilde{\alpha}(u) \quad &:= \quad \textstyle\int_Y p(dy)[r(y)\mathbf{R}_0 r(y)u \\ &- \quad (r(y)\mathbf{R}_0 \tfrac{r(y)}{K(y)} + \tfrac{r(y)}{K(y)}\mathbf{R}_0 r(y))u^2 + (2\tfrac{r(y)}{K(y)}\mathbf{R}_0 \tfrac{r(y)}{K(y)})u^3]/m, \end{aligned}$$

and

$$\begin{aligned} \tilde{\beta}(u) \quad &:= \quad 2\textstyle\int_Y p(dy)[r(y)\mathbf{R}_0 r(y) + 1/2\tfrac{r^2(y)}{K^2(y)}u^2 \\ &+ \quad (r(y)\mathbf{R}_0 \tfrac{r(y)}{K(y)} + \tfrac{r(y)}{K(y)}\mathbf{R}_0 r(y)\tfrac{r(y)}{K(y)})u^3 \\ &+ \quad (\tfrac{r(y)}{K(y)}\mathbf{R}_0 \tfrac{r(y)}{K(y)}) + 1/2\tfrac{r^2(y)}{K^2(y)})u^4]/m, \end{aligned}$$

$w(t)$ is a standard Wiener process.

7.6. Normal Deviations of Logistic Growth Model in Random Media.

Consider the logistic growth model with discrete-time in random media:

$$\begin{cases} N^\varepsilon_{n+1} - N^\varepsilon_n = \varepsilon r(y_{n+1})N^\varepsilon_n \cdot (1 - N^\varepsilon_n/K(y_{n+1})), \\ N^\varepsilon_0 = N_0, \end{cases} \tag{7.24}$$

and the difference scheme for averaged logistic growth model:

$$\begin{cases} \tilde{N}^\varepsilon_{n+1} - \tilde{N}^\varepsilon_n = \varepsilon \tilde{R}(N^\varepsilon_n), \\ \tilde{N}^\varepsilon_0 = N_0, \end{cases}$$

where

$$\tilde{R}(x) := \int_Y p(dy)\frac{r(y)}{K(y)}(K(y) - x)x/m,$$

$r(y)$ and $K(y)$ are bounded and measurable functions on Y.

Let us define the following deviated process:

$$N^\varepsilon(t) := \sum_{n=0}^{+\infty} N^\varepsilon(n)\mathbf{1}\{\tau_n \le t/\varepsilon < \tau_{n+1}\}, \tag{7.25}$$

where

$$N^\varepsilon(n) := [N_n^\varepsilon - \tilde{N}_n^\varepsilon]/\sqrt{\varepsilon}$$

Applying the normal deviation theorem from Subsection 2.4 of Chapter 2, we conclude that the process $N^\varepsilon(t)$ converges weakly as $\varepsilon \to 0$ to the process $\tilde{N}(t)$ such that

$$\tilde{N}(t) = \int_0^t \int_Y p(dy)\frac{r(y)}{K(y)}(K(y) - 2\tilde{N}(s))\tilde{N}(s)ds + \int_0^t \sigma(\tilde{N}(s))dw(s),$$

where

$$\sigma^2(u) := \int_Y p(dy)(R(u,y) - \tilde{R}(u))\mathbf{R}_0(R(u,y) - \tilde{R}(u)) + (R(u,y) - \tilde{R}(u)^2/2]/m,$$

$$R(u,y) := \frac{r(y)}{K(y)} \cdot u \cdot (K(y) - u),$$

$\tilde{N}(s)$ is defined from the averaged equation:

$$\frac{d\tilde{N}(t)}{dt} \int_Y p(dy)\frac{r(y)}{K(y)}\tilde{N}(t)(K(y) - \tilde{N}(t))/m := \tilde{R}(\tilde{N}_t),$$

$w(t)$ is a standard Wiener process.

7.7. Stochastic Stability of Logistic Growth Model in Averaging and Diffusion Approximation Schemes.

7.7.1. STABILITY OF LGM IN AVERAGING SCHEME.

Let us consider the LGM in series scheme in the form of difference equation:

$$N^\epsilon_{\nu(t/\epsilon)+1} - N_{\nu(t/\epsilon)} = \varepsilon r(y_{\nu(t/\epsilon)+1})N^\epsilon \cdot (1 - \frac{N^\epsilon}{K(y_{\nu(t/\epsilon)})}), \tag{7.26}$$

where $r(y)$ and $K(y)$ are positive bounded measurable functions on Y.

In Subsection 7.3 we have stated that under averaging conditions the process $N^\epsilon_{\nu(t/\epsilon)}$ in (7.26) converges weakly as $\epsilon \to 0$ to the averaged process $\tilde{N}_t$ such that

$$\frac{d\tilde{N}_t}{dt} = \tilde{r} \cdot \tilde{N}_t \cdot (1 - \tilde{N}_t/\tilde{K}), \tag{7.27}$$

where

$$\tilde{r} := \int_Y p(dy)r(y)/m,$$

$$\tilde{K} := \Big(\int_Y p(dy)\frac{r(y)}{K(y)}\int_Y p(dy)r(y)\Big). \tag{7.28}$$

We know study the stability of the average model in (7.27)-(7.28). There are two standard steady states or equilibrium states for (7.27), namely $\tilde{N}=0$ and $\tilde{N}=\tilde{K}$, that is where $\frac{d\tilde{N}_t}{dt}=0$. $\tilde{N}=0$ is unstable since linearization about it (that is, $\tilde{N}^2$ is neglected compared with $\tilde{N}$) gives $\frac{d\tilde{N}}{dt}=\tilde{r}\tilde{N}$, and so $\tilde{N}$ grows exponentially from any initial value. The other equilibrium $\tilde{N}=\tilde{K}$ is stable and linearization about it (that is, $(\tilde{N}-\tilde{K})^2$ is neglected compared with $|\tilde{N}-\tilde{K}|$) gives $\frac{d(\tilde{N}-\tilde{K})}{dt}\simeq -\tilde{r}\cdot(\tilde{N}-\tilde{K})$ and so $\tilde{N}\to\tilde{K}$ as $t\to+\infty$. The averaged carrying capacity $\tilde{K}$ determines the average size of the stable steady state population, while $\tilde{r}$ is an average measure of the averaged rate at which it is reached, that is, it is an average measure of the dynamics. Also, as observed in Section 7.1.2., if $\tilde{N}_0=N_0>0$, then the solution of (7.27) is

$$\tilde{N}_t=\frac{N_0\tilde{K}e^{\tilde{r}t}}{[\tilde{K}+N_0(e^{\tilde{r}t}-1)]}\to\tilde{K}\text{as } t\to+\infty. \tag{7.29}$$

If $N_0<\tilde{K},\tilde{N}_t$ simply increases monotically to $\tilde{K}$, while $N_0>\tilde{K}$ decreases monotonically to $\tilde{K}$. In the former case, there is a qualitative difference depending on whether $N_0>\tilde{K}/2$ or $N_0<\tilde{K}/2$; with $N_0<\tilde{K}/2$ the form has a typical sigmoid character.

We are now interested in the stability of the difference equation (7.26), in random environment in series scheme.

It means that we are interested in function $V(x)$, satisfying the conditions i)-iii) (Subsection 4.7).

By Theorem 7 (Subsection 2.6.1) process $N^{\epsilon}_{\nu(t/\epsilon)}$ in (7.26) is stochastically asymptotically stable one, namely,

$$P_{N_0,y}\{\lim_{t\to+\infty}|N^{\epsilon}_{\nu(t/\epsilon)}|=\tilde{K}\}=1.$$

In this case the Lyapunov function is equal to $V(x)=(x-\tilde{K})$, and inequality $\tilde{r}x(\tilde{K}-x)V(x)\le\beta V(x)$ holds for $\beta<2\tilde{r}N_0/\tilde{K}$.

7.7.2. STABILITY OF LGM IN DIFFUSION APPROXIMATION SCHEME.

As we have seen in subsection 7.1 the equation (7.27) has two steady states: $\tilde{N}=0$ and $\tilde{N}=\tilde{K}$. Under these conditions the solution of (7.27) behaves deterministically: gwos, exponentially from any initial value or tends to $\tilde{K}$ as $t\to+\infty$, respectively.

There is the third case, where the solution of (7.27) behaves stochastically as some diffusion process. It is

$$\tilde{N}_0=\tilde{K}, \tag{7.30}$$

but we do not neglect the term $(\tilde{N}-\tilde{K})^2$.

Condition (7.30) is a balance condition and we can apply diffusion approximation scheme for the following difference equation:

$$N^{\epsilon}_{\nu(t/\epsilon)+1}-N^{\epsilon}_{\nu(t/\epsilon^2)}=\epsilon r(y_{\nu(t/\epsilon^2)})\cdot N^{\epsilon}\cdot(1-N^{\epsilon}_{\nu(t/\epsilon^2)}/K(y_{\nu(t/\epsilon^2)})) \tag{7.31}$$

The function $g(x, y) = r(y)x(1 - x/K(y))$ satisfies all the conditions of Theorem 8 in Chapter 4. Process $N^{\epsilon}_{\nu(t/\epsilon^2)}$ converges weakly as $\epsilon \to 0$ to the diffusion process $\tilde{N}(t)$:

$$d\tilde{N}(t) = \tilde{\alpha}(\tilde{N}(t))dt + \tilde{\beta}(\tilde{N}(t))dw(t), \tag{7.32}$$

where

$$\tilde{\alpha}(x) := \tilde{\alpha}_1 x - \tilde{\alpha}_2 x^2 + \tilde{\alpha}_3 x^3, \tag{7.33}$$

$$\tilde{\alpha}_1 := \int_Y p(dy)(r(y)\mathbf{R}_0 r(y))/m,$$

$$\tilde{\alpha}_2 := \int_Y p(dy)(r(y)\mathbf{R}_0 \frac{r(y)}{K(y)} + \frac{r(y)}{K(y)}\mathbf{R}_0 r(y))/m,$$

$$\tilde{\alpha}_3 := \int_Y p(dy)(2\frac{r(y)}{K(y)}\mathbf{R}_0 \frac{r(y)}{K(y)})/m,$$

and

$$\tilde{\beta}^2(x) := 2(\tilde{\beta}_1 \cdot x^2 + \tilde{\beta}_2 \cdot x^3 + \tilde{\beta}_3 \cdot x^4) \tag{7.34}$$

$$\tilde{\beta}_1 := \int_Y p(dy)(r(y)\mathbf{R}_0 r(y) + 1/2\frac{r^2(y)}{K^2(y)})/m,$$

$$\tilde{\beta}_2 := \int_Y p(dy)(r(y) \cdot \mathbf{R}_0 \frac{r(y)}{K(y)} + \frac{r(y)}{K(y)}\mathbf{R}_0 \cdot r(y) + \frac{r(y)}{K(y)})/m,$$

$$\tilde{\beta}_3 := \int_Y p(dy)(\frac{r(y)}{K(y)}\mathbf{R}_0 \frac{r(y)}{K(y)} + 1/2\frac{r^2(y)}{K^2(y)})/m,$$

$w(t)$ is a standard Wiener process.

Let us take the function $W(x) = x^2$. Then stability condition (subsection 2.7, Chapter 2) takes the following form:

$$\tilde{N}^2[(\tilde{\alpha}_1 + 2\tilde{\beta}_1 + \gamma) + (2\tilde{\beta}_2 - \tilde{\alpha}_2)\tilde{N} + (2\tilde{\alpha}_3 + 2\tilde{\beta}_3)\tilde{N}^2] \leq 0,$$

or

$$(\tilde{\alpha}_1 + 2\tilde{\beta}_1 + \gamma) + (2\tilde{\beta}_2 - \tilde{\alpha}_2)\tilde{N} + (2\tilde{\alpha}_3 + \tilde{\beta}_3)\tilde{N}^2 \leq 0. \tag{7.35}$$

Since $\tilde{\alpha}_3 > 0$ and $\tilde{\beta}_3 > 0$, then $(\tilde{\alpha}_3 + \tilde{\beta}_3) > 0$ and the inequality (7.35) has a solution iff

$$D := (2\tilde{\beta}_2 - \tilde{\alpha}_2)^2 - 8(\tilde{\alpha}_3 + \tilde{\beta}_3)(\tilde{\alpha}_1 + 2\tilde{\beta}_1 + \gamma) > 0,$$

or

$$\gamma < \frac{(2\tilde{\beta}_2 - \tilde{\alpha}_2)^2 - 8(\tilde{\alpha}_1 + 2\tilde{\beta}_1)(\tilde{\alpha}_3 + \tilde{\beta}_3)}{8(\tilde{\alpha}_3 + \tilde{\beta}_3)} \tag{7.36}$$

where $\tilde{\alpha}_i, \tilde{\beta}_i$ are defined in (7.33) and (7.34), respectively, $i = \overline{1,3}$.

Let $\tilde{N}_1$ and $\tilde{N}_2$ be the roots of respected quadratic equation in (7.35), $\tilde{N}_1 \leq \tilde{N}_2$. The inequality (7.33) is fulfilled if

$$\tilde{N}_1 < \tilde{N} < \tilde{N}_2. \tag{7.37}$$

Since $\tilde{N}$ depends on γ we have to choose γ from (7.36) and inquality (7.37).

It means, by Theorem 8, Chapter 2, that process $N^{\epsilon}_{\nu(t/\epsilon^2}$ in (7.36) is stochastically asymptotically stable process.

Remarks. Stochastic stability of stochastic differential equations (SDE) has been studied in [70], including asymptotic and global stability.

Asymptotic stability of linear stochastic systems was studied in [45]. Asymptotic stability of SDE with jumps has been studied in [31, p.325]. Asymptotic stochastic stability of stochastic systems with wide-band noise disturbances using martingale approach was studied in [12].

CHAPTER 8: PREDATOR-PREY MODELS

8.1. Deterministic Predator-Prey Model

When species interact the population dynamics of each species is affected. In general, there is a whole web of interacting species, called *trophic web*, which makes for structurally complex communities. We consider here systems involving two species. There are three main types of interaction:

(i) When the growth rate of one population is decreased, the other increased. In this case, we have a *predator-prey* system;

(ii) When the growth rate of one population is decreased, then the other is increased. We have the *competition*;

(iii) The other population's growth rate is enhanced by the growth of one population. We have the so-called *mutualism* or *symbiosis*.

Here, we are mainly concerned with *predator-prey models*. The book by Hassel [35] deals with predator-prey models. Beddington et al. [8] present some results on the dynamic complexity of coupled predator-prey systems. The book by Gurowski and Mira [33] deals generally with the mathematics of coupled systems, but also includs some interesting numerically computed results. The book by Murray [64] contains detailed analysis of continuous and discrete predator-prey models as well.

8.1.1. DISCRETE PREDATOR-PREY MODEL.

We consider a simple model in which predators simply search over a constant area and have unlimited capacity for consuming the prey. This can be described by the system:

$$\begin{cases} N_{t+1} = r \cdot N_t \cdot \exp[-aP_t], \\ P_{t+1} = N_t \cdot \{1 - \exp[-aP_t]\}, \quad a > 0, \quad r > 0, \end{cases} \tag{8.1}$$

where N_t is the prey population and P_t is the predator population at time t. Here $r > 0$ is the net linear rate of increase of the prey and $a > 0$ is the net linear rate of decrease of the predator.

The equilibrium values (N^*, P^*) of (8.1) are give by

$$N^* = 0, \quad P^* = 0$$

or $1 = r \cdot \exp[-a \cdot P^*], P^* = N^* \cdot (1 - \exp[-aP^*]$ and so positive steady state populations are

$$P^* = \frac{1}{a}\ln r, \quad N^* = \frac{r}{a(r-1)}\ln r, \quad r > 1. \tag{8.2}$$

The linear stability of the equilibria can be determined in the usual way by writing

$$N_t = N^* + n_t, \quad P_t = P^* + p_t, \quad |\frac{n_t}{N^*}| \ll 1, \quad |\frac{p_t}{p^*}| \ll 1,$$

substituting into (8.1) and retaining only linear terms. For the steady state (0,0) the analysis is particularly simple since

$$n_{t+1} = rn_t, \qquad p_{t+1} = 0,$$

and so it is stable for $r < 1$ since $n_t \to 0$ as $t \to +\infty$, and unstable for $r > 1$ (that is the range of r when the positive steady state (8.2) exists). For more detailed analysis of (8.1), see Murray [59].

8.1.2. CONTINUOUS PREDATOR-PREY MODEL.

Volterra (1926) [87] first proposed a simple model for the predation of one species by another to explain the oscillatory levels of certain fish catches in the Adriatic. We now discuss his model. If $N(t)$ is the prey population and $P(t)$ that of the predator at time t, then Volterra's model is

$$\begin{cases} \frac{dN}{dt} = N(a - bP), \\ \frac{dP}{dt} = P(cN - d), \end{cases} \tag{8.3}$$

where a, b, c and d are positive constants.

This model is based on the following assumptions:

(i) the prey in the absence of any predation grows unboundedly in a Malthusian way, thus is the aN term in (8.3);

(ii) the effect of the predation is to reduce the prey's per capita growth rate by a term proportional to the prey and predator populations (this is the $(-bNP)$ term);

(iii) in the absence of any prey for sustenance the predator's death rate results in exponential decay (that is the $(-d \cdot P)$ term in (8.3));

(iv) the prey's contribution to the predators' growth rate is cNP (that is, it is proportional to the available prey as well as to the size of the predator population). The NP terms can be thought of as representing the conversion of energy from one source to another: bNP is taken from the prey and $c - N \cdot P$ accrues to the predators.

The model (8.3) is known as the *Lotka-Volterra model* since the same equation were also derived by Lotka (1920) [55] (see also Lotka (1925) [56]) from a hypothetical chemical reaction which he claimed could exhibit periodic behaviour in the chemical concentrations.

Usually, a first step in analyzing the Lotka-Volterra model is to non-dimensionalize the system by writing

$$u(\tau) = \frac{cN(t)}{d}, \quad v(\tau) = \frac{bP(t)}{a}, \quad \tau = \alpha t, \quad \alpha = d/a$$

to obtain

$$\frac{du}{d\tau} = u(1-v), \quad \frac{dv}{d\tau} = \alpha v(u-1). \tag{8.4}$$

In the (u, v) phase plane, we get

$$\frac{dv}{du} = \alpha \cdot \frac{v(u-1)}{u \cdot (1-v)}, \tag{8.5}$$

which has singular points at $u = 0 = v$ and $u = v = 1$. Integrating (8.5), we then get the phase trajectories

$$\alpha \cdot u + v - \ln u^{\alpha} \cdot v = H, \tag{8.6}$$

where $H > H_{\min}$ is a constant: $H_{\min} = 1 + \alpha$ is the minimum of H over all (u, v) and it occurs at $u = v = 1$.

A closed trajectory in the u, v plane implies periodic solutions in τ for u and v in (8.4). The initial conditions, $u(0)$ and $v(0)$, determine the constant H in (8.6) and hence the phase trajectory.

8.2. Stochastic Predator-Prey Model (PPM in RM).

Here we consider *discrete-time and continuous-time predator-prey model*, introduced in Sections 8.1, in random environment.

We introduce several predator-prey models, discrete-time and continuous-time, in Markov and semi-Markov random environment.

8.2.1. DISCRETE PREDATOR-PREY MODEL IN MARKOV RANDOM MEDIA

Let $r(y)$ and $a(y)$ be bounded measurable positive functions on Y and $(y_n)_{n \in Z_+}$ be a homogeneous Markov chain in Y. We consider

$$\begin{cases} N_{r+1} = r(y_{n+1}) N_n \exp\left[-a(y_{n+1}) P_n\right], \\ P_{n+1} = N_n \cdot \{1 - \exp\left[-a(y_{n+1}) P_n\right]\}, \end{cases} \tag{8.7}$$

where N_n and P_n are the prey and predator populations, respectively, at time n. Note that the net linear rate $r(y)$ of increase of the prey and the net linear rate $a(y)$ of decrease of the predator are dependent on the state of Markov random environment.

Of course, functions N_n and P_n in (8.7) are random processes.

8.2.2. CONTINUOUS PREDATOR-PREY MODEL IN MARKOV RENEWAL RANDOM MEDIA.

Let $r(y)$ and $a(y)$ be the functions defined above, and $\nu(t) := \max\{n; \tau_n \le t\}$ be a counting process. We consider,

$$\begin{cases} N_{\nu(t)+1} &= r(y_{\nu(t)+1})N_{\nu(t)}\exp\left[-a(y_{\nu(t)+1})P_{\nu(t)}\right], \\ P_{\nu(t)+1} &= N_{\nu(t)}\{1 - \exp\left[-a(y_{\nu(t)+1})P_{\nu(t)}\right]\}. \end{cases} \tag{8.8}$$

Clearly, functions $N_{\nu(t)}$ and $P_{\nu(t)}$ are random processes.

We note that if $t\epsilon[\tau_n, \tau_{n+1}]$, then $\nu(t) = n$, and system (8.8) coincides with system (8.7).

8.2.3. CONTINUOUS PREDATOR-PREY MODEL IN SEMI-MARKOV RANDOM MEDIA

This model is defined as

$$\begin{cases} \frac{dP(t)}{dt} &= N(t)\cdot(a(y(t)) - b(y(t))\cdot P(t)), \\ \frac{dP(t)}{dt} &= P(t)\cdot(c(y(t))\cdot N(t) - d(y(t))), \end{cases} \tag{8.9}$$

where $a(y), b(y), c(y)$ and $d(y)$ are bounded measurable positive functions on Y and $y(t) := y_{\nu(t)}$ is a semi-Markov process.

Here, functions $N(t)$ and $P(t)$ are random functions. We can see that prey in the absence of any predation growth $a(y)N$, effect of the predaction is to reduce the prey's per capita growth rate $(-b(y)NP)$, the $(-d(y)P)$ term as the predator's death rate in the absence of any prey results in exponential decay, and the prey's contribution to the predators' growth rate term $c(y)NP$ are dependent on the state of semi-Markov random environment.

The measurability of predator-prey models in random media is described as follows.

The discrete-time predator-prey model in Markov random enviroment in (8.7) is $\mathcal{F}_{n+1}/\mathcal{R}$-measurable. This means that N_{n+1} and P_{n+1} are $\mathcal{F}^Y_{n+1}/\mathcal{R}$-measurable, as $r(y_{n+1})$ and $a(y_{n+1})$ are $\mathcal{F}^Y_{n+1}/\mathcal{R}$-measurable, and P_n and N_n are $\mathcal{F}^Y_n/\mathcal{R}$- measurable, and $\mathcal{F}^Y_n \subset \mathcal{F}^Y_{n+1} \subset \mathcal{F}_{n+1}$.

Continuous-time predator-prey model in markov renewal random environment in (8.8) is $\mathcal{G}_t/\mathcal{R}$-measurable, as such are $r(y_{\nu(t)+1})$, $a(y_{\nu(t)+1})$, $P_{\nu(t)}$ and $N_{\nu(t)}$ are $\mathcal{H}_t/\mathcal{R}$-measurable and $\mathcal{H}_t \subset \mathcal{G}_t$.

Continuous-time predator-prey model in semi-Markov rfandom environment in (8.9) is $\mathcal{H}_t/\mathcal{R}$-measurable. In particular, as $a(y(t))$ and $b(y(t)), c(y(t))$ and $d(y(t))$ in (8.9) are $\mathcal{H}_t/\mathcal{R}$-measurable, and $y(t) = y_{\nu(t)}$.

8.3. Averaging of Predator-Prey Model in Random Media.

Here, we apply the results from Section 2.7.1 of Chapter 2, namely, Theorems 9-10, to the predator-prey models in random media. We consider the *discrete predator-prey model in Markov renewal random environment* in series scheme:

$$\begin{cases} N^{\varepsilon}_{\nu(t/\varepsilon)+1} - N^{\varepsilon}_{\nu(t/\varepsilon)} &= N^{\varepsilon}_{\nu(t/\varepsilon)} \cdot (a(y^{\varepsilon}_{\nu(t/\varepsilon)+1}) - b(y_{\nu(t/\varepsilon)+1}) \cdot P^{\varepsilon}_{\nu(t/\varepsilon)}), \\ P^{\varepsilon}_{\nu(t/\varepsilon)+1} - P^{\varepsilon}_{\nu(t/\varepsilon)} &= P^{\varepsilon}_{\nu(t/\varepsilon)} \cdot (c(y_{\nu(t/\varepsilon)+1}) N^{\varepsilon}_{\nu(t/\varepsilon)} - d(y_{\nu(t/\varepsilon)+1})), \\ N^{\varepsilon}_0 &= N_0, \quad P^{\varepsilon}_0 = P_0, \end{cases} \tag{8.10}$$

where $a(y), b(y), c(y)$ and $d(y)$ are bounded measurable positive functions on Y and $\nu(t)$ is a counting process.

Here $X = R^2$, and $\mathbf{x} = [N, P]^T$. Using the function $\mathbf{g}(x, y)$ as the vector-function $\mathbf{g}([N, P]^T, y) = ((N(a(y) - b(y)P), (P(c(y)N - d(y)))^T$ in Theorem 10 we obtain that

$$(N^{\varepsilon}_{\nu(t/\varepsilon)}, P^{\varepsilon}_{\nu(t/\varepsilon)}) \rightarrow_{\varepsilon\to 0} (\tilde{N}_t, \tilde{P}_t),$$

where

$$\begin{cases} \frac{d\tilde{N}_t}{dt} &= \tilde{N}_t(\tilde{a} - \tilde{b} \cdot \tilde{P}_t), \\ \frac{d\tilde{P}_t}{dt} &= \tilde{P}_t \cdot (\tilde{c}\tilde{N}_t - \tilde{d}), \\ \tilde{N}_0 &= N_0, \quad \tilde{P}_0 = P_0, \end{cases} \tag{8.11}$$

and

$$\begin{aligned} \tilde{a} &:= \int_Y p(dy)a(y)/m, \\ \tilde{b} &:= \int_Y p(dy)b(y)/m, \\ \tilde{c} &:= \int_Y p(dy)c(y)/m, \\ \tilde{d} &:= \int_Y p(dy)d(y)/m, \\ m &:= \int_Y p(dy)m(y). \end{aligned} \tag{8.12}$$

Also, for the following *discrete predator-prey model in Markov random environment* in series scheme

$$\begin{cases} N^{\varepsilon}_{[t/\varepsilon]+1} - N_{[t/\varepsilon]} &= N^{\varepsilon}_{[t/\varepsilon]} \cdot (a(y_{[t/\varepsilon]+1}) - b(y_{[t/\varepsilon]+1}) \cdot P^{\varepsilon}_{[t/\varepsilon]}), \\ P^{\varepsilon}_{[t/\varepsilon]+1} - P^{\varepsilon}_{[t/\varepsilon]} &= P^{\varepsilon}_{[t/\varepsilon]} \cdot (c(y_{[t/\varepsilon]+1}) N^{\varepsilon}_{[t/\varepsilon]} - d(y_{[t/\varepsilon]+1})), \\ N^{\varepsilon}_0 &= N_0, \quad P^{\varepsilon}_0 = P_0. \end{cases}$$

We can apply Theorem 9, in Subsection 2.7.1 of Chapter 2 to obtain that

$$(N^{\varepsilon}_{[t/\varepsilon]}, P^{\varepsilon}_{[t/\varepsilon]}) \to_{\varepsilon \to 0} (\tilde{N}_t, \hat{P}_t),$$

where

$$\begin{cases} \frac{d\hat{N}_t}{dt} &= \hat{N}_t(\hat{a} - \hat{b} \cdot \hat{P}_t), \\ \frac{d\hat{P}_t}{dt} &= \hat{P}_t \cdot (\hat{c}\hat{N}_t - \hat{d}), \\ \hat{P}_0 &= P_0, \quad \hat{N}_0 = N_0, \end{cases}$$

and $\hat{a} := \tilde{a} \cdot m, \hat{b} := \tilde{b} \cdot m, \hat{c} := \tilde{c} \cdot m, \hat{d} := \tilde{d} \cdot m.$

We remark that the factor m^{-1} in (8.12) characterizes the mean stationary intensity of the jumps of random environment in the predator-prey model.

We shall also consider the following continuous-time predator-prey model in semi-Markov random environment in series scheme:

$$\begin{cases} \frac{dN^{\varepsilon}(t)}{dt} &= N^{\varepsilon}(t) \cdot (a(y(t/\varepsilon)) - b(y(t/\varepsilon)) \cdot P^{\varepsilon}(t), \\ \frac{dP^{\varepsilon}(t)}{dt} &= P^{\varepsilon}(t) \cdot (c(y(t/\varepsilon)) \cdot N^{\varepsilon}(t) - d((y(t/\varepsilon)), \\ P^{\varepsilon}(0) &= P_0, \quad N^{\varepsilon}(0) = N_0, \end{cases} \tag{8.13}$$

where $y(t) := y_{\nu(t)}$ is a semi-Markov process.

From the general theory of random evolutions [49,81,82] it follows that

$$(N^{\varepsilon}(t), P^{\varepsilon}(t)) \to_{\varepsilon \to 0} (\bar{N}(t), \bar{P}(t)),$$

where

$$\begin{cases} \frac{d\bar{N}(t)}{dt} &= \bar{N}(t) \cdot (\bar{a} - \bar{b} \cdot \bar{P}(t)), \\ \frac{d\bar{P}(t)}{dt} &= \bar{P}(t) \cdot (\bar{c} \cdot \bar{N}(t) - \bar{d}, \\ \bar{N}(0) &= N_0, \quad \bar{P}(0) = P_0, \end{cases} \tag{8.14}$$

and

$$
\begin{aligned}
\bar{a} &:= \int_Y p(dy)a(y)m(y)/m, \\
\bar{b} &:= \int_Y p(dy)b(y)m(y)/m, \\
\bar{c} &:= \int_Y p(dy)c(y)m(y)/m, \\
\bar{d} &:= \int_Y p(dy)d(y)m(y)/m, \\
m(y) &:= \int_0^{+\infty} tG_y(dt), \\
m &:= \int_Y p(dy)m(y).
\end{aligned}
\tag{8.15}
$$

Here, $m(y)$ is the lifetime in the state y of the continuous-time predator-prey model in semi-Markov random environment.

As a first step in analysing the above averaged (Lotka-Volterra) model, we first non-dimensionalize system (8.11), by writing

$$\tilde{u}(\tau) := \frac{\tau \tilde{N}_t}{\tilde{d}}, \quad \tilde{v}(\tau) := \frac{\tilde{b}\tilde{P}_t}{\tilde{a}}, \tau := \tilde{a}t, \quad \tilde{\alpha} := \frac{\tilde{d}}{\tilde{a}}.$$

We obtain

$$\frac{d\tilde{u}}{d\tau} = \tilde{u}(1-\tilde{v}), \quad \frac{d\tilde{v}}{d\tau} = \tilde{\alpha}\tilde{v}(\tilde{u}-1).$$

In the $(\tilde{u}, \tilde{v})$ phase place, we have

$$\frac{d\tilde{v}}{d\tilde{u}} = \tilde{\alpha}\frac{\tilde{v}(\tilde{u}-1)}{\tilde{u}(1-\tilde{v})}, \tag{8.16}$$

which has singular points at $\tilde{u} = \tilde{v} = 0$ and $\tilde{u} = \tilde{v} = 1$. We can integrate (8.16) to get the phase trajectories

$$\tilde{\alpha}\tilde{u} + \tilde{v} - \ln \tilde{u} \cdot \tilde{v} = \tilde{H},$$

where $\tilde{H} > \tilde{H}_{\min}$ is a constant: $\tilde{H}_{\min} = 1 + \tilde{\alpha}$ is the minimum of $\tilde{H}$ over all $(\tilde{u}, \tilde{v})$ and it occurs at $\tilde{u} = \tilde{v} = 1$, and $\tilde{H}$ is determined by the initial condition $(\tilde{u}(0), \tilde{v}(0))$.

We now derive the averaging of the continuous-time predator-prey model in semi-Markov random environment (8.13) by another method using ergodicity of semi-Markov process $y(t)$. We first rewrite system (8.13) in the form:

$$
\begin{cases}
\frac{\ln N^\varepsilon(t)}{N_0^\varepsilon} = \varepsilon \cdot \int_0^{t/\varepsilon}(a(y(s)) - b(y(s))P^\varepsilon(s))ds, \\
\frac{\ln P^\varepsilon(t)}{P_0^\varepsilon} = \varepsilon \cdot \int_0^{t/\varepsilon}(c(y(s))N^\varepsilon(s) - d(y(s))ds.
\end{cases}
\tag{8.17}
$$

Also, the system (8.17) can be rewritten as

$$
\begin{cases}
\frac{\ln \bar{N}(t)}{N_0} = \int_0^t(\bar{a} - \bar{b}\bar{P}(s))ds, \\
\frac{\ln \bar{P}(t)}{P_0} = \int_0^t(\bar{c} \cdot \bar{N}(s) - \bar{d})ds.
\end{cases}
\tag{8.18}
$$

Now, we prove that $N^\varepsilon_{(t)}$ in (8.17) converges weakly to $\bar{N}(t)$ in (8.18) as $\varepsilon \to 0$. Subtracting the first equation in (8.18) from the first equation in (8.17) we obtain

$$\begin{aligned}
\ln \frac{N^\varepsilon_{(t)}}{\bar{N}(t)} &= \varepsilon \cdot \int_0^{t/\varepsilon}(a(y(s)) - b(y(s))P^\varepsilon(s))ds - \int_0^t(\bar{a} - \bar{b}\cdot \bar{P}(s))ds \\
&= \varepsilon \cdot \int_0^{t/\varepsilon}(a(y(s)) - \bar{a})ds - \varepsilon \cdot \int_0^{t/\varepsilon}(b(y(s))P^\varepsilon(s) - \bar{b}\cdot \bar{P}(s)/ds \\
&= \varepsilon \cdot \int_0^{t/\varepsilon}(a(y(s)) - \bar{a})ds - \varepsilon \cdot \int_0^{t/\varepsilon}(b(y(s)) - \bar{b})\bar{P}(s)ds \\
&- \varepsilon \cdot \int_0^{t/\varepsilon} b(y(s))(P^\varepsilon(s) - \bar{P}(s))ds.
\end{aligned} \tag{8.19}$$

The first term in the right hand-side of (8.19) converges weakly to zero as $\varepsilon \to 0$, since $y(t)$ is an ergodic semi-Markov process and

$$\varepsilon \int_0^{t/\varepsilon} a(y(s))ds \to_{\varepsilon\to 0} \bar{a}t, \tag{8.20}$$

As $\bar{P}(s)$ is bounded, the second term on the right hand-side of (8.19) converges weakly as $\varepsilon \to 0$ to zero, again due to the ergodic result (8.20) for the semi-Markov process with function $b(y(t))$. Further, since $P^\varepsilon(t)$ converges weakly to $\bar{P}(t)$ as $\varepsilon \to 0$ and $b(y)$ is bounded, the third term on the right hand-side of (8.19) converges weakly to zero as $\varepsilon \to 0$. Therefore,

$$\ln \frac{N^\varepsilon(t)}{\bar{N}(t)} \to_{\varepsilon\to 0} 0, \quad \forall t \epsilon R_+,$$

and, hence

$$N^\varepsilon(t) \to_{\varepsilon\to 0} \bar{N}(t).$$

A similar result is true for the second equation in (8.13). Namely,

$$\ln \frac{P^\varepsilon(t)}{\bar{P}(t)} \to_{\varepsilon\to 0} 0, \quad \forall t \epsilon R_+,$$

and

$$P^\varepsilon(t) \to_{\varepsilon\to 0} \bar{P}(t).$$

8.4. Merging of Predator-Prey Model.

Here, we apply the results from Section 2.7.4 of Chapter 2 (namely, Theorem 14) to the predator-prey models in random media. We consider the *discrete predator-prey model in Markov renewal random environment* in series scheme:

$$\begin{cases}
N^\varepsilon_{\nu(t/\varepsilon)+1} - N^\varepsilon_{\nu(t/\varepsilon)} = N^\varepsilon_{\nu(t/\varepsilon)} \cdot (a(y^\varepsilon_{\nu(t/\varepsilon)+1}) - b(y_{\nu(t/\varepsilon)+1}) \cdot P^\varepsilon_{\nu(t/\varepsilon)}), \\
P^\varepsilon_{\nu(t/\varepsilon)+1} - P^\varepsilon_{\nu(t/\varepsilon)} = P^\varepsilon_{\nu(t/\varepsilon)} \cdot (c(y_{\nu(t/\varepsilon)+1})N^\varepsilon_{\nu(t/\varepsilon)} - d(y_{\nu(t/\varepsilon)+1})), \\
N^\varepsilon_0 = N_0, \quad P^\varepsilon_0 = P_0,
\end{cases}$$

where $a(y), b(y), c(y)$ and $d(y)$ are bounded measurable positive functions on Y and $\nu(t)$ is a counting process.

Here $X = R^2$, and $\mathbf{x} = [N, P]^T$. Using the function $\mathbf{g}(x, y)$ as the vector-function $\mathbf{g}([N, P]^T, y) = ((N(a(y) - b(y)P), (P(c(y)N - d(y)))^T$ in Theorem 14 (Section 2.7.4 of Chapter 2) we obtain that

$$(N^{\varepsilon}_{\nu(t/\varepsilon)}, P^{\varepsilon}_{\nu(t/\varepsilon)}) \to_{\varepsilon\to 0} (\hat{N}_t, \hat{P}_t),$$

where

$$\begin{cases} \frac{d\hat{N}_t}{dt} &= \hat{N}_t(\hat{a}(\hat{y}(t)) - \hat{b}(\hat{y}(t)) \cdot \hat{P}_t), \\ \frac{d\hat{P}_t}{dt} &= \hat{P}_t \cdot (\hat{c}(\hat{y}(t))\hat{N}_t - \hat{d}(\hat{y}(t))), \\ \hat{N}_0 &= N_0, \quad \hat{P}_0 = P_0, \end{cases}$$

and

$$\begin{aligned} \hat{a}(v) &:= \textstyle\int_{Y_v} p(dy)a(y)/m_v, \\ \hat{b}(v) &:= \textstyle\int_{Y_v} p(dy)b(y)/m_v, \\ \hat{c}(v) &:= \textstyle\int_{Y_v} p(dy)c(y)/m_v, \\ \hat{d}(v) &:= \textstyle\int_{Y_v} p(dy)d(y)/m_v, \\ m_v &:= \textstyle\int_{Y_v} p(dy)m(y). \end{aligned}$$

We now take another approach to obtain the analogous result for the system

$$\begin{cases} \frac{dN^{\varepsilon}(t)}{dt} &= N^{\varepsilon}(t)(a(y^{\varepsilon}(t)) - b(y^{\varepsilon}(t))P^{\varepsilon}(t)), \\ \frac{dP^{\varepsilon}(t)}{dt} &= P^{\varepsilon}(t)(c(y^{\varepsilon}(t))N^{\varepsilon}(t) - d(y^{\varepsilon}t)), \\ N^{\varepsilon}(0) &= N_0, \quad P^{\varepsilon}(0) = P_0, \end{cases} \tag{8.21}$$

where $y^{\varepsilon}(t) := y^{\varepsilon}(\nu(t/\varepsilon))$. The idea is similar to that used in Subsection 8.3. First, we apply the following famous result for merging of integral functional of a perturbed semi-Markov process (see Chapter 1, Subsection 1.9):

$$\varepsilon \int_0^{t/\varepsilon} a(y^{\varepsilon}(s/\varepsilon))ds \to_{\varepsilon\to 0} \int_0^t \tilde{a}(\hat{y}(s))ds,$$

where $\hat{y}(t)$ is a merged Markov process. In this way, we obtain that $(N^{\varepsilon}, P^{\varepsilon})$ converges weakly as $\varepsilon \to 0$ to the pair $(\tilde{N}, \tilde{P})$ which satisfies the following system:

$$\begin{cases} \frac{d\bar{N}(t)}{dt} = \bar{N}(t)(\bar{a}(\hat{y}(t)) - \bar{b}(\hat{y}(t))\bar{P}(t)), \\ \frac{d\bar{P}(t)}{dt} = \bar{P}(t)\cdot(\bar{c}(\hat{y}(t)) - \bar{d}(\hat{y}(t)\bar{N}(t)), \\ \bar{P}(0) = P_0, \quad \bar{N}(0) = N_0, \end{cases} \tag{8.22}$$

$$\begin{aligned} \bar{a}(v) &:= \int_{Y_v} p_v(dy)a(y)/m_v, \\ \bar{b}(v) &:= \int_{Y_v} p_v(dy)b(y)/m_v, \\ \bar{c}(v) &:= \int_{Y_v} p_v(dy)c(y)/m_v, \\ \bar{d}(v) &:= \int_{Y_v} p_v(dy)d(y)/m_v, \\ \bar{d}(v) &:= \int_{Y_v} p_v(dy)m(y)d(y)/m_v, \\ m(y) &:= \int_0^\infty tG_y(dt), \\ m_v &:= \int_{Y_v} p_v(dy)m(y). \end{aligned}$$

8.5. Diffusion Approximation of Predator-Prey Model.

Here, we apply the results from Section 2.7.2 of Chapter 2 (namely, Theorem 12) to the predator-prey models in random media.

Let us suppose that $\tilde{a} = \tilde{b} = \tilde{c} = \tilde{d} = 0$ (see (8.12)) and hence , *the balance condition is satisfied for predator-prey models* (8.10) and (8.13).

We consider the system (8.10) in scale of time t/ε^2 :

$$\begin{cases} N^\varepsilon_{\nu(t/\varepsilon^2)+1} - N^\varepsilon_{\nu(t/\varepsilon^2)} = N^\varepsilon_{\nu(t/\varepsilon^2)}(a(y_{\nu(t/\varepsilon^2)+1}) - b(y_{\nu(t/\varepsilon^2)+1})P^\varepsilon_{\nu(t/\varepsilon^2)}), \\ P^\varepsilon_{\nu(t/\varepsilon^2)+1} - P^\varepsilon_{\nu(t/\varepsilon^2)} = P^\varepsilon_{\nu(t/\varepsilon^2)}(c(y_{\nu(t/\varepsilon^2)+1})N^\varepsilon_{\nu(t/\varepsilon^2)} - d(y_{\nu(t/\varepsilon^2)+1})), \\ N^\varepsilon_0 = N_0, \quad P^\varepsilon_0 = P_0, \end{cases} \tag{8.23}$$

where $a(y), b(y), c(y)$ and $d(y)$ are bounded measurable positive functions on Y, and $\nu(t)$ is a counting process.

Here, $X = R^2$, function $\mathbf{g}(\mathbf{x}, y)$ is the vector-function

$$\mathbf{g}([N,P]^T, y) = [(N(a(y) - b(y)P), (P(c(y)N - d(y))]^T,$$

and $\mathbf{x} = [N, P]^T$.

Applying Theorem 12 to this predator-prey model with the above vector-function $\mathbf{g}([N,P]^T,y)$, we obtain that drift coefficient $\tilde{\alpha}$ and diffusion matrix $\tilde{\beta}$ are given by

$$\tilde{\alpha} = [\alpha_1, \alpha_2]^T,$$

where

$$\begin{aligned}\alpha_1 &= \textstyle\int_Y p(dy)[N(a(y)-b(y)P)R_0(a(y)-b(y)P) - P^2(c(y)N-d(y))R_0c(y)]/m,\\ \alpha_2 &= \textstyle\int_Y p(dy)[N^2(b(y)-a(y)P)R_0b(y) + P(c(y)N-d(y))R_0(c(y)N-d(y))]/m,\end{aligned} \tag{8.24}$$

and

$$\tilde{\beta}([N,P]^T) = \begin{pmatrix} \beta_{11} & \beta_{12} \\ \beta_{21} & \beta_{22} \end{pmatrix},$$

where

$$\begin{aligned}\beta_{11} &= 2\textstyle\int_Y p(dy)[N^2(a(y)-b(y)P)R_0(a(y)-b(y)P)\\ &+ 2^{-1}N^2(a(y)-b(y)P)^2,\\ \beta_{12} &= 2\textstyle\int_Y p(dy)[PN(a(y)-b(y)P)R_0(c(y)N-d(y))\\ &+ 2^{-1}N(a(y)-b(y)P)(c(y)N-d(y))^2,\\ \beta_{21} &= 2\textstyle\int_Y p(dy)[PN(c(y)N-d(y))R_0(a(y)-b(y)P)\\ &+ 2^{-1}NP(a(y)-b(y)P)(c(y)N-d(y))^2,\\ \beta_{22} &= 2\textstyle\int_Y p(dy)[P^2(c(y)N-d(y))R_0(c(y)N-d(y))\\ &+ 2^{-1}P^2(c(y)N-d(y))^2.\end{aligned} \tag{8.25}$$

We obtain that in diffusion approximation the system (8.23) has the following limiting system as $\varepsilon \to 0$

$$\begin{cases} d\hat{N}(t) = \alpha_1(N,P)dt + \beta_{11}dw_1(t) + \beta_{12}dw_2(t),\\ d\hat{P}(t) = \alpha_2(N,P)dt + \beta_{21}dw_1(t) + \beta_{22}dw_2(t),\end{cases} \tag{8.26}$$

where $\alpha_i := \alpha_i(N,P)$, $i = 1,2$, and $\beta_{ij} := \beta_{ij}(N,P)$, $i,j = 1,2$, are defined in (8.24) and (8.25), respectively.

8.6. Normal Deviations of Predator-Prey Model in Random Media.

Here, we apply the results from Section 2.7.3 of Chapter 2 (namely, Theorem 13) to the predator-prey models in random media. We consider the *discrete predator-prey model in Markov renewal random environment* in series scheme:

$$\begin{cases} N^{\varepsilon}_{\nu(t/\varepsilon)+1} - N^{\varepsilon}_{\nu(t/\varepsilon)} &= N^{\varepsilon}_{\nu(t/\varepsilon)} \cdot (a(y^{\varepsilon}_{\nu(t/\varepsilon)+1}) - b(y_{\nu(t/\varepsilon)+1}) \cdot P^{\varepsilon}_{\nu(t/\varepsilon)}), \\ P^{\varepsilon}_{\nu(t/\varepsilon)+1} - P^{\varepsilon}_{\nu(t/\varepsilon)} &= P^{\varepsilon}_{\nu(t/\varepsilon)} \cdot (c(y_{\nu(t/\varepsilon)+1}) N^{\varepsilon}_{\nu(t/\varepsilon)} - d(y_{\nu(t/\varepsilon)+1})), \\ N^{\varepsilon}_0 &= N_0, \quad P^{\varepsilon}_0 = P_0, \end{cases} \tag{8.27}$$

where $a(y), b(y), c(y)$ and $d(y)$ are bounded measurable positive functions on Y, and $\nu(t)$ is a counting process.

Here, $X = R^2$, and $\mathbf{x} = (N, P)^T$. Using the function $\mathbf{g}(x, y)$ as the vector-function $\mathbf{g}((N, P)^T, y) = ((N(a(y) - b(y)P), (P(c(y)N - d(y)))^T$ in Theorem 10 (see Section 8.4 and Section 2.7.3) we obtain that

$$(N^{\varepsilon}_{\nu(t/\varepsilon)}, P^{\varepsilon}_{\nu(t/\varepsilon)}) \to_{\varepsilon\to 0} (\hat{N}_t, \hat{P}_t),$$

where

$$\begin{cases} \frac{d\hat{N}_t}{dt} &= \hat{N}_t(\hat{a} - \hat{b}\hat{P}_t), \\ \frac{d\hat{P}_t}{dt} &= \hat{P}_t(\hat{c}\hat{N}_t - \hat{d}), \\ \hat{N}_0 &= N_0, \quad \hat{P}_0 = P_0, \end{cases} \tag{8.28}$$

and

$$\begin{aligned} \hat{a} &:= \textstyle\int_Y p(dy)a(y)/m, \\ \hat{b} &:= \textstyle\int_Y p(dy)b(y)/m, \\ \hat{c} &:= \textstyle\int_Y p(dy)c(y)/m, \\ \hat{d} &:= \textstyle\int_Y p(dy)d(y)/m, \\ m &:= \textstyle\int_Y p(dy)m(y). \end{aligned}$$

Applying the Theorem 13 (see Section 2.7.3 of Chapter 2) to this predator-prey model with the above vector-function $\mathbf{g}((N, P)^T, y)$ we obtain that the drift coefficient $\tilde{\alpha}(\mathbf{x}) := \tilde{\mathbf{g}}_{\mathbf{x}}(\hat{\mathbf{x}})\tilde{\mathbf{Z}}$ and diffusion matrix $\beta(\hat{\mathbf{x}})$ in (23) are given by (here, $\hat{\mathbf{x}} := [\hat{N}, \hat{P}]^T$ and $\tilde{\mathbf{Z}} := (\tilde{N}, \tilde{P})^T$)

$$\tilde{\alpha}(\mathbf{x}) = (\alpha_1(\hat{N}, \hat{P}, \tilde{N}, \tilde{P}), \alpha_2(\hat{N}, \hat{P}, \tilde{N}, \tilde{P}))^T,$$

where

$$\begin{aligned} \alpha_1(\hat{N}, \hat{P}, \tilde{N}, \tilde{P}) &= \textstyle\int_Y p(dy)[\tilde{N}(a(y) - b(y)\hat{P}) - b(y)\hat{N}\tilde{P}]/m, \\ \alpha_2(\hat{N}, \hat{P}, \tilde{N}, \tilde{P}) &= \textstyle\int_Y p(dy)[c(y)\hat{P}\tilde{N} + \tilde{P}(c(y)\hat{N} - d(y))]/m, \end{aligned} \tag{8.30}$$

and

$$\tilde{\beta}(\hat{N},\hat{P}) = \begin{pmatrix} \beta_{11}(\hat{N},\hat{P}) & \beta_{12}(\hat{N},\hat{P}) \\ \beta_{21}(\hat{N},\hat{P}) & \beta_{22}(\hat{N},\hat{P}) \end{pmatrix},$$

where

$$\begin{aligned}
\beta_{11}(\hat{N},\hat{P}) = & \; 2 \quad \hat{N}^2 \textstyle\int_Y p(dy)[(a(y)-\hat{a})R_0(a(y)-\hat{a}) + \hat{P}(a(y)-\hat{a})R_0(\hat{b}-b(y)) \\
& + \quad \hat{P}(\hat{b}-b(y))R_0(a(y)-\hat{a}) + \hat{P}^2(\hat{b}-b(y))R_0(\hat{b}-b(y)) \\
& + \quad (a(y)-\hat{a})^2 + 2\hat{P}(a(y)-\hat{a})(\hat{b}-b(y)) + \hat{P}^2(\hat{b}-b(y))^2]/m, \\
\beta_{12}(\hat{N},\hat{P}) = & \; 2 \quad \hat{N}\hat{P} \textstyle\int_Y p(dy)[\hat{N}(a(y)-\hat{a})R_0(c(y)-\hat{c}) + (a(y)-\hat{a})R_0(\hat{d}-d(y)) \\
& + \quad \hat{N}\hat{P}(\hat{b}-b(y))R_0(c(y)-\hat{c}) + \hat{P}^2(b(y)-\hat{b})R_0(d(y)-\hat{d}) \\
& + \quad \hat{N}(a(y)-\hat{a})(c(y)-\hat{c}) + (a(y)-\hat{a})(\hat{d}-d(y)) \\
& + \quad \hat{N}\hat{P}(\hat{b}-b(y))(c(y)-\hat{c}) + \hat{P}^2(b(y)-\hat{b})(d(y)-\hat{d})]/m, \\
\beta_{21}(\hat{N},\hat{P}) = & \; 2 \quad \hat{N}\hat{P} \textstyle\int_Y p(dy)[\hat{N}(c(y)-\hat{c})R_0(a(y)-\hat{a}) + (d(y)-\hat{d})R_0(\hat{a}-a(y)) \\
& + \quad \hat{N}\hat{P}(\hat{c}-c(y))R_0(b(y)-\hat{b}) + \hat{P}^2(d(y)-\hat{d})R_0(b(y)-\hat{b}) \\
& + \quad \hat{N}(c(y)-\hat{c})(a(y)-\hat{a}) + (d(y)-\hat{d})(\hat{a}-a(y)) \\
& + \quad \hat{N}\hat{P}(\hat{c}-c(y))(b(y)-\hat{b}) + \hat{P}^2(d(y)-\hat{d})(b(y)-\hat{b}]/m, \\
\beta_{22}(\hat{N},\hat{P}) = & \; 2 \quad \hat{P}^2 \textstyle\int_Y p(dy)[\hat{N}^2(c(y)-\hat{c})R_0(c(y)-\hat{c}) + \hat{N}(c(y)-\hat{c})R_0(\hat{d}-d(y)) \\
& + \quad \hat{N}(\hat{d}-d(y))R_0(c(y)-\hat{c}) + (\hat{d}-d(y))R_0(\hat{d}-d(y)) \\
& + \quad \hat{N}^2(c(y)-\hat{c})^2 + 2\hat{N}(c(y)-\hat{c})(\hat{d}-d(y)) + (\hat{d}-d(y))^2]/m.
\end{aligned} \tag{8.31}$$

We conclude that $\mathbf{Z}_t^\epsilon$, where $\mathbf{Z}_n^\epsilon := [\mathbf{X}_n^\epsilon - \hat{\mathbf{X}}_n^\epsilon]/\sqrt{\epsilon} = [N^\varepsilon_{\nu(t/\varepsilon)} - \hat{N}^\varepsilon_{\nu(t/\varepsilon)}, P^\varepsilon_{\nu(t/\varepsilon)} - \hat{P}^\varepsilon_{\nu(t/\varepsilon)}]/\sqrt{\epsilon}$ converges weakly as $\varepsilon \to 0$ to the processes $\tilde{\mathbf{Z}}_t := (\tilde{N}(t), \tilde{P}(t))$ that satisfies the following stochastic differential equations:

$$\begin{cases} d\tilde{N}(t) = \alpha_1(\hat{N},\hat{P},\tilde{N},\tilde{P})dt + \beta_{11}(\hat{N},\hat{P})dw_1(t) + \beta_{12}(\hat{N},\hat{P})dw_2(t), \\ d\tilde{P}(t) = \alpha_2(\hat{N},\hat{P},\tilde{N},\tilde{P})dt + \beta_{21}(\hat{N},\hat{P})dw_1(t) + \beta_{22}(\hat{N},\hat{P})dw_2(t), \end{cases} \tag{8.32}$$

where $\alpha_i := \alpha_i(\hat{N}, \hat{P}, \tilde{N}, \tilde{P})$, $i = 1, 2$, and $\beta_{ij} := \beta_{ij}(\hat{N}, \hat{P})$, $i, j = 1, 2$, are defined in (8.30) and (8.31), respectively.

8.7. Stochastic Stability of Predator-prey Model in Averaging and Diffusion Approximation Schemes.

8.7.1. STOCHASTIC STABILITY OF PREDATOR-PREY MODEL IN AVERAGING SCHEME.

Let us consider the predator-prey model in series scheme

$$\begin{cases} N^{\varepsilon}_{\nu(t/\varepsilon)+1} - N^{\varepsilon}_{\nu(t/\varepsilon} &= N^{\varepsilon}_{\nu(t/\varepsilon)}(a(y_{\nu(t/\varepsilon)+1}) - b(y_{\nu(t/\varepsilon)+1}) \cdot P^{\varepsilon}_{\nu(t/\varepsilon)}), \\ P^{\varepsilon}_{\nu(t/\varepsilon)+1} - P^{\varepsilon}_{\nu(t/\varepsilon)} &= P^{\varepsilon}_{\nu(t/\varepsilon)} \cdot (C(y_{\nu(t/\varepsilon)+1}) N^{\varepsilon}_{\nu(t/\varepsilon)} - d(y_{\nu(t/\varepsilon)+1})), \\ N^{\varepsilon}_0 &= N_0, \quad P^{\varepsilon}_0 = P_0, \end{cases} \tag{8.33}$$

and the averaged predator-prey model

$$\begin{cases} d\tilde{N}_t &= \tilde{N}_t \cdot (\tilde{a} - \tilde{b} \cdot \tilde{P}_t), \\ d\tilde{P}_t &= \tilde{P}_t \cdot (\tilde{c} \cdot \tilde{N}_t - \tilde{d}), \\ \tilde{N}_0 &= N_0, \quad \tilde{P}_0 = P_0, \end{cases} \tag{8.34}$$

where

$$\tilde{a} := \int_Y p(dy)a(y)/m,$$

$$\tilde{b} := \int_Y p(dy)b(y)/m,$$

$$\tilde{c} := \int_Y p(dy)c(y)/m,$$

$$\tilde{d} := \int_Y p(dy)d(y)/m.$$

In Subsection 8.3, we have studied some stability property of *averaged Lotka-Volterra model* (8.11). Also, for the continuous-time predator-prey model in semi-Markov random media $y(t/\varepsilon)$ in series scheme

$$\begin{cases} \frac{dN^{\varepsilon}(t)}{dt} &= N^{\varepsilon}(t)(a(y(t/\varepsilon)) - b(y(t/\varepsilon))P^{\varepsilon}(t)), \\ \frac{dP^{\varepsilon}(t)}{dt} &= P^{\varepsilon}(t) \cdot (c(y(t/\varepsilon))N^{\varepsilon}(t) - d(y(t/\varepsilon)), \\ N^{\varepsilon}(0) &= N_0, \quad P^{\varepsilon}(0) = P_0, \end{cases} \tag{8.35}$$

we have shown that $(N^\varepsilon(t), P^\varepsilon(t))$ converges weakly as $\varepsilon \to 0$ to the couple $(\bar{N}(t), \bar{P}(t))$ with

$$\begin{cases} \frac{d\bar{N}(t)}{dt} &= \bar{N}(t)\cdot(\bar{a}-\bar{b}\cdot\bar{P}(t)), \\ \frac{d\bar{P}(t)}{dt} &= \bar{P}(t)\cdot(\bar{c}\cdot\bar{N}(t)-\bar{d}), \\ \bar{N}(0) &= N_0, \quad \bar{P}(0)=P_0 \end{cases} \tag{8.36}$$

where $\bar{a}, \bar{b}, \bar{c}$ and $\bar{d}$ are defined in (8.15).

Let us analyze the stability of the perturbed system (8.35) via the analysis of averaged system (8.36).

Let us non-dimensionalize the averaged system (8.36) by writing:

$$\bar{u}(\tau) := \frac{\bar{c}\bar{N}(t)}{\bar{d}}, \quad \bar{v}(\tau) := \frac{\bar{b}\cdot\bar{P}(t)}{\bar{a}}, \quad \tau := \bar{a}\cdot t, \bar{\alpha} := \frac{\bar{d}}{\bar{a}}, \tag{8.37}$$

and rewriting (8.36) as

$$\frac{d\bar{u}}{d\tau} = \bar{u}\cdot(1-\bar{v}), \quad \frac{d\bar{v}}{d\tau} = \bar{\alpha}\cdot\bar{v}\cdot(\bar{u}-1). \tag{8.38}$$

In the $(\bar{u},\bar{v})$ phase plane, we get

$$\frac{d\bar{v}}{d\bar{u}} = \frac{\bar{\alpha}\cdot\bar{v}(\bar{u}-1)}{\bar{u}\cdot(1-\bar{v})}, \tag{8.39}$$

which has singular points at $\bar{u}=\bar{v}=0$ and $\bar{u}=\bar{v}=1$. We can integrate (8.39) to get the phase trajectories

$$\bar{\alpha}\cdot\bar{u}+\bar{v}-\ln\bar{u}^{\bar{\alpha}}\cdot\bar{v} = \bar{H}, \tag{8.8.40}$$

where $\bar{H} > \bar{H}_{\min}$ is a constant; $\bar{H}_{\min} := 1+\bar{\alpha}$ is the minimum of $\bar{H}$ over all $(\bar{u},\bar{v})$ and it occurs at $\bar{u}=\bar{v}=1$. The solutions $\bar{u}(\tau)$ and $\bar{v}(\tau)$ are periodic functions.

From the analysis of $N^\varepsilon(t)$ and $\bar{N}(t)$ in Subsection 8.3, we have obtained the following expression:

$$\begin{aligned} \ln\frac{N^\varepsilon(t)}{\bar{N}(t)} &= \varepsilon\cdot\int_0^{t/\varepsilon}(a(y(s)-\bar{a})ds \\ &-\varepsilon\cdot\int_0^{t/\varepsilon}(b(y(s))-\bar{b})\bar{P}(s)ds \\ &-\varepsilon\cdot\int_0^{t/\varepsilon}b(y(s))(P^\varepsilon(s)-\bar{P}(s))ds \end{aligned} \tag{8.41}$$

for the prey population and the following expression for the predator population

$$\begin{aligned} \ln\frac{P^\varepsilon(t)}{\bar{P}(t)} &= \varepsilon\int_0^{t/\varepsilon}c(y(s))(N^\varepsilon(s)-\bar{N}(s))ds \\ &-\varepsilon\int_0^{t/\varepsilon}(c(y(s))-\bar{c})\bar{N}(s)ds-\varepsilon\int_0^{t/\varepsilon}d(y(s))-\bar{d})ds. \end{aligned} \tag{8.42}$$

We note that the prey and predator populations are bounded. Therefore,

$$|N^{\varepsilon}(t)| \leq N, \tag{8.43}$$
$$|P^{\varepsilon}(t)| \leq P,$$

uniformly by $t \epsilon R_+$, and

$$|\bar{N}(t)| \leq \bar{N}, \quad |\bar{P}(t)| \leq \bar{P}, \text{ uniformly} \forall t \in R_+. \tag{8.44}$$

Now, we fix ε such that $0 \leq \varepsilon \leq \varepsilon_0$, where ε_0 is a sufficiently small positive number.

From the rates of convergence for RE, it follows that [80]:

(a)

$$|P^{\varepsilon}(s) - \bar{P}(s)| \leq \varepsilon \cdot P_T, \quad \forall s \epsilon [0, T], \text{ a.s.}, \tag{8.45}$$

where the positive constant $P_T \equiv P_T(P_0)$ depends only on T and $P_0 > 0$;

(b)

$$|N^{\varepsilon}(s) - \bar{N}(s)| \leq \varepsilon \cdot N_T, \quad \forall s \epsilon [0, T], \tag{8.46}$$

where $N_T \equiv N_T(N_0)$ depends only on T and $N_0 > 0$. From renewal theorem (Chapter 1) it follows the following rates of convergence:

$$\begin{aligned}
|\varepsilon \textstyle\int_0^{t/\varepsilon}(a(y(s)) - \bar{a})ds| \leq & \ \varepsilon a_T, \\
|\varepsilon \textstyle\int_0^{t/\varepsilon}(b(y(s)) - \bar{b})ds| \leq & \ \varepsilon b_T, \\
|\varepsilon \textstyle\int_0^{t/\varepsilon}(c(y(s)) - \bar{c})ds| \leq & \ \varepsilon c_T, \\
|\varepsilon \textstyle\int_0^{t/\varepsilon}(d(y(s)) - \bar{d})ds| \leq & \ \varepsilon d_T,
\end{aligned} \tag{8.47}$$

for all $t \epsilon [0, T]$, where constants $a_T, \quad b_T, \quad c_T, \quad d_T$ depend only on T.

From (8.43)-(8.44) and (8.47) it follows that

$$\ln \frac{N^{\varepsilon}(t)}{\bar{N}(t)} \simeq \varepsilon \cdot (a_T + \bar{P} \cdot b_t + P_T \cdot b),$$

where

$$b := \sup_{y \epsilon Y} b(y),$$

and

$$\ln \frac{P^{\varepsilon}(t)}{\bar{P}(t)} \simeq \varepsilon \cdot (c \cdot N_T + \bar{N} \cdot c_T + d_T),$$

with

$$c := \sum_{y \epsilon Y} c(y).$$

In this way, we get

$$\bar{N}(t) \simeq N^{\varepsilon}(t) \cdot e^{-\varepsilon \cdot A_T} \tag{8.48}$$

and

$$\bar{P}(t) \simeq P^{\varepsilon}(t)e^{\varepsilon B_T}, \tag{8.49}$$

where

$$\begin{aligned} A_T \; &:= a_T + \bar{P} \cdot b_T + P_t \cdot b, \\ B_T \; &:= d_T + \bar{N} \cdot c_T + N_T \cdot c. \end{aligned} \tag{8.50}$$

We now substitute the expression for $\bar{N}(t)$ and $\bar{P}(t)$ in (8.48)-(8.49) into the equality (8.48) to get

$$\bar{\alpha} \cdot N^{\varepsilon}(t)d^{-\varepsilon A_T} + P^{\varepsilon}(t) \cdot e^{-\varepsilon B_T} - \ln(N^{\varepsilon}(t))^{\bar{\alpha}-\varepsilon\bar{\alpha}\cdot A_T} P^{\varepsilon}(t) \cdot d^{-\varepsilon B_T} = \bar{H},$$

or

$$\bar{\alpha}N^{\varepsilon}(t)e^{-\varepsilon A_T} + P^{\varepsilon}(t)e^{-\varepsilon B_T} - \ln(N^{\varepsilon}(t))^{\bar{\alpha}} \cdot P^{\varepsilon}(t) - \varepsilon(\bar{\alpha}A_T + B_T) = \bar{H}.$$

In the $(u^{\varepsilon}, v^{\varepsilon})$ phase plane these give the phase trajectories:

$$\bar{\alpha}u^{\varepsilon}e^{-\varepsilon A_T} + v^{\varepsilon} \cdot e^{-\varepsilon B_T} - \ln(u^{\varepsilon})^{\bar{\alpha}} \cdot v^{\varepsilon} - \varepsilon(\bar{\alpha}A_T + B_T) = \bar{H}. \tag{8.51}$$

Compare it with phase trajectories in (8.40) for the $\bar{u}, \bar{v}$ phase plane we obtain:

$$\bar{\alpha}(\bar{u} - u^{\varepsilon}e^{-\varepsilon A_T}) + \bar{v} - v^{\varepsilon}d^{-\varepsilon B_T}) + (\ln \bar{u}^{\bar{\alpha}} \cdot \bar{v} - \ln(u^{\varepsilon})^{\bar{\alpha}} \cdot v^{\varepsilon}) = \varepsilon(\bar{\alpha}B_T + B_T). \tag{8.52}$$

Therefore, for a small fixed ε with $0 \leq \varepsilon \leq \varepsilon_0$, the phase trajectories $(u^{\varepsilon}, v^{\varepsilon})$ are very closed to the phase trajectories $(\bar{u}, \bar{v})$. Let us return to the form (8.38):

$$\frac{d\bar{u}}{d\tau} = \bar{u} \cdot (1 - \bar{v}); \quad \frac{d\bar{v}}{d\tau} = \bar{\alpha} \cdot \bar{v} \cdot (\bar{u} - 1), \tag{8.53}$$

where

$$\bar{u}(\tau) := \frac{\bar{c} \cdot \bar{N}(t)}{\bar{d}}; \quad \bar{v}(\tau) := \frac{\bar{b} \cdot \bar{P}(t)}{\bar{a}}, \quad \tau = \bar{a} \cdot t, \quad \bar{\alpha} := \bar{d}/\bar{a}.$$

The linearization about the singular points determines the type of singularity and the stability of the steady states.

Let us consider the steady state $(\bar{u}, \bar{v}) = (0, 0)$. Let $\bar{x}$ and $\bar{y}$ be small perturbations about $(0, 0)$. If we keep only linear terms, (8.53) becomes

$$\begin{pmatrix} \frac{d\bar{x}}{d\tau} \\ \frac{d\bar{y}}{d\tau} \end{pmatrix} \simeq \begin{pmatrix} 1 & 0 \\ 0 & -\bar{\alpha} \end{pmatrix} \begin{pmatrix} \bar{x} \\ \bar{y} \end{pmatrix} := \bar{A} \cdot \begin{pmatrix} x \\ y \end{pmatrix}. \tag{8.54}$$

The solution is of the form

$$\begin{pmatrix} \bar{x}(\tau) \\ \bar{y}(\tau) \end{pmatrix} = Be^{\bar{\lambda}\tau},$$

where B is an arbitrary constant column vector and the eigenvalues $\bar{\lambda}$ are given by solving the characteristic polynomial of the matrix $\bar{A}$:

$$|\bar{A} - \bar{\lambda} \cdot I| = \begin{vmatrix} 1 - \bar{\lambda} & 0 \\ 0 & -\bar{\alpha} - \bar{\lambda} \end{vmatrix} = 0.$$

In particular, we have

$$\bar{\lambda}_1 = 1, \quad \bar{\lambda}_2 = -\bar{\alpha} = \frac{\bar{d}}{\bar{a}}.$$

Since at least one eigenvalue, $\bar{\lambda}_2 > 0$, $\bar{x}(\tau)$ and $\bar{y}(\tau)$ grow exponentially and so $\bar{u} = \bar{v} = 0$ is *linearly unstable.*

Since $\bar{\lambda}_1 > 0$ and $\bar{\lambda}_2 < 0$, $(0,0)$ is an *averaged saddle point.*

Linearizing about the steady state $\bar{u} = \bar{v} = 1$ by setting $\bar{u} = 1 + \bar{x}, \bar{v} = 1 + \bar{y}$ with $|\bar{x}|$ and $|\bar{y}|$ small, (8.53) becomes

$$\begin{pmatrix} \frac{d\bar{x}}{d\tau} \\ \frac{d\bar{y}}{d\tau} \end{pmatrix} = \bar{A} \cdot \begin{pmatrix} \bar{x} \\ \bar{y} \end{pmatrix}, \quad \bar{A} = \begin{pmatrix} 0 & -1 \\ \bar{\alpha} & 0 \end{pmatrix} \tag{8.55}$$

with eigenvalues $\bar{\lambda}$ given from the following determinant:

$$\begin{vmatrix} -\bar{\lambda} & -1 \\ \bar{\alpha} & -\bar{\lambda} \end{vmatrix} = 0.$$

Namely,

$$\bar{\lambda}_1, \bar{\lambda}_2 = \pm i \cdot \sqrt{\bar{\alpha}} = \pm i \sqrt{\frac{\bar{d}}{\bar{a}}}.$$

Thus $\bar{u} = \bar{v} = 1$ is an *averaged center* since the eigenvalues are purely imaginary. Since $Re\bar{\lambda} = 0$, the steady state is *neutrally stable.* The solution of (8.55) is of the form:

$$\begin{pmatrix} \bar{x}(\tau) \\ \bar{y}(\tau) \end{pmatrix} = l e^{i\sqrt{\bar{\alpha}}\tau} + m e^{-i\sqrt{\bar{\alpha}}\tau},$$

where l and m are eigenvectors.

So, the solutions in the neighbourhood of the singular point $\bar{u} = \bar{v} = 1$ are periodic in τ with period $2\pi/\sqrt{\bar{\alpha}}$. In dimensional terms from (8.37), this period is $\bar{T} = 2\pi(\bar{a}/\bar{d})^{1/2}$. That is, this period is proportional to the square root of the ratio of the linear growth averaged rate $\bar{a}$ of the prey to the death averaged rate $\bar{d}$ of the predator.

In the ecological context, the matrix $\bar{A}$ in the linear equations (8.54) and (8.55) is called the *community matrix*, and its eigenvalues $\bar{\lambda}$ determine the stability of the steady states. If $Re\bar{\lambda} > 0$ then the steady state is unstable, while if both $Re\bar{\lambda} < 0$ it is stable. The critical case $Re\bar{\lambda} = 0$ is termed as *neutral stability.*

8.7.2. STABILITY OF PREDATOR-PREY MODEL IN DIFFUSION APPROXIMATION SCHEME.

Let us suppose that we have predator-prey model in series scheme t/ε^2 with balance condition: $\hat{a} = \hat{b} = \hat{c} = \hat{d} = 0$ (see Subsection 8.5):

$$\begin{cases} \ln \frac{N^\varepsilon(t)}{N_0} &= \varepsilon \cdot \int_0^{t/\varepsilon^2} (a(y(s)) - b(y(s))P^\varepsilon(s))ds, \\ \ln \frac{P^\varepsilon(t)}{P_0} &= \varepsilon \cdot \int_0^{t/\varepsilon^2} (c(y(s))N^\varepsilon(s) - d(y(s)))ds. \end{cases} \tag{8.56}$$

Te limiting predator-prey model is then

$$\begin{cases} d\hat{N}(t) &= \alpha_1(N,P)dt + \beta_{11}dw_1(t) + \beta_{12}dw_2(t), \\ d\hat{P}(t) &= \alpha_2(N,P)dt + \beta_{21}dw_1(t) + \beta_{22}dw_2(t), \end{cases} \tag{8.57}$$

where $\alpha_i := \alpha_i(N,P), \quad i = 1,2,$ and $\beta_{ij} := \beta_{ij}(N,P), \quad i,j = 1,2,$ are defined in (8.24) and (8.25), respectively.

We will study the stability of mean value of the predator-prey model in diffusion approximation scheme, namely, the behaviour of $E\hat{N}(t)$ and $E\hat{P}(t)$. For this purpose we will make the linearization of the system (8.57) near the point $(0,0)$. Let us calculate the first derivatives of the funactions $\alpha_i, \quad i = 1,2,$ which are defined in (8.24):

$$\begin{aligned} d\alpha_1/dN|_{(0,0)} &= \int_Y p(dy)a(y)R_0a(y)/m, \\ d\alpha_1/dP|_{(0,0)} &= 0, \\ d\alpha_2/dN|_{(0,0)} &= 0, \\ d\alpha_2/dP|_{(0,0)} &= \int_Y p(dy)d(y)R_0d(y)/m. \end{aligned}$$

From there and (8.57), we obtain the following system of equations for $E\hat{N}(t)$ and $E\hat{P}(t)$:

$$\begin{cases} dE\hat{N}(t) &= \int_Y p(dy)a(y)R_0a(y)/mE\hat{N}(t), \\ dE\hat{P}(t) &= \int_Y p(dy)d(y)R_0d(y)/mE\hat{P}(t). \end{cases} \tag{8.58}$$

We note that the error of this approximation is $O(r^2)$, where $r^2 := \sqrt{N^2 + P^2}$.

From (8.58), we have the following behaviour of the mean values $E\hat{N}(t)$ and $E\hat{P}(t)$:

$$E\hat{N}(t) \to +\infty$$

and

$$E\hat{P}(t) \to +\infty$$

as $t \to +\infty$.

We note that since $N^\varepsilon_{\nu(t/\varepsilon^2)}$ and $P^\varepsilon_{\nu(t/\varepsilon^2)}$ converges weakly to the pair $(\hat{N}(t), \hat{P}(t))$, then the behaviour of the mean value of predator-prey model in diffusion approximation scheme is completely defined by the behaviour of the mean value of the limiting predator-prey model (8.58).

8.7.3. STABILITY OF PREDATOR-PREY MODEL IN NORMAL DEVIATIONS SCHEME.

Let us study the behaviour of the mean value of the predator-prey model in normal deviations scheme, namely, the behaviour of $E\tilde{N}(t)$ and $E\tilde{P}(t)$, where $\tilde{N}(t)$ and $\tilde{P}(t)$ are defined in (8.32):

$$\begin{cases} d\tilde{N}(t) = \alpha_1(\hat{N},\hat{P},\tilde{N},\tilde{P})dt + \beta_{11}(\hat{N},\hat{P})dw_1(t) + \beta_{12}(\hat{N},\hat{P})dw_2(t), \\ d\tilde{P}(t) = \alpha_2(\hat{N},\hat{P},\tilde{N},\tilde{P})dt + \beta_{21}(\hat{N},\hat{P})dw_1(t) + \beta_{22}(\hat{N},\hat{P})dw_2(t), \end{cases} \tag{8.59}$$

where $\alpha_i := \alpha_i(\hat{N},\hat{P},\tilde{N},\tilde{P})$, $\quad i = 1,2$, and $\beta_{ij} := \beta_{ij}(\hat{N},\hat{P})$, $\quad i,j = 1,2$, are defined in (8.30) and (8.31), respectively. In particular, we have

$$\begin{aligned} \alpha_1(\hat{N},\hat{P},\tilde{N},\tilde{P}) &= \textstyle\int_Y p(dy)[\tilde{N}(a(y) - b(y)\hat{P}) - b(y)\hat{N}\tilde{P}]/m, \\ \alpha_2(\hat{N},\hat{P},\tilde{N},\tilde{P}) &= \textstyle\int_Y p(dy)[c(y)\hat{P}\tilde{N} + \tilde{P}(c(y)\hat{N} - d(y))]/m. \end{aligned} \tag{8.60}$$

Let us make a linearization of the system (8.59) near the point $(0,0)$. For this purpose, we calculate the first derivatives of the coefficients α_i, $\quad i = 1,2$, in (8.60):

$$\begin{aligned} d\alpha_1/dN|_{(0,0)} &= (a(y) - b(y)\hat{P}), \\ d\alpha_1/dP|_{(0,0)} &= -b(y)\hat{N}, \\ d\alpha_2/dN|_{(0,0)} &= c(y)\hat{P}, \\ d\alpha_2/dP|_{(0,0)} &= c(y)\hat{N} - d(y). \end{aligned}$$

From there and (8.59), we obtain the following system of equations for the mean values $E\tilde{N}(t)$ and $E\tilde{P}(t)$:

$$\begin{cases} dE\tilde{N}(t) = \int_Y p(dy)(a(y) - b(y)/m\hat{P}(t))E\tilde{N}(t) - \int_Y p(dy)b(y)/m\hat{N}(t)E\tilde{P}(t), \\ dE\tilde{P}(t) = \int_Y p(dy)c(y)/m\hat{P}(t)E\tilde{N}(t) + \int_Y p(dy)(c(y)\hat{N}(t) - d(y))/mE\tilde{P}(t). \end{cases} \tag{8.61}$$

We note that the error of this approximation is $O(r^2)$, where $r := \sqrt{N^2 + P^2}$.

The stability in mean of normal deviated predator-prey model is completely determined by the behaviour of the system (8.61), which contains only averaged predator-prey model components $(\hat{N},\hat{P})$ and its averaged coefficients.

We note that since

$$\mathbf{X}^\epsilon(t) \approx \hat{\mathbf{X}}(t) + \sqrt{\epsilon}\tilde{\mathbf{X}}(t),$$

where

$$\mathbf{X}^{\epsilon}(t) := [N^{\varepsilon}_{\nu(t/\varepsilon)}, P^{\varepsilon}_{\nu(t/\varepsilon)}]^T,$$

$$\hat{\mathbf{X}}(t) := (\hat{N}(t), \hat{P}(t))^T,$$

$$\tilde{\mathbf{X}}(t) := (\tilde{N}(t), \tilde{P}(t))^T,$$

the stability in mean of the predator-prey model in normal deviations scheme is defined by the following relation

$$E\mathbf{X}^{\epsilon}(t) \approx \hat{\mathbf{X}}(t) + \sqrt{\epsilon}E\tilde{\mathbf{X}}(t),$$

where $E\tilde{\mathbf{X}}(t)$ is defined in (8.61).

In this way, we finally obtain the following relations with respect to the behaviour of mean values of predator-prey model in normal deviations scheme:

$$EN^{\epsilon}(t) \approx \hat{N}(t) + \sqrt{\epsilon}E\tilde{N}(t),$$

$$EP^{\epsilon}(t) \approx \hat{P}(t) + \sqrt{\epsilon}E\tilde{P}(t).$$

BIBLIOGRAPHY

[1] **Anderson, R. M.**(ed.) Population Dynamics of Infectious Diseases: Theory and Applications. London: Chapman and Hall, 1982.

[2] **Anderson, R. M and May, R. M.** Vaccination and hers immunity to infectious diseases. Nature 318, 1985, 323-329.

[3] **Anderson, R. M and May, R. M.** The invasion, persistence and spread of infectious diseases within animal and plant communities. Phil. Trans. Roy. Soc. (London), B324, 1986, 533-570.

[4] **Arnold, L.** Stochastic Differential Equations: Theory and Applications, J. Wiley & Sons Inc., 1974, 228 p.

[5] **Athreya, K. B. and Ney, P. E.** Branching Processes, Springer-Verlag, Berlin, 1972.

[6] **Bailey, N. T. J.** The mathematical theory of Epidemics. Charles Griffin, London, 1957.

[7] **Bailey, N. T. J.** The Mathematical Theory of Infectious Diseases, 2nd edn. London: Griffin, 1975.

[8] **Beddington, J. F., Free, C. A. and Lawton J. H.** Dynamic complexity in predator-prey models found in difference equations. nature 255, 1985, 58-60.

[9] **Bellman, R. and Harris, T. E.** On age-dependent binary branching processes, Am. of Math. 55, 952, 280-295.

[10] **Bernardelli, H.** Population waves. F. Burma Res. Soc., 37, 1962, 1-18.

[11] **Bernoulli, D.** Essai d'une nouvelle analyse de la mortalite causée par la petite vérole, et des avantages de l'inoculation pour la poéveuir. Réstoite de l'Acad. Roy. Sci. (Paris) avec Mém. des Math. et Phys., Mém., 1760, 1-45.

[12] **Blankenship, G. and Papanicolaou, G.** Stability and control of stochastic systems with wide-band noise disturbances. I., SIAM J. Appl. Math., 34, No3, 1978, 437-476.

[13] **Capasso, V. and Paveri-Fontana, S. L.** A mathematical model for the 1973 cholera epidemic in the european mediterranean region. Rev. Epidém. et Sauté Publ. 27, 1979, 121-132.

[14] **Cavalli-Sforza, L. L. and Bodmer, W. A.** The Genetics of Human Populations. W.H. Freeman, San Francisco, 1971.

[15] **Chung, K. L.** Markov Chains with Stationary Transition Probabilities, Berlin, 1960.

[16] **Cogburn, R.** Markov chains in random environment, the case of Markovian environment, Ann. of Probab., 8, 1980, 989-1016.

[17] **Cogburn, R. and Bourgin, R. O.** On determining Absorbtion probabilities for Markov chains in random environment, Adv. in Appl. Probab., 13, 1981, 369-387.

[18] **Cogburn, R. and Torres, W.** Birth and death processes with random environments in continuous time, J. Appl. Probab., 18, 1981, 19-30.

[19] **Cogburn, R.** Recurrence transience for spatially inhomogeneous birth and death processes in random environment, Z. Wahrsch. Verv. Gebiete, 16, 1982, 153-160.

[20] **Cogburn, R.**The ergodic theory of Markov chains in random environment, Z. Wahrsch. Verv. Gebiete, 66, 1984, 109-128.

[21] **Crow, J. F. and Kimura, M.** An Introduction to Population Genetics Theory. Harper & Row, NY. 1970.

[22] **Doeblin, W.** Sur les properties asymptotiques de mouvement regis par certain types de chaines simples. Ball. Huer. Soc. Roum. Sci, 39, 1937, No. 1, 57-115; No. 2, 3-61.

[23] **Doob, J. I.** Stochastic processes, John Wiley, New York, 1953.

[24] **Dynkin, E. G.** Markov processes, Fizmatgiz, Moscow, 1963 (In Russian).

[25] **Ewens, W. J.** Population Genetics. Methuen, London, 1969.

[26] **Feller, W.** An Introduction to Probability Theory and its Applications. Wiley & Sons Inc., New York, 1, 1966.

[27] **Feller, W.** An Introduction to Probability Theory and Its Applications, Wiley & Sons Inc., New York, 2, 1971.

[28] **Fibonacci, L.** Tipographia della Scienze Mathematichee Fischehe, Roma, 1702.

[29] **Fisher, R. A.** The genetical theory of natural selection. The wave of advance of an advantageous gene. Ann. Eugen. Dover, NY, 7, 1936, 335-369.

[30] **Freidlin, M. I. and Wentzel, A. D.** Random Perturbations of Dynamical Systems,, Springer-Verlag, Berlin, 1984.

[31] **Gikhman, I. and Skorokhod, A.** Stochastic Differential Equations, Springer-Verlag, Berlin, 1972.

[32] **Griego, R. and Hersh, R.** Random evolutions, Markov chains, and systems of partial differential equations, Proc. Nat. Acad. Sci., USA, 62, 1969, 305-308.

[33] **Gurowski, I. and Mira, C.** Dynamique Chaotique. Toulouse: Collection Nabla, Cepadue Edition 1980.

[34] **Harris, T. E.** The Theory of Branching Processes, Springer-Verlag, Berlin, 1963.

[35] **Hassell, M. P.** The Dynamics of Arthropod Predator-Prey systems. Princeton: Princeton Univ. Press, 1978.

[36] **Hersh, R.** Random evolutions: a survey of results and problems, Rocky Mount. Math. J., 4, 1974, 443-477.

[37] **Hoppensteadt, F. C.** Mathematical Theories of Populations: Demographics, Genetics and Epidemics, CBMS, SIAM, Philadelphia, 20, 1975.

[38] **Hoppensteadt, F. C. and Minanker W.** Multi-time methods for difference equations. Stud. Appl. Math., 56, 1977, 273-289.

[39] **Hoppensteadt, F. C.** Mathematical Methods of Population Biology. Cambridge Univ. press, 1982.

[40] **Hoppensteadt, F. C. and Peskin C. S.** Modeling and Simulation in Medicine and Life Science. Springer-Verlag, New-York, 2001.

[41] **Hoppensteadt, F. C., Salehi H. and Skorokhod A.** Discrete time semigroup transformation with random perturbations. J. of Dynamics & Diff. Eq. v. 9, N. 3, 1997.

[42] **Hoppensteadt, F. C. and Peskin, C. S.** Mathematics in Medicine and the Life Sciences, Springer-Verlag, 1992.

[43] **Jolly, C. and Brett, F. L.** J. Med. Primatol, 1973.

[44] **Kermack, W. O. and McKendrick, A. G.** A contribution to the theory of epidemics, I,II. Proc. Roy. Soc. Lond. Sec. A, 115, 1927, 700-21; 139, 1932, 55-83.

[45] **Khasminskii, R.** Necessary and sufficient conditions for the asymptotic stability of linear stochastic systems, Theory Probab. Appl., 12, 1967, 144-147.

[46] **Kertz, R.** Discontinuous Random Evolutions, Ann. Probab., 2, 1974, No. 6.

[47] **Keyfitz, N. and Flieger, W.** Population: Facts and Methods of Demography, W.H. Freeman, San Francisco, 1971.

[48] **Kolmogorov, A. N.** A fongsgrunde der Markoffschen Ketten mit unendlich vieleu moglichen Zustangen, Rec. Math. Moscow (Mat. Sb.)1 (43), 1936, 607-610.

[49] **Korolyuk, V. S. and Swishchuk, A. V.** Evolution of Systems in Random Media, CRC Press, Boca Raton, U.S.A., 1995.

[50] **Korolyuk, V. S. and Swishchuk, A. V.** Semi-Markov Random Evolutions. Kluwer AP, Dordrecht, The Netherlands, 1995.

[51] **Kushner, H.** Stochastic Stability and Control, Academic Press, New York, 1967.

[52] **Leslie, P. H.** Biometrika. 33, 1945, 183-212; 35, 1948, 213-243.

[53] **Levy, P.** Processes semi-Markovians, in: Proceed. of the 3rd Internat. Cong. of Math., 1954, 416-426.

[54] **Limnios, N. and Oprisan, G.** Semi-Markov Processes and Reliability. Birkhauser, Boston, 2001.

[55] **Lotka, A. J.** Undaped oscillations derived from the law of mass action. J.Amer. Chem. Soc. 42, 1920, 1595-1599.

[56] **Lotka, A. J.** Elements of Physical Biology. Williams and Wilkins: Baltimore 1925.

[57] **Ludwig, D. A.** Stochastic population theories. Lect. Notes Biomath., Springer-Verlag, 3, 1979.

[58] **Malthus, T. R.** An essay on the Principal of Population. 1798 [Penguin Books, 1970].

[59] **Markov, A. A.** Investigation of remarkable case of dependent trials, Izvestiya Rosiyskoi Akademic Nauk, v. 1, 1907.

[60] **McKusick, V. A.** Human Genetics, 2nd. edi:, Prentice-Hall, Englewood Cliffs, New Jersey, 1969.

[61] **May, R. M. and Oster, G. F.** Bifurcations and dynamic complexity in simple ecological models. Amer. Nature, 110, 1976, 573-599.

[62] **Mendel, G. J.** Versuche über Pflanzeu-Hybriden. Verh. Naturforsch. Ver. Brunn, 19, 1865.

[63] **Moran, P. A. P.** The Statistical Process of Evolutionary Theory. Clarendon

Press, Oxford, 1962.
[64] **Murray, J. D.** Mathematical Biology. springer-Verlag, Sec. ed., Biomath. Series., 19, 1993.
[65] **Norris, J. K.** Markov Chains. Cambridge Series in Statistical and Probabilistical Mathematics. Cambridge Univ. Press, 1997.
[66] **Nummelin, E.** General Irreducible Markov chains and Non-negative Operators, Mir, Moscow, 1989 (In Russian).
[67] **Papanicolaou, G.** Asymptotic analysis of transport processes. BAMS, 81, 1975, 330-392.
[68] **Papanicolaou, G.** Random Media, Springer-Verlag, Berlin, 1987.
[69] **Papanicolaou, G., Kohler, W. and White, B.** Random Media, Lectures in Applied Mathematics, 27, SIAM, Philadelphia, 1991.
[70] **Pinsky, M.** Stochastic stability and the Dirichlet problem, Comm. Res. Appl. Math., 27, 1974, 311-350.
[71] **Pinsky, M.** Lectures on Random Evolution. World Scientific Publ., Singapore, 1991.
[72] **Revuz, D.** Markov chains, North-Holland, Amsterdam, 1975.
[73] **Rosenblatt, M.** Random Processes, Springer-Verlag, Berlin, 1974, 228p.
[74] **Sevastyanov, B. A.** Branching processes. Moscow: Nanka, 1971 (In Russian).
[75] **Shiryaev, A. N.** Probability, Nanka, Moscow, 1980 (In Russian).
[76] **Skorokhod, A., Hoppensteadt, F. and Salehi, H.** Random Perturbation Methods with Applications to Science and Engineering,Springer-Verlag, Berlin, 2002.
[77] **Smith, W. L.** Regenerative stochastic processes. Proc. Roy. Soc. London, A232, 1956, 6-31.
[78] **Sobolev S.** Some Applications of Functional Analysis in Mathematical Physics. Nanka, Moscow, 1988 (In Russian).
[79] **Swishchuk A. V.** Random Evolutions and Their Applications, Kluwer AP, Dordrecht, The Netherlands, 1997.
[80] **Swishchuk, A. V.** Random Evolutions and their Applications. New Thrends, Kluwer AP, Dordrecht, The Netherlands, 2000.
[81] **Swishchuk, A. and Wu, J.-H.** Averaging and diffusion approximation of vector difference equations in random media with applications to biological systems. J. Diff. Eqns. and Dyna. Systems, 2003, in press.
[82] **Swishchuk, A. and Wu, J.-H.** Limit theorems for difference equations in random media with applications to biological systems, Random Operators and Stoch. Equat., 11:1, 2003, 24-75.
[83] **Swishchuk, A. and Wu, J.-H.** Stability of difference equations in random media in averaging and diffusion approximation schemes. Technical Report N5, LIAM, Dept. of Math. & Stat., York University, 2003.
[84] **Takacs, L.** On secondary processes generated by recurrent processes. Arch. Math. 7, 1956, 17-29.
[85] **Verhulst, P. F.** Recherches mathematiques sur la loi d'accroissement de la population. Meae. Acad. Roy., Belgium, 18, 1845, 1-38.
[86] **Verhulst, P. F.** Notice sur la loque la population suit dans sou accroissement.

Corr. math. et Phys. 10, 1838, 113-121.

[87] **Volterra, V.** Variziomie fluttuaziom del numero d'individici in specie animali conviventi. Mem. Acad. Lincee. 2, 1926, 31-113. (Variations and fluctuations of a number of individuals in animal species living together. Translation In: R.N. Chapman: Animal Ecology. New York: McGraw Hill, 1931, 409-448).

[88] **Waltman, P.** Deterministic Threshold Models in the Theory of Epidemics. Lect. Notes in Biomath., 1, Springer, 1974.

[89] **Watkins, J.** Consistency and fluctuation theorems for discrete time structured population models having demographic stochasticity, J. of Mathem. Biology, 41, 2000, 253-271.

[90] **Watkins, J.** A central limit theorems for radnom evolutions, Ann. of Probab., 12, 1984, 480-513.

[91] **Watson, H. W. and Galton, F.** On the probability of the extinction of families. J. Anthropol. Inst. Great Br. and Ireland, 4, 138-144, 1874.

[92] **Wright, S.** Evolution of Mendelian genetics. Genetics, 16, 1931, 97-159.

[93] **Wickwire, K. H.** Mathematical models for the control of pests and infectious diseases: a survey. Theor. Pop. Biol., 11, 1977, 182-283.

[94] **Wu, J.-H.** Theory and Applications of Partial Differential Equations, Springer-Verlag, 1996.

Index

MATHEMATICAL MODELLING: Theory and Applications

---

1. M. Křížek and P. Neittaanmäki: *Mathematical and Numerical Modelling in Electrical Engineering.* Theory and Applications. 1996
ISBN 0-7923-4249-6
2. M.A. van Wyk and W.-H. Steeb: *Chaos in Electronics.* 1997
ISBN 0-7923-4576-2
3. A. Halanay and J. Samuel: *Differential Equations, Discrete Systems and Control.* Economic Models. 1997 ISBN 0-7923-4675-0
4. N. Meskens and M. Roubens (eds.): *Advances in Decision Analysis.* 1999
ISBN 0-7923-5563-6
5. R.J.M.M. Does, K.C.B. Roes and A. Trip: *Statistical Process Control in Industry.* Implementation and Assurance of SPC. 1999
ISBN 0-7923-5570-9
6. J. Caldwell and Y.M. Ram: *Mathematical Modelling.* Concepts and Case Studies. 1999 ISBN 0-7923-5820-1
7. 1. R. Haber and L. Keviczky: *Nonlinear System Identification - Input-Output Modeling Approach.* Volume 1: Nonlinear System Parameter Identification. 1999 ISBN 0-7923-5856-2; ISBN 0-7923-5858-9 Set

   2. R. Haber and L. Keviczky: *Nonlinear System Identification - Input-Output Modeling Approach.* Volume 2: Nonlinear System Structure Identification. 1999 ISBN 0-7923-5857-0; ISBN 0-7923-5858-9 Set
8. M.C. Bustos, F. Concha, R. Bürger and E.M. Tory: *Sedimentation and Thickening.* Phenomenological Foundation and Mathematical Theory. 1999
ISBN 0-7923-5960-7
9. A.P. Wierzbicki, M. Makowski and J. Wessels (eds.): *Model-Based Decision Support Methodology with Environmental Applications.* 2000
ISBN 0-7923-6327-2
10. C. Rocşoreanu, A. Georgescu and N. Giurgiţeanu: *The FitzHugh-Nagumo Model.* Bifurcation and Dynamics. 2000 ISBN 0-7923-6427-9
11. S. Aniţa: *Analysis and Control of Age-Dependent Population Dynamics.* 2000
ISBN 0-7923-6639-5
12. S. Dominich: *Mathematical Foundations of Informal Retrieval.* 2001
ISBN 0-7923-6861-4

13. H.A.K. Mastebroek and J.E. Vos (eds.): *Plausible Neural Networks for Biological Modelling*. 2001 ISBN 0-7923-7192-5
14. A.K. Gupta and T. Varga: *An Introduction to Actuarial Mathematics*. 2002 ISBN 1-4020-0460-5
15. H. Sedaghat: *Nonlinear Difference Equations*. Theory with Applications to Social Science Models. 2003 ISBN 1-4020-1116-4
16. A. Slavova: *Cellular Neural Networks: Dynamics and Modelling*. 2003 ISBN 1-4020-1192-X
17. J.L. Bueso, J.Gómez-Torrecillas and A. Verschoren: *Algorithmic Methods in Non-Commutative Algebra*. Applications to Quantum Groups. 2003 ISBN 1-4020-1402-3

KLUWER ACADEMIC PUBLISHERS – DORDRECHT / BOSTON / LONDON